Dmitri N. Akhiezer

Lie Group Actions in Complex Analysis

Aspects of Mathematics

Edited by Klas Diederich

*A Publication of the Max-Planck-Institut für Mathematik, Bonn

Volumes of the German-language subseries "Aspekte der Mathematik" are listed at the end of the book.

Dmitri N. Akhiezer

Lie Group Actions in Complex Analysis

vieweg

Dmitri N. Akhiezer
129010 Moskau
B. Spasskaja 33, KV 33
Russia

CIP-Codierung angefordert

Softcover reprint of the hardcover 1st edition 1995

Vieweg is a subsidiary company of the Bertelsmann Professional Information.

Cover design: Wolfgang Nieger, Wiesbaden

Printed on acid-free paper

ISSN 0179-2156
ISBN-13:978-3-322-80269-9 e-ISBN-13:978-3-322-80267-5
DOI: 10.1007/978-3-322-80267-5

Preface

This book was planned as an introduction to a vast area, where many contributions have been made in recent years. The choice of material is based on my understanding of the role of Lie groups in complex analysis. On the one hand, they appear as the automorphism groups of certain complex spaces, e.g., bounded domains in $\mathbb{C}^n$ or compact spaces, and are therefore important as being one of their invariants. On the other hand, complex Lie groups and, more generally, homogeneous complex manifolds, serve as a proving ground, where it is often possible to accomplish a task and get an explicit answer. One good example of this kind is the theory of homogeneous vector bundles over flag manifolds. Another example is the way the global analytic properties of homogeneous manifolds are translated into algebraic language.

It is my pleasant duty to thank A.L.Onishchik, who first introduced me to the theory of Lie groups more than 25 years ago. I am greatly indebted to him and to E.B.Vinberg for the help and advice they have given me for years.

I would like to express my gratitude to M.Brion, B.Gilligan, P.Heinzner, A.Hukleberry, and E.Oeljeklaus for valuable discussions of various subjects treated here. A part of this book was written during my stay at the Ruhr-Universität Bochum in 1993. I thank the Deutsche Forschungsgemeinschaft for its research support and the colleagues in Bochum for their hospitality.

Moscow, June 1994 D.N.Akhiezer

Contents

Introduction

The aim of this book is the study of Lie group actions on complex spaces. For the same reason as in the general Lie theory, it is natural to begin with the local properties of such actions, and so one is led to the foundations and, in particular, to Hilbert's fifth problem.

In this problem Hilbert considers a local transformation group, i.e., a system of continuous functions $f_i(x;a)$, $x = (x_1, \dots, x_n), a = (a_1, \dots, a_d)$, defined in a neighborhood of $(0,0) \in \mathbb{R}^n \times \mathbb{R}^d$, such that

$$f_i(f_1(x;a), \dots, f_n(x;a); b_1, \dots, b_d) = f_i(x_1, \dots, x_n; c_1, \dots, c_d),$$

where $c_i = c_i(a,b)$. The question is whether one can make f_i differentiable by a continuous change of local coordinates in $\mathbb{R}^n$ and in $\mathbb{R}^d$. The general answer is negative, though the similar problem for abstract locally Euclidean groups has been solved positively. Moreover, after changing the local parameters in $\mathbb{R}^d$ one may suppose that the functions $c_i(a,b)$ are real analytic. Assuming this has been done, let $x \in \mathbb{C}^n$ and consider the case when $f_i(x;a)$ are holomorphic in x for every a. Then it turns out that, without any further coordinate changes, $f_i(x;a)$ are also real analytic. This theorem goes back to H.Cartan (1935).

In Chapter 1 we prove the theorem of H.Cartan, using ideas of W.Kaup, for arbitrary complex spaces. As a result we define the Lie homomorphism for a local Lie group action. We discuss the properties of this homomorphism and give some conditions sufficient for the corresponding local action to extend to a global one.

From the viewpoint of the theory of transformation groups, there is an important distinction between real and complex analysis. Namely, there exist complex manifolds whose automorphism groups are Lie groups, whereas the diffeomorphism group of a real manifold is always infinite dimensional. S.Bochner and D.Montgomery proved in 1947 that the automorphism group of a compact complex manifold, endowed with the compact-open topology, is a Lie group. In Chapter 2 we introduce a natural topology into the automorphism group of an arbitrary complex space and prove this theorem for complex spaces. An important result used in the proof is the local linearization theorem for compact transformation groups having a fixed point. Another classical theorem proved here is the theorem of H.Cartan stating that the automorphism group of a bounded domain in $\mathbb{C}^n$ is a Lie group. As an example we give a characterization of the unit ball in terms of its automorphism group.

In Chapter 3 we consider compact homogeneous complex manifolds. The main tool in studying their geometry is the Tits fibration, whose base and fiber are well-understood. Namely, the base is a flag manifold of a semisimple complex Lie group and the fiber is complex parallelizable, i.e., has the form G/Γ, where Γ is a discrete uniform subgroup of a connected complex Lie group G. Under certain assumptions on the fundamental group of the manifold we can say more about the fiber and also about the algebraic properties of transitive transformation groups. By using this fibration we prove that the dimension of a transitive transformation group is bounded by a function depending on the dimension of the manifold. Another application is the structure theorem for compact homogeneous Kähler manifolds due to A.Borel and R.Remmert.

In Chapter 4 we consider the representations of complex Lie groups on the cohomology spaces of homogeneous vector bundles over flag manifolds. These manifolds are of the form G/P, where G is a semisimple complex Lie group, $P \subset G$ a parabolic subgroup. Our main objective is the theorem of R.Bott, showing how to determine the representations of G induced by irreducible holomorphic representations of P. The proof in the text is due to M.Demazure. Some applications to complex geometry are given. In particular, we compute the cohomology of the tangent sheaf of G/P, which makes it possible to list all transformation groups transitive on flag manifolds.

In Chapter 5 we study holomorphic functions in K-invariant domains of homogeneous manifolds G/H, where K is a connected compact Lie group, G its complexification, and H a closed complex Lie subgroup of G. The exposition is based on the theorem of Harish-Chandra about the representations of compact Lie groups on Fréchet vector spaces. As an application, we prove the algebraicity of H in the case when G/H is holomorphically separable. It turns out that G/H is holomorphically separable if and only if G/H is an orbit in a finite dimensional rational G-module. The isotropy subgroups which appear in this situation are called observable. Using various tools from the geometric invariant theory, we describe observable subgroups of arbitrary connected linear algebraic groups. This is done in two steps. Firstly, we define special observable subgroups called qusiparabolic. Secondly, we prove that an algebraic subgroup $H \subset G$ is observable if and only if H is embedded in a quasiparabolic subgroup $Q \subset G$ in such a way that the unipotent radical of H is contained in the unipotent radical of Q. Returning to reductive groups, we give a characterization of Stein homogeneous manifolds in terms of their isotropy subgroups. At the end we discuss the connection between geodesic convexity on $K\backslash G$ and the plurisubharmonicity of K-invariant functions on G/H.

The concluding remarks present some modern trends in the theory. These remarks can also serve as an explanatory note to the bibliography.

1 Lie Theory

In this chapter we define local and global Lie group actions on complex spaces. It is shown that a local action of a Lie group G on a complex space X is real analytic. Such an action gives rise to the Lie homomorphism, which is a map from the Lie algebra of G into the Lie algebra of vector fields on X. The second fundamental theorem of S.Lie states that the local action can be recovered from this homomorphism. We prove this theorem and give some sufficient conditions for a local action to extend to a global one.

1.1 Complex spaces

In this section we recall some basic definitions and results of the theory of complex spaces in the form convenient for the future exposition. The proofs can be found in [GR3].

A *$\mathbb{C}$-ringed space* is a topological space with a sheaf of local $\mathbb{C}$-algebras. A *complex space* $(X, \mathcal{O}_X)$ is a Hausdorff $\mathbb{C}$-ringed space, which is locally isomorphic to a *complex model space*. The latter is defined as follows. Let D be a domain in $\mathbb{C}^n$ and let $\mathcal{O}_D$ be the sheaf of germs of holomorphic functions on D. Consider an ideal sheaf $\mathfrak{J} \subset \mathcal{O}_D$, which is locally generated by a finite number of sections. Denote by Y the *support* of the quotient sheaf $\mathcal{O}_D/\mathfrak{J}$, i.e., the set of all points $z \in D$ where the stalk $(\mathcal{O}_D/\mathfrak{J})_z$ is a non-zero algebra. Locally, in a neighborhood of any point $z \in D$, the subset $Y = \operatorname{supp} \mathcal{O}_D/\mathfrak{J} \subset D$ is the zero set of a finite number of holomorphic functions. The $\mathbb{C}$-ringed space $(Y, \mathcal{O}_Y)$, where $\mathcal{O}_Y := (\mathcal{O}_D/\mathfrak{J})|Y$, is called a complex model space defined by the ideal sheaf $\mathfrak{J} \subset \mathcal{O}_D$.

The sheaf $\mathcal{O} = \mathcal{O}_X$ is called the *structure sheaf* of a complex space $(X, \mathcal{O}_X)$. One of the most important theorems of complex analysis goes back to Oka and says that the structure sheaf of a complex space is *coherent*.

We shall often denote a complex space $(X, \mathcal{O}_X)$ merely by X.

Let x be a point of a complex space X. Then X is said to be *smooth at* x, if there is an open neighborhood U of $x \in X$ such that the $\mathbb{C}$-ringed space $(U, \mathcal{O}_X|U)$ is isomorphic to $(D, \mathcal{O}_D)$, where D is a domain in $\mathbb{C}^n$. Such a point $x \in X$ is also called *regular* or *simple.* If X is smooth at every point then, by definition, X is a *complex manifold.* Points of X at which X is not smooth are called *singular* points. X is said to be *irreducible* (resp. *reduced*) *at* $x \in X$, if the stalk $\mathcal{O}_x$ has no zero divisors (resp. no nilpotent elements $\neq 0$). If this is true for all $x \in X$ then X is called *locally irreducible* (resp. *reduced*).

Let $\mathfrak{m}_x \subset \mathcal{O}_x$ be the maximal ideal, $\mathfrak{n}_x \subset \mathcal{O}_x$ the nilradical, i.e., the ideal of all nilpotent elements of $\mathcal{O}_x$, and $\nu_x : \mathcal{O}_x \to \mathbb{C}$ the canonical epimorphism of $\mathbb{C}$-algebras. Then Ker $(\nu_x) = \mathfrak{m}_x$. Let U be an open subset in X. Any section $f \in \mathcal{O}(U)$ gives rise to a continuous function $[f] : U \to \mathbb{C}$. Namely, $[f]\ (x) := \nu_x(f_x)$, where f_x is the germ of f at $x \in U$. Thus we obtain a sheaf homomorphism $\mathcal{O} \to \mathcal{C}$, where $\mathcal{C}(= \mathcal{C}_X)$ is the sheaf of germs of continuous complex functions on X. The kernel is the ideal sheaf

$$\mathcal{N} := \bigcup_{x \in X} \mathfrak{n}_x \subset \mathcal{O},$$

which is called the *nilradical* of $\mathcal{O}$. The quotient sheaf $\mathcal{O}/\mathcal{N}$ is denoted by red $\mathcal{O}$. The $\mathbb{C}$-ringed space red $X := (X, \text{red}\, \mathcal{O})$ is a reduced complex space which is called the *reduction* of X.

The above construction of a complex model space admits the following natural generalization. Let $(X, \mathcal{O}_X)$ be a complex space, $\mathcal{J} \subset \mathcal{O}_X$ a coherent ideal sheaf, $Y := \text{supp}\, \mathcal{O}_X/\mathcal{J}$, and $\mathcal{O}_Y := (\mathcal{O}_X/\mathcal{J})|Y$. Then the $\mathbb{C}$-ringed space $(Y, \mathcal{O}_Y)$ is a complex space. This complex space is called a *closed complex subspace* of $(X, \mathcal{O}_X)$ defined by the ideal sheaf $\mathcal{J}$. For example, if $\mathcal{J} = \mathcal{N}$ then $Y = X$ and $\mathcal{O}_Y = \text{red}\, \mathcal{O}_X$, so that red X is a closed complex subspace of X.

Let A be a subset of a complex space X and $x \in X$. The subset A is said to be *analytic at* x if there exists a neighborhood U of x and a finite number of sections $f_1, \ldots, f_k \in \mathcal{O}(U)$ such that

$$A \cap U = \{x \in U \mid [f_i](x) = 0, \quad i = 1, \ldots, k\}.$$

The set A is called *analytic in* X if A is analytic at every point of X. Note that, by definition, an analytic set $A \subset X$ is closed. The support of a coherent $\mathcal{O}_X$-sheaf is analytic in X. Thus, the underlying topological space of a closed complex subspace of X is an analytic set. Conversely, for any analytic set $A \subset X$ one can define the *ideal sheaf of* A, denoted by $\mathcal{J}_A$. Namely, $\mathcal{J}_{A,x}$ is the collection of all germs f_x, where $f \in \mathcal{O}(U)$, $U \ni x$, and $[f]|_{U \cap A} = 0$. The Oka-Cartan theorem states that the sheaf $\mathcal{J}_A$ is coherent, so that $(A, (\mathcal{O}_X/\mathcal{J}_A)|A)$ is a closed complex subspace of $(X, \mathcal{O}_X)$.

The set of all points, where a complex space X is not reduced, is analytic. Another example of an analytic set is the set $S(X)$ of all singular points of X. A reduced complex space is called *(globally) irreducible* if there are no proper analytic sets A_1 and A_2 in X such that $X = A_1 \cup A_2$. This is equivalent to the connectedness of the complex manifold $X - S(X)$.

A *holomorphic mapping* between two complex spaces $(X, \mathcal{O}_X)$ and $(Y, \mathcal{O}_Y)$ is defined to be a morphism in the category of $\mathbb{C}$-ringed spaces. It is given by a continuous mapping $\varphi : X \to Y$ along with a homomorphism

$$\tilde{\varphi} : \mathcal{O}_Y \to \varphi_* \mathcal{O}_X$$

of sheafs of rings on Y.

Let $x \in X$ and $y := \varphi(x) \in Y$. Denote by $\tilde{\varphi}_x$ the composition of

$$\tilde{\varphi}_y : \mathcal{O}_{Y,y} \to (\varphi_* \mathcal{O}_X)_y$$

with the natural map

$$(\varphi_* \mathcal{O}_X)_y \to \mathcal{O}_{X,x}.$$

The stalk map $\tilde{\varphi}_x : \mathcal{O}_{Y,\varphi(x)} \to \mathcal{O}_{X,x}$ is a local $\mathbb{C}$-algebra homomorphism. Any section $f \in \mathcal{O}_Y(V)$ over an open set $V \subset Y$ gives rise to a section

$$f \circ \varphi \in \mathcal{O}_X(\varphi^{-1}(V)),$$

which is defined by

$$(f \circ \varphi)_x = \tilde{\varphi}_x (f_{\varphi(x)}),$$

where $x \in \varphi^{-1}(V)$.

A sheaf of modules over $\mathcal{O}_X$ is called an *analytic sheaf* on X. Let $\mathcal{F}$ be a coherent analytic sheaf on X. For every $x \in X$ the local algebra $\mathcal{O}_x$ and the $\mathcal{O}_x$-module $\mathcal{F}_x$ carry the so called *sequence topology.* More precisely, $\mathcal{O}_x$ is a topological $\mathbb{C}$-algebra and $\mathcal{F}_x$ is a topological module over $\mathcal{O}_x$. In the simplest case when $\mathcal{F} = \mathcal{O}$ and X is reduced the convergence in this topology means the following. A sequence of germs $\{f_{m,x}\}$ has the limit $f_x \in \mathcal{O}_x$ if and only if there is a neighborhood U of x, such that all f_m, $m = 1, 2, \ldots$, and f are defined in U and the sequence of functions $\{f_m\}$ converges to f uniformly on compact sets in U. We refer the reader to [GR1], Ch.2, §1, for the definition of the sequence topology in the general case.

We assume throughout the book that all complex spaces under consideration have *countable topology.* For any open set $U \subset X$ the vector space $\mathcal{F}(U)$ can be made into Fréchet space so that all restriction maps $\mathcal{F}(U) \to \mathcal{F}_x$ are continuous, where $x \in U$ and $\mathcal{F}_x$ is considered with the sequence topology. This uniquely determined Fréchet topology on $\mathcal{F}(U)$ is called *canonical.* If $\mathcal{F} = \mathcal{O}$ and X is reduced, then the canonical topology coincides with the topology of compact convergence, see [GR2], Ch.5, §6. Furthermore, the canonical Fréchet topology has the following important property.

Let $(\varphi, \tilde{\varphi}) : (X, \mathcal{O}_X) \to (Y, \mathcal{O}_Y)$ be a holomorphic mapping. For any open $V \subset Y$ the induced linear mapping of Fréchet spaces

$$\mathcal{O}_Y(V) \to \mathcal{O}_X(\varphi^{-1}(V)), \quad f \mapsto f \circ \varphi,$$

is continuous.

Proof By the Banach open mapping theorem it is enough to prove that the graph Γ of our linear mapping is closed. We have

$$\Gamma = \bigcap_{x \in \varphi^{-1}(V)} \Gamma_x ,$$

where

$$\Gamma_x := \{ (f, g) \in \mathcal{O}_Y(V) \times \mathcal{O}_X(\varphi^{-1}(V)) \mid g_x = \tilde{\varphi}_x(f_{\varphi(x)}) \}.$$

Since all restriction mappings to stalks and all homomorphisms $\tilde{\varphi}_x$ are continuous in the sequence topology, the sets Γ_x are closed and so is Γ. □

The composition $(\chi, \tilde{\chi}) : (X, \mathcal{O}_X) \to (Z, \mathcal{O}_Z)$ of two holomorphic mappings $(\varphi, \tilde{\varphi}) : (X, \mathcal{O}_X) \to (Y, \mathcal{O}_Y)$ and $(\psi, \tilde{\psi}) : (Y, \mathcal{O}_Y) \to (Z, \mathcal{O}_Z)$ is the composition of two morphisms of $\mathbb{C}$-ringed spaces. The underlying mapping $\chi : X \to Z$ is, of course, $\psi \circ \varphi$. The second component $\tilde{\chi} : \mathcal{O}_Z \to \chi_* \mathcal{O}_X$ is constructed as follows. The homomorphism of sheaves on Y

$$\tilde{\varphi} : \mathcal{O}_Y \to \varphi_* \mathcal{O}_X$$

induces the homomorphism of their direct images on Z

$$\psi_*(\tilde{\varphi}) : \psi_* \mathcal{O}_Y \to \psi_* \varphi_* \mathcal{O}_X = \chi_* \mathcal{O}_X,$$

and the composition $\psi_*(\tilde{\varphi}) \circ \tilde{\psi}$ is $\tilde{\chi}$.

An *isomorphism* or a *biholomorphic mapping* between two complex spaces is a holomorphic mapping $(\varphi, \tilde{\varphi}) : (X, \mathcal{O}_X) \to (Y, \mathcal{O}_Y)$ for which there exists a holomorphic mapping $(\psi, \tilde{\psi}) : (Y, \mathcal{O}_Y) \to (X, \mathcal{O}_X)$, such that

$$\psi \circ \varphi = \mathrm{id}_X, \quad \varphi \circ \psi = \mathrm{id}_Y, \quad \psi_*(\tilde{\varphi}) \circ \tilde{\psi} = \mathrm{id}_{\mathcal{O}_X} \quad \text{and} \quad \varphi_*(\tilde{\psi}) \circ \tilde{\varphi} = \mathrm{id}_{\mathcal{O}_Y}.$$

Clearly, $(\varphi, \tilde{\varphi})$ is biholomorphic if and only if φ is a homeomorphism and $\tilde{\varphi}_x$ is an isomorphism for every $x \in X$.

More generally, a holomorphic mapping $(\varphi, \tilde{\varphi}) : (X, \mathcal{O}_X) \to (Y, \mathcal{O}_Y)$ is called a *closed embedding* if $(\varphi, \tilde{\varphi})$ is a composition of a biholomorphic mapping onto a closed complex subspace $(Y', \mathcal{O}_{Y'})$ of $(Y, \mathcal{O}_Y)$ and the injection mapping $(\iota, \tilde{\iota}) : (Y', \mathcal{O}_{Y'}) \to (Y, \mathcal{O}_Y)$. (By definition, $\tilde{\iota}$ is the natural sheaf epimorphism $\mathcal{O}_Y \to \iota_*(\mathcal{O}_{Y'}) = \mathcal{O}_Y/\mathcal{J}$, where $\mathcal{J} \subset \mathcal{O}_Y$ is the ideal sheaf defining $(Y', \mathcal{O}_{Y'})$).

A holomorphic mapping of reduced complex spaces $(\varphi, \tilde{\varphi}) : (X, \mathcal{O}_X) \to (Y, \mathcal{O}_Y)$ is uniquely determined by the underlying mapping $\varphi : X \to Y$. Namely, $\tilde{\varphi}$ is the restriction of the canonical homomorphism $\mathcal{C}_Y \to \varphi_*\mathcal{C}_X$ to the subsheaf $\mathcal{O}_Y$. In order to simplify the notation, we shall quite often write X instead of $(X, \mathcal{O}_X)$ and $\varphi : X \to Y$ instead of $(\varphi, \tilde{\varphi}) : (X, \mathcal{O}_X) \to (Y, \mathcal{O}_Y)$, even if the complex spaces in question are not reduced. So, for example, we have a closed embedding red $X \to X$.

The set of all holomorphic mappings from X to Y is denoted by $\mathrm{Hol}(X, Y)$. There is a natural bijection between $\mathrm{Hol}(X, \mathbb{C}^n)$ and $\mathcal{O}(X)^n$. For this reason, sections of the structure sheaf $\mathcal{O}_X$ are often called *holomorphic functions* .

If $\varphi \in \mathrm{Hol}(X, Y)$ and $\psi \in \mathrm{Hol}(Y, Z)$, then the definitions show that

$$(f \circ \psi) \circ \varphi = f \circ (\psi \circ \varphi) \quad \text{for all} \quad f \in \mathcal{O}(Z).$$

An *automorphism* of a complex space X is an isomorphism of X onto X. The set of all automorphisms is denoted by $\mathrm{Aut}(X)$. Composition of holomorphic mappings gives a group structure to $\mathrm{Aut}(X)$. Let $\varphi \in \mathrm{Aut}(X)$. As we have seen above, $f \mapsto f \circ \varphi$ is a continuous linear operator in the Fréchet space $\mathcal{O}(X)$. It follows that $\mathrm{Aut}(X)$ has a natural representation as a group of invertible linear operators of $\mathcal{O}(X)$. The operator corresponding to $\varphi \in \mathrm{Aut}(X)$ is given by $f \mapsto f \circ \varphi^{-1}$.

The group $\mathrm{Aut}(X)$ is of central importance for this book. In Chapter 2 we provide $\mathrm{Aut}(X)$ with a natural topology (see §2.1) and study in detail those complex spaces, for which $\mathrm{Aut}(X)$ is a Lie group (see §§2.3, 2.6).

1.2 Lie group actions

Let X be a complex space with countable topology. Topological transformation groups of X, which are defined below, were first considered by W. Kaup [Ka1]. We start with an abstract group G. Given a homomorphism $\Phi : G \to \mathrm{Aut}(X)$, we shall often denote the automorphism $\Phi(g)$ of X simply by g. In particular, we shall write $f \circ g$ instead of $f \circ \Phi(g)$, where $f \in \mathcal{O}(V)$, $V \subset X$. One says that Φ defines an *action of* G (or a G*-action*) on X. The pair (G, Φ) is also called a *transformation group* of X. The normal subgroup $I := \mathrm{Ker}\, \Phi \lhd G$ is called the *ineffectivity kernel* of a G-action. A G-action and the corresponding transformation group (G, Φ) are said to be *effective* if and only if $I = \{e\}$. A G-action on a complex space $(X, \mathcal{O}_X)$ induces an action on the underlying topological space X. Since we do not assume that the complex space is reduced, the latter action may have a non-trivial ineffectivity kernel, even if the former action is effective. Suppose now that G is a topological group. The G-action defined by Φ is said to be *continuous* if for each open relatively compact $U \subset\subset X$ and for each open $V \subset X$:

(i) $W := W_{C,V} := \{g \in G \mid g \cdot C \subset V\}$, where $C = \overline{U}$, is open in G;

(ii) for any $f \in \mathcal{O}(V)$ the mapping

$$W \to \mathcal{O}(U), \qquad g \mapsto f \circ g|_U, \tag{*}$$

is continuous ($\mathcal{O}(U)$ carries the canonical Fréchet topology).

If these conditions are fulfilled, (G, Φ) is called a *topological transformation group* of the complex space X. Note that (i) means that the mapping of topological spaces $G \times X \to X$, $(g, x) \mapsto gx$, is continuous. For a continuous G-action I is a closed normal subgroup in G. The action is said to be *locally effective* if I is discrete.

Recall that a mapping α from a real analytic (resp. complex) manifold W to a Fréchet space F over $\mathbb{C}$ is called real analytic (resp. holomorphic) if for each point $w_0 \in W$ there exists an open coordinate neighborhood $N = N(w_0)$ and a real analytic (resp. holomorphic) coordinate system $t_1, \dots, t_n$ in N such that $t_i(w_0) = 0$ and for all $w \in N$ we have

$$\alpha(w) = \sum a_{i_1,\dots,i_n} t_1^{i_1}(w) \cdot \ldots \cdot t_n^{i_n}(w), \qquad \text{where} \qquad a_{i_1,\dots,i_n} \in F$$

and the convergence is absolute with respect to any continuous seminorm on F.

Let G be a real (resp. complex) Lie group. The G-action defined by Φ is said to be *real analytic* (resp. *holomorphic*) if it is continuous and, in addition, the mapping (*) in (ii) is a real analytic (resp. a holomorphic) mapping from W to the Fréchet space $\mathcal{O}(V)$. In this situation (G, Φ) is called a *real* (resp. *complex*) *(Lie) transformation group* of the complex space X. In fact, we shall prove in §1.6 that a continuous action of a Lie group on a complex space is always real analytic.

For a real analytic G-action I is a closed normal Lie subgroup in G. If G is a complex Lie group acting on X holomorphically then, in addition, I is a complex Lie subgroup (see §1.7 for the proof). In both cases the induced action of G/I on X is effective and belongs to the same category.

Since any holomorphic mapping between complex spaces determines in a canonical way a holomorphic mapping between their reductions, a G-action on a complex space X induces a G-action on red X. Furthermore, if the action on X is continuous, real analytic, or holomorphic then the induced action on red X enjoys the same property. For reduced complex spaces the above definitions can be simplified. Namely, a G-action on a reduced complex space X is just a mapping of sets $G \times X \to X$, $(g, x) \mapsto gx$, such that

$$g_1(g_2 x) = (g_1 g_2)x \qquad \text{for all} \qquad x \in X,\ g_1, g_2 \in G,$$

$$ex = x \qquad \text{for all} \qquad x \in X,$$

and gx depends holomorphically on x for any fixed $g \in G$. One can easily show that a G-action on a reduced complex space X is continuous (resp. real analytic, resp. holomorphic) in the sense of the general definition if and only if G is a topological group (resp. a real Lie group, resp. a complex Lie group) and the mapping $G \times X \to X$, $(g, x) \mapsto gx$, is continuous (resp. real analytic, resp. holomorphic).

Suppose we are given two G-actions of the same type (i.e. both continuous, real analytic or holomorphic) on complex spaces X and Y. Let $\Phi : G \to \mathrm{Aut}(X)$ and $\Psi : G \to \mathrm{Aut}(Y)$ be the corresponding group homomorphisms. A holomorphic mapping $\varphi : X \to Y$ is called a *morphism of G-actions* or a *G-equivariant mapping* if for every $g \in G$ the diagram

$$\begin{array}{ccc} X & \xrightarrow{\Phi(g)} & X \\ \varphi\downarrow & & \downarrow\varphi \\ Y & \xrightarrow{\Psi(g)} & Y \end{array}$$

commutes.

As we noted above, a G-action on X induces a G-action on red X. The identity mapping of the underlying spaces together with the residue epimorphism $\mathcal{O} \to \mathcal{O}/\mathcal{N} = \mathrm{red}\,\mathcal{O}$ defines a holomorphic mapping $\mathrm{red}\,X = (X, \mathrm{red}\,\mathcal{O}) \to X = (X, \mathcal{O})$, which is obviously a morphism of G-actions.

A biholomorphic G-equivariant mapping is also called an *isomorphism of G-actions*. Two G-actions, for which such a mapping exists, are said to be *isomorphic*.

Given a G-action on X, the subset $Gx := \{gx | g \in G\} \subset X$ is called the *orbit of $x \in X$*. The subgroup $G_x := \{g \in G | gx = x\}$ is called the *isotropy subgroup* of $x \in X$. Define a mapping of sets $\theta_x : G/G_x \to X$ by $\theta_x(g \cdot G_x) = gx$. Clearly, θ_x determines a bijection between G/G_x and the orbit Gx. A G-action is called *transitive* if $Gx = X$ for some $x \in X$ (and then, of course, for any $x \in X$). A reduced complex space X with a transitive G-action is a complex manifold, which is said to be *homogeneous with respect to G* or *G-homogeneous*. A complex manifold X is called *homogeneous* if it is homogeneous with respect to $\mathrm{Aut}(X)$. It should be noted that a homogeneous complex manifold may not admit a transitive Lie transformation group (see Concluding remarks).

If G is a topological group acting continuously on X then G_x is a closed subgroup and the quotient topology on G/G_x is Hausdorff. By the definition of this topology θ_x is a continuous mapping. Furthermore, if G is a Lie group and the action is real analytic, then G/G_x has a structure of a real analytic manifold and θ_x is a real analytic mapping to red X, viewed as a real analytic space.

If G is a complex Lie group and the action is holomorphic, then G_x is a complex Lie subgroup, G/G_x a complex manifold and θ_x a holomorphic mapping to red X. Moreover, G acts holomorphically on G/G_x and $\theta_x : G/G_x \to \mathrm{red}\,X$ is a morphism of actions. In particular, if X is a G-homogeneous complex manifold, then $\theta_x : G/G_x \to X$ is a biholomorphic G-equivariant mapping. With the help of this mapping one can identify X with the coset space G/H, where $H := G_x$ is the isotropy subgroup of some point $x \in X$. A representation $X = G/H$ is often called a *Klein form* of a homogeneous manifold X.

It is useful to generalize the notion of a homogeneous complex manifold in the following way. Let X be an irreducible complex space and G a complex Lie group acting holomorphically on X. One says that X is *almost homogeneous* with respect to G, if G has an open orbit $\Omega \subset X$. We shall see in §1.7 that an open orbit is unique and that $E := X - \Omega$ is a proper analytic subset in X.

Example 1 If G is a real Lie group acting on X with an open orbit, then such an orbit is, in general, not unique. For example, $G = \mathrm{SL}(2, \mathbb{R})$ acts on $\mathbb{P}_1 = \mathbb{C} \cup \{\infty\}$

by

$$(g, z) \mapsto \frac{az+b}{cz+d}, \qquad \text{where } z \in \mathbb{P}_1 \,, \; g = \begin{pmatrix} a & b \\ c & d \end{pmatrix} \in \mathrm{SL}(2, \mathbb{R}),$$

and has three orbits, two of them being open.

Example 2 In contrast with the theory of algebraic transformation groups, an orbit of a holomorphic group action is not necessarily open in its closure. Let Γ be a lattice of rank $2n$ in $\mathbb{C}^n$, $n > 1$, and $T = \mathbb{C}^n/\Gamma$ the complex torus defined by Γ. Fix $a \in \mathbb{C}^n$. The action of $G = \mathbb{C}$ on $\mathbb{C}^n$

$$(\zeta, z) \mapsto z + \zeta a, \qquad \text{where } \zeta \in \mathbb{C} \,, \; z \in \mathbb{C}^n,$$

induces a G-action on T. Let $a = \sum_{k=1}^{2n} \lambda_k \gamma_k$, where $\lambda_k \in \mathbb{R}$ and $\{\gamma_k\}$ is a basis of Γ. Suppose that the set of real numbers $\{\lambda_1, \ldots, \lambda_{2n}, 1\}$ is linearly independent over $\mathbb{Q}$. Then, by a theorem of Kronecker, the additive group $\Gamma + \mathbb{Z}a$ is dense in $\mathbb{C}^n$. It follows that each G-orbit on T is dense in T. On the other hand, any such orbit is an immersed manifold of real dimension 2.

The group actions considered above are sometimes called *global*. Later on, we shall be also concerned with the so called *local* actions, which we now define. Let G be a topological group and X a complex space. Denote by Π_X the collection of all pairs $\pi = (U_\pi, V_\pi)$, where U_π and V_π are open subsets in X such that $U_\pi \subset\subset V_\pi$. Suppose that for each $\pi \in \Pi_X$ we have an open neighborhood G_π of $e \in G$ and a mapping $\Phi_\pi : G_\pi \to \mathrm{Hol}(U_\pi, V_\pi)$. One says that the system $\{\Phi_\pi\}$ defines a *local (continuous) G-action* on X and calls $(G, \{\Phi_\pi\})$ a *local (topological) transformation group* of X if the following conditions are fulfilled:

(a) for all $g, h \in G_\pi$ such that $k := gh \in G_\pi$ we have

$$\Phi_\pi(g) \circ \Phi_\pi(h)|_{U_{\pi,h}} = \Phi_\pi(k)|_{U_{\pi,h}},$$

where $U_{\pi,h} := \{x \in U_\pi | \Phi_\pi(h)x \in U_\pi\}$;

(b) $\Phi_\pi(e) = \mathrm{id}$;

(c) for all $\pi, \rho \in \Pi_X$ and $g \in G_\pi \cap G_\rho$ we have

$$\Phi_\pi(g)|_{U_\pi \cap U_\rho} = \Phi_\rho(g)|_{U_\pi \cap U_\rho},$$

so that $gx := \Phi_\pi(g)x$ is independent of the choice of π with $x \in U_\pi, g \in G_\pi$;

(d) for any two open sets $U \subset\subset U_\pi$ and $V \subset V_\pi$ the set

$$W := W_{C,V} := \{g \in G_\pi | g \cdot C \subset V\}, \qquad \text{where} \qquad C = \overline{U},$$

is open in G_π and the mapping (*) is continuous for all $f \in \mathcal{O}(V)$.

If G is a real (resp. complex) Lie group and (*) is real analytic (resp. holomorphic), then the local G-action is called *real analytic* (resp. *holomorphic*). The local transformation group in question is then called a *local real* (resp. *complex*) *(Lie) transformation group* of the complex space X. We shall prove in §1.6 that any local action of a Lie group on a complex space is real analytic.

Two local actions $\{\Phi_\pi\}$, $\{\Phi'_\pi\}$ of the same group G are said to be *equivalent* if, for all $\pi \in \Pi_X$, the mappings $\Phi_\pi : G_\pi \to \mathrm{Hol}(U_\pi, V_\pi)$ and $\Phi'_\pi : G'_\pi \to \mathrm{Hol}(U_\pi, V_\pi)$

coincide on a subdomain of $G_\pi \cap G'_\pi$ containing e. In fact, (a) implies that in this case they coincide on the corresponding connected component of $G_\pi \cap G'_\pi$. Any global continuous action can be viewed as a local action with $G_\pi := \{g \in G | g \cdot \overline{U_\pi} \subset V_\pi\}$, but a local action is not necessarily equivalent to a global one. For example, a G-action on X defines a local G-action on any open subspace $U \subset X$. If G is connected then this local action is equivalent to a global one if and only if U is *G-invariant*, i.e., $g \cdot U = U$ for all $g \in G$.

To construct a local action one need not consider all possible pairs $\pi = (U_\pi, V_\pi)$ in Π_X. Let $\{\pi_i\}_{i \in J}$ be a collection of pairs $\pi_i = (U_i, V_i)$, $U_i \subset\subset V_i$, such that $\{U_i\}_{i \in J}$ is a covering of X. Assume that there are neighborhoods $G_i = G_{\pi_i}$ and mappings $\Phi_i = \Phi_{\pi_i}$ satisfying (a) - (d). In particular, (c) should be true for any two $\pi_i, \pi_j, i, j \in J$. Then the system $\{\Phi_i\}$ gives rise to a local G-action on X.

Suppose we have a continuous G-action on X, defined by a homomorphism $\Phi : G \to \mathrm{Aut}(X)$. Then any continuous homomorphism of topological groups $\phi : H \to G$ induces a continuous H-action on X given by $\Phi \circ \phi$. The same is true for local actions. Namely, if a local G-action on X is defined by a system of mappings $\Phi_\pi : G_\pi \to \mathrm{Hol}(U_\pi, V_\pi)$ then the induced local H-action is defined by $\Psi_\pi : H_\pi \to \mathrm{Hol}(U_\pi, V_\pi)$, where $H_\pi = \phi^{-1}(G_\pi)$, $\Psi_\pi = \Phi_\pi \circ \phi$. Note that if G and H are real Lie groups and the (local) G-action is real analytic, then the induced (local) H-action is also real analytic. If G and H are complex Lie groups, $\phi : G \to H$ is holomorphic and G acts on X holomorphically, then the induced (local) action of H is also holomorphic. The following observation is useful for the proof of the analyticity of a local G-action.

Let G be a real (resp. complex) Lie group. A local G-action on a complex space X is real analytic (resp. holomorphic) if and only if for each pair $U \subset\subset V$ and each $f \in \mathcal{O}(V)$ the mapping (*) is real analytic (resp. holomorphic) at $e \in G$. In particular, if two local actions $\{\Phi_\pi\}$ and $\{\Phi'_\pi\}$ are equivalent and $\{\Phi_\pi\}$ is real analytic (holomorphic) then $\{\Phi'_\pi\}$ has the same property.

Proof Let U, V, W be as in (d), $f \in \mathcal{O}(V)$ and $g_0 \in W$. If $g \in W$ is sufficiently close to g_0 then $g(\overline{U})$ is contained in an open subset $U' \subset\subset V$ and $f \circ g|_U = (f \circ h|_{U'}) \circ g_0|_U$, where $h = g g_0^{-1}$. By our hypothesis the mapping $h \mapsto f \circ h|_{U'} \in \mathcal{O}(U')$ is real analytic (resp. holomorphic) at e. Thus the mapping (*) is real analytic (resp. holomorphic) at g_0. □

1.3 One-parameter transformation groups

Suppose we have a local action of $G = \mathbb{R}$ on a complex space X. We retain the notation of the previous section. For $t \in G_\pi \subset \mathbb{R}$ and $x \in U_\pi \subset X$ we write $\gamma_t(x)$ instead of $\Phi_\pi(t)x$. A local transformation group $(\mathbb{R}, \{\Phi_\pi\})$, usually denoted by $\{\gamma_t\}$, is called a *local one-parameter transformation group*.

Proposition *Let $\{\gamma_t\}$ be a local one-parameter transformation group of a complex space X. Given two open subsets $U \subset\subset V$ in X, choose a positive ϵ so that $\gamma_t(\overline{U}) \subset V$ for all t, $|t| < \epsilon$. Then, for any $f \in \mathcal{O}(V)$, the curve*

$$(-\epsilon, \epsilon) \ni t \mapsto f \circ \gamma_t|_U \in \mathcal{O}(U)$$

is differentiable at $t = 0$.

Remark For X non-singular the result goes back to H.Cartan, see [Ca3]. The generalization to the singular case is due to W.Kaup, see [Ka1]. Our proof is based on the work of S.Bochner and D.Montgomery [BM1].

For the proof we shall need the following lemma.

Lemma *Let $D \subset D'$ be two domains in $\mathbb{C}^n$ and $\mathcal{J} \subset \mathcal{O}_{D'}$ a coherent ideal sheaf. Suppose we have a continuous mapping $(-\epsilon, \epsilon) \times D \to D'$,*

$$(t, z) \mapsto \gamma_t(z) = (f_1(t, z), \ldots, f_n(t, z)),$$

where $\gamma_0 = \mathrm{id}$ and all f_i are holomorphic functions of $z \in D$. Assume that in each domain $\Omega \subset\subset D$

$$f_i(t, \gamma_s(z)) \equiv f_i(s + t, z) \qquad \mod \mathcal{J}(\Omega)$$

for all i and for all sufficiently small s, t. Then, given $f \in \mathcal{O}(D')$, we have in each $\Omega \subset\subset D$

$$f(\gamma_t(z)) - f(z) \equiv t\big(p(z) + q(t, z)\big) \qquad \mod \mathcal{J}(\Omega),$$

where p, $q(t, \cdot) \in \mathcal{O}(\Omega)$ and $q(t, z) \to 0$ uniformly for $z \in \Omega$, as $t \to 0$.

Proof Let $z_1, .., z_n$ be coordinates in $\mathbb{C}^n$. From Cauchy's integral formula it follows that all derivatives $f_{ij} := \frac{\partial f_i}{\partial z_j}$ are continuous in $(-\epsilon, \epsilon) \times D$. Define

$$T_i(t, z) := \int_0^t f_i(\tau, z)\, d\tau \qquad (|t| < \epsilon,\ z \in D).$$

Suppose that the line segment $[z, w]$ is contained in D. Then

$$T_i(t, w) - T_i(t, z) = \sum_j h_{ij}(t, z, w)(w_j - z_j), \tag{1}$$

where

$$h_{ij}(t, z, w) = \int_0^1 \frac{\partial T_i}{\partial z_j}(t, z + \theta(w - z))\, d\theta = \int_0^1 d\theta \int_0^t f_{ij}\big(\tau, z + \theta(w - z)\big)\, d\tau \tag{2}$$

are holomorphic functions of z and w. On the other hand, since $\mathcal{J}(\Omega)$ is closed in $\mathcal{O}(\Omega)$ (see [GR2], Ch.5, §6), we have

$$T_i(t, \gamma_s(z)) - T_i(t, z) = \int_0^t f_i(\tau, \gamma_s(z))\, d\tau - \int_0^t f_i(\tau, z)\, d\tau \overset{\mod \mathcal{J}(\Omega)}{\equiv}$$

$$\equiv \int_0^t f_i(s + \tau, z)\, d\tau - \int_0^t f_i(\tau, z)\, d\tau = \int_s^{s+t} f_i(\tau, z)\, d\tau - \int_0^t f_i(\tau, z)\, d\tau =$$

$$= \int_t^{s+t} f_i(\tau, z)\, d\tau - \int_0^s f_i(\tau, z)\, d\tau = \int_0^s [f_i(\tau + t, z) - f_i(\tau, z)]\, d\tau, \tag{3}$$

where $z \in \Omega$ and s, t are sufficiently small. More precisely, we can find $\delta = \delta(\Omega) \in (0, \epsilon)$ so that for all $z \in \overline{\Omega}$, $|s| \leq \delta$, the points z and $\gamma_s(z)$ are in an open ball, which is contained in D. Now, take $w := \gamma_s(z)$ in (1) and compare the result with (3). We obtain a congruence which, after dividing by t, becomes:

$$\frac{1}{t}\int_0^s [f_i(\tau+t,z) - f_i(\tau,z)]\, d\tau \equiv \sum_j a_{ij}(s,t,z)(f_j(s,z)-z_j) \qquad \text{mod } \mathfrak{J}(\Omega), \quad (4)$$

where

$$a_{ij}(s,t,z) := \frac{1}{t}\, h_{ij}(t,z,\gamma_s(z)).$$

Since f_{ij} are continuous and $f_{ij}(0,z) = \delta_{ij}$, it follows from (2) that

$$\lim\nolimits_{t\to 0} a_{ij}(s,t,z) = \delta_{ij}.$$

Observe that for all $z \in \overline{\Omega}$, $\theta \in [0,1]$ and $|s| \le \delta$ the point $z + \theta(\gamma_s(z) - z)$ remains in a compact set in D. Thus, again by virtue of (2), we can find $t_0 > 0$ so that for all $z \in \overline{\Omega}$, $s \in [-\delta,\delta]$ the matrix $(a_{ij}(s,t_0,z))$ is uniformly close to (δ_{ij}) and, in particular, invertible. Therefore, (4) yields:

$$f_i(s,z) - z_i \equiv \sum_j \frac{1}{t_0} b_{ij}(s,t_0,z) \int_0^s [f_j(\tau+t_0,z) - f_j(\tau,z)]\, d\tau \qquad \text{mod } \mathfrak{J}(\Omega),$$

where (b_{ij}) is the inverse matrix for (a_{ij}). This relation can be rewritten in the form

$$f_i(s,z) - z_i \equiv s\big(p_i(z) + q_i(s,z)\big) \qquad \text{mod } \mathfrak{J}(\Omega),$$

where

$$p_i(z) := \frac{1}{t_0}\sum_j b_{ij}(0,t_0,z)\big(f_j(t_0,z) - z_j\big)$$

and

$$q_i(s,z) := -p_i(z) + \frac{1}{t_0}\sum_j b_{ij}(s,t_0,z)\frac{1}{s}\int_0^s [f_j(\tau+t_0,z) - f_j(\tau,z)]\, d\tau \quad \to 0$$

uniformly for $z \in \Omega$ as $s \to 0$. This proves the lemma for $f = z_i$.

Now, for any $f \in \mathcal{O}(D')$ we have

$$f(w) - f(z) = \sum_j k_j(z,w)(w_j - z_j), \qquad (5)$$

where

$$k_j(z,w) = \int_0^1 \frac{\partial f}{\partial z_j}\big(z + \theta(w-z)\big)\, d\theta,$$

provided the line segment $[z,w]$ is contained in D'. Substituting $w = \gamma_s(z), z \in \Omega, |s| < \delta$ into (5), we obtain

$$f(\gamma_s(z)) - f(z) = \sum_j k_j(z,\gamma_s(z))(f_j(s,z) - z_j) \equiv s\big(p(z) + q(s,z)\big) \qquad \text{mod } \mathfrak{J}(\Omega),$$

where

$$p(z) := \sum_j \frac{\partial f}{\partial z_j}(z)\, p_j(z)$$

and

$$q(z,s) := \sum_j k_j(z,\gamma_s(z))q_j(s,z) + \sum_j [k_j(z,\gamma_s(z)) - \frac{\partial f}{\partial z_j}]p_j(z).$$

Suppose $s \to 0$. Then $q_j(z,s) \to 0$ and $k_j(z,\gamma_s(z)) \to \frac{\partial f}{\partial z_j}$ uniformly for $z \in \Omega$. Since $p_j(z)$ are bounded, it follows that $q(z,s) \to 0$ uniformly for $z \in \Omega$. □

Proof of the proposition Since $\overline{U}$ is compact, there exist two finite open coverings $\{U_i\}, \{V_i\}$, $i = 1,\dots,d$, of $\overline{U}$ such that $U_i \subset\subset V_i \subset V$ for all i. We can always pick a smaller ϵ so that $\gamma_t(\overline{U_i}) \subset V_i$ for all i and for all t, $|t| < \epsilon$. Without loss of generality we may assume that for each i the open subspace V_i is a complex model space defined by an ideal sheaf in a polydisk. We may also assume that U_i is a complex model space defined by the restriction of the same sheaf to a compactly embedded polydisk. Let $U' := \cup_{i=1}^d U_i$. The commutative diagram

$$\begin{array}{ccc} \mathcal{O}(V) & \longrightarrow & \mathcal{O}(U') \\ \downarrow & & \downarrow \iota \\ \oplus_{i=1}^d \mathcal{O}(V_i) & \longrightarrow & \oplus_{i=1}^d \mathcal{O}(U_i), \end{array}$$

where the arrows denote the restriction mappings, shows that for each $f \in \mathcal{O}(V)$

$$\iota(f \circ \gamma_t|_{U'}) = \Big(f|_{V_1} \circ (\gamma_t|_{U_1}), \dots, f|_{V_d} \circ (\gamma_t|_{U_d})\Big).$$

Suppose we know that each curve $t \mapsto f|_{V_i} \circ (\gamma_t|_{U_i}) \in \mathcal{O}(U_i)$ is differentiable at $t = 0$. Then, since ι is a closed embedding (see [GR2], Ch.5, §6), the same is true for the curve $t \mapsto f \circ \gamma_t|U' \in \mathcal{O}(U')$. Finally, the curve $t \mapsto f \circ \gamma_t|U \in \mathcal{O}(U)$ is the composition of the previous one with the restriction mapping $\mathcal{O}(U') \to \mathcal{O}(U)$. Therefore, it is enough to prove our statement for each pair (U_i, V_i).

Let $D \subset\subset D'$ be two Stein domains in $\mathbb{C}^n$, Y' a complex model space in D' defined by an ideal sheaf $\mathfrak{J} \subset \mathcal{O}_{D'}$, and Y a complex model space in D, defined by $\mathfrak{J}|_D$. The above argument reduces our proposition to the case $X = V = Y'$, $U = Y$. We shall write γ_t^* instead of $\gamma_t|_Y$ and view γ_t^* as a holomorphic mapping from Y to $D' \subset \mathbb{C}^n$. Denote by $f_i^*(t,\cdot) \in \mathcal{O}_Y(Y)$ the functions defining γ_t^*.

We want to extend f_i^* holomorphically to D, so that the extended functions depend continuously on t. For this we need the following result.

Selection Theorem (see [BarGr], [Mi]) *Let E, F be Fréchet spaces and $\phi : E \to F$ a continuous linear mapping of E onto F. Then there exists a continuous (in general, non-linear) mapping $\psi : F \to E$ such that $\psi(f) \in \phi^{-1}(f)$ for every $f \in F$.*

In our case $E := \mathcal{O}_D(D)$, $F := \mathcal{O}_Y(Y)$, and ϕ is the restriction mapping. In order to show that $\phi : E \to F$ is onto we can apply Theorem B of H.Cartan, see e.g. [GR2], Ch.4, §1. Namely, we have the exact sequence of sheaves

$$0 \to \mathfrak{J} \to \mathcal{O}_D \to \mathcal{O}_D/\mathfrak{J} \to 0$$

and ϕ is the mapping

$$\mathcal{O}_D(D) \to (\mathcal{O}_D/\mathfrak{J})\,(D) = \mathcal{O}_Y(Y)$$

from the corresponding exact cohomology sequence. Since $H^1(D, \mathcal{J}) = 0$, this mapping is onto. Therefore, by the Selection Theorem, there exist continuous curves

$$t \mapsto f_i(t, z) \in \mathcal{O}_D(D)$$

such that $f_i|_Y = f_i^*$ for all t. Since $\gamma_0^* = \mathrm{id}$ we can choose f_i so that $f_i(0, z) = z_i$. Denote again by $\gamma_t : D \to \mathbb{C}^n$ the mapping defined by $f_i(t, z)$. Clearly, $\gamma_t|_Y = \gamma_t^*$. We can replace D by a (not necessarily Stein) neighborhood of Y and take a smaller ϵ, so that $\gamma_t(D) \subset D'$ if $|t| < \epsilon$. Then, by our construction, $f \in \mathcal{J}(D')$ implies $f \circ \gamma_t \in \mathcal{J}(D)$. The group property of $\{\gamma_t|_Y\}$ shows that, for $\Omega \subset\subset D$ and for small s, t,

$$f_i(t, \gamma_s(z)) \equiv f_i(s + t, z) \mod \mathcal{J}(\Omega).$$

It follows from the lemma that, for any $f \in \mathcal{O}_{D'}(D')$, the curve

$$t \mapsto (f|_{Y'}) \circ (\gamma_t|_Y) = (f \circ \gamma_t)|_Y \in \mathcal{O}_Y(Y)$$

is differentiable at $t = 0$. Since the restriction mapping $\mathcal{O}_{D'}(D') \to \mathcal{O}_{Y'}(Y')$, $f \mapsto f|_{Y'}$, is onto, the proof is complete. □

1.4 Vector fields

Let R be a commutative algebra with identity element over a field, M a unitary R-module. A *derivation of R with values in M* is a linear mapping $\partial : R \to M$, satisfying the Leibniz rule

$$\partial(fg) = f\partial(g) + g\partial(f) \qquad \text{for all } f, g \in R.$$

The set of all derivations of R with values in M will be denoted by $\mathrm{Der}(R, M)$. If $\partial \in \mathrm{Der}(R, M)$ and $h \in R$, then the mapping $R \to M$, $f \mapsto h\partial(f)$, is again a derivation, i.e., $\mathrm{Der}(R, M)$ is an R-module. For $M = R$ we write $\mathrm{Der}(R)$ instead of $\mathrm{Der}(R, R)$. Elements of this R-module are called *derivations of R*. If $\partial, \partial' \in \mathrm{Der}(R)$ then the bracket

$$[\partial, \partial'] := \partial \circ \partial' - \partial' \circ \partial$$

is again a derivation of R. Considered with this operation, $\mathrm{Der}(R)$ is a Lie algebra.

Let X be a complex space with the structure sheaf $\mathcal{O}(= \mathcal{O}_X)$ and let $\mathcal{F}$ be a coherent analytic sheaf on X. A *derivation on X with values in $\mathcal{F}$* is a continuous mapping of sheaves $A : \mathcal{O} \to \mathcal{F}$ such that A_x is a derivation of $\mathcal{O}_x$ with values in $\mathcal{F}_x$ for all $x \in X$. For any subset $Y \subset X$ the induced mapping $A_Y : \mathcal{O}(Y) \to \mathcal{F}(Y)$ belongs to $\mathrm{Der}(\mathcal{O}(Y), \mathcal{F}(Y))$. A derivation on X with values in $\mathcal{O}_X$ is called a *derivation of $\mathcal{O}_X$* or a *(holomorphic) vector field on X*.

For domains in $\mathbb{C}^n$ this definition gives us classical objects.

Given n holomorphic functions $a_1(z), \ldots, a_n(z)$ in a domain $D \subset \mathbb{C}^n$, there exists a unique vector field A in D, such that $A_D(z_i) = a_i(z)$. The induced mapping $A_D : \mathcal{O}(D) \to \mathcal{O}(D)$ is of the form

$$A_D = \sum_j a_j(z) \frac{\partial}{\partial z_j}.$$

Proof Define $A : \mathcal{O} \to \mathcal{O}$ by

$$f_x \mapsto \sum_j a_{j,x} \left(\frac{\partial f}{\partial z_j}\right)_x ,$$

where f is a holomorphic function in a neighborhood of $x \in D$. It is easy to check that A is a vector field with the required property.

On the other hand, assume that for a vector field A we have $A_D(z_i) = 0$. Then $A_x(z_{i,x}) = 0$ at any point $x \in D$ and $A_Y(z|_Y) = 0$ for any $Y \subset D$. Let $\Delta \subset D$ be a polydisk and $f \in \mathcal{O}(\Delta)$. Fix a point $x \in \Delta$. Then

$$f = f(x) + \sum_j f_j \cdot (z_j - z_j(x)),$$

where

$$f_j(z) = \int_0^1 \frac{\partial f}{\partial z_j}(x + \theta(z - x))d\theta.$$

From this we obtain that $A_\Delta(f)\ (x) = 0$. Since x is an arbitrary point of Δ, it follows that $A_\Delta = 0$. Since Δ is an arbitrary polydisk in D, we conclude that $A = 0$. □

Let $U \subset X$ be an open subset. Denote by $\mathrm{D}_U(\mathcal{O}, \mathcal{F})$ the $\mathcal{O}(U)$-module of derivations on U with values in $\mathcal{F}|_U$. Note that $\mathrm{D}_U(\mathcal{O}) := \mathrm{D}_U(\mathcal{O}, \mathcal{O})$ is a Lie algebra. If $V \subset U$ is another open subset, we have the natural restriction mapping

$$r_V^U : \mathrm{D}_U(\mathcal{O}, \mathcal{F}) \to \mathrm{D}_V(\mathcal{O}, \mathcal{F}).$$

The system $\{\mathrm{D}_U(\mathcal{O}, \mathcal{F}),\ r_V^U\}$ is a presheaf satisfying the sheaf conditions of Serre, see e.g. [GR3], Annex, §1.3. The associated sheaf is denoted by $\mathcal{D}er(\mathcal{O}, \mathcal{F})$ and is called the *sheaf of germs of derivations on X with values in* $\mathcal{F}$. In case $\mathcal{F} = \mathcal{O}$ we obtain a sheaf of Lie algebras, which is denoted by $\mathcal{D}er(\mathcal{O})$ or $\mathcal{T}(= \mathcal{T}_X)$ and is called the *sheaf of germs of derivations of* $\mathcal{O}$ or the *sheaf of germs of local (holomorphic) vector fields on X.*

If X is a manifold, then $\mathcal{T}_X$ is a locally free sheaf and the corresponding vector bundle is the holomorphic tangent bundle. For this reason, $\mathcal{T}_X$ is often called the *tangent sheaf* (even if X has singularities). One can also construct the *sheaf* $\Omega(= \Omega_X)$ *of germs of local holomorphic 1-forms on X* together with the canonical derivation $d : \mathcal{O} \to \Omega$, called the *exterior derivative.* The pair (Ω, d) is uniquely determined up to an isomorphism by the following property: for any coherent sheaf $\mathcal{F}$ on X the homomorphism

$$\mathcal{H}om(\Omega, \mathcal{F}) \to \mathcal{D}er(\mathcal{O}, \mathcal{F}), \quad \xi \mapsto \xi \circ d,$$

is bijective. Let $(Y, \mathcal{O}_D/\mathcal{J}|_Y)$ be a complex model space in a domain $D \subset \mathbb{C}^n$ defined by an ideal sheaf $\mathcal{J} = f_1\mathcal{O}_D + \ldots + f_s\mathcal{O}_D$. Then

$$\Omega_Y := \mathcal{O}_D^n/\mathcal{L}|_Y,$$

where

$$\mathcal{L} = \sum_{k=1}^s \left(\frac{\partial f_k}{\partial z_1}, \ldots, \frac{\partial f_k}{\partial z_n}\right) \mathcal{O}_D + \sum_{k=1}^s f_k\mathcal{O}_D^n,$$

and the exterior derivative is given by

$$df_x := (\frac{\partial f_x}{\partial z_1}, \dots, \frac{\partial f_x}{\partial z_n}) \bmod \mathcal{L}_x, \quad f_x \in \mathcal{O}_{D,x}, \ x \in D$$

(see [GR1], Ch.3, §4). Since any complex space $(X, \mathcal{O}_X)$ is locally isomorphic to a complex model space, the sheaf Ω is coherent. The isomorphism $\mathcal{T} \simeq \mathcal{H}om(\Omega, \mathcal{O})$ shows that the sheaf $\mathcal{T}$ is also coherent.

Now we can prove the following important lemma.

Extension Lemma *Let $D \subset \mathbb{C}^n$ be a Stein domain and let Y be a complex model space in D defined by a coherent ideal sheaf $\mathcal{J} \subset \mathcal{O}_D$. Given $A \in \mathcal{T}_Y(Y)$, there exists $A' \in \mathcal{T}_D(D)$ such that the diagram*

$$\begin{array}{ccc} \mathcal{O}_D|_Y & \xrightarrow{A'|_Y} & \mathcal{O}_D|_Y \\ \downarrow & & \downarrow \\ \mathcal{O}_Y & \xrightarrow{A} & \mathcal{O}_Y \end{array}$$

commutes (and, in particular, $A(\mathcal{J}) \subset \mathcal{J}$). *Proof* First of all, consider the situation at one point. Let I be an ideal in $R = \mathcal{O}_{\mathbb{C}^n,0}$ and $\alpha : R \to R/I$ the natural epimorphism. Then, given $\partial \in \mathrm{Der}(R/I)$ there exists $\partial' \in \mathrm{Der}(R)$ such that $\partial \circ \alpha = \alpha \circ \partial'$. In fact, it is enough to define ∂' on the coordinate functions z_i. This can be done arbitrarily, the only condition on $w_i := \partial'(z_i)$ being $\alpha(w_i) = \partial(\alpha(z_i))$. Observe that ∂' automatically preserves I. Conversely, denote by $\mathrm{Der}_I(R)$ the set of all derivations of R preserving I. Then each $\partial' \in \mathrm{Der}_I(R)$ induces an element of $\mathrm{Der}(R/I)$. The induced derivation is equal to zero if and only if $\partial' \in I \cdot \mathrm{Der}(R) \subset \mathrm{Der}_I(R)$. Therefore, we have an exact sequence

$$0 \to I \cdot \mathrm{Der}(R) \to \mathrm{Der}_I(R) \to \mathrm{Der}(R/I) \to 0.$$

Denote by $\mathcal{T}_{D,Y}$ the subsheaf of $\mathcal{T}_D$, consisting of those germs of derivations of $\mathcal{O}_D$, which preserve $\mathcal{J}$. Geometrically speaking, $\mathcal{T}_{D,Y}$ is the sheaf of germs of holomorphic vector fields, which are tangent to Y. Then the above exact sequence implies that

$$\mathcal{T}_{D,Y}/(\mathcal{J} \cdot \mathcal{T}_D)\,|_Y \ = \ \mathcal{T}_Y.$$

Since D is a Stein domain, it follows from Theorem B that the mapping

$$\mathcal{T}_{D,Y}(D) \to \mathcal{T}_{D,Y}/(\mathcal{J} \cdot \mathcal{T}_D)(D) = \mathcal{T}_Y(Y)$$

is an epimorphism. □

1.5 Infinitesimal transformations

As we have shown in §1.3, a local one-parameter transformation group $\{\gamma_t\}$ of a complex space X is differentiable. More precisely, for any two open subsets $U \subset\subset V \subset X$ and any $f \in \mathcal{O}(V)$ the derivative $\frac{d}{dt}\big|_{t=0} (f \circ \gamma_t|_U)$ is an element of $\mathcal{O}(U)$.

Now, let $f_x \in \mathcal{O}_x$, $x \in X$. Take two open subsets $U \subset\subset V$ so that $x \in U$ and f_x is represented by a holomorphic function $f \in \mathcal{O}(V)$.

The mapping

$$f_x \mapsto \left\{ \frac{d}{dt}\bigg|_{t=0} (f \circ \gamma_t|_U) \right\}_x \tag{1}$$

is a well-defined derivation of $\mathcal{O}_x$. The resulting sheaf mapping $\mathcal{O}_X \to \mathcal{O}_X$ is a holomorphic vector field on X.

Proof Let V_1, V_2 and $U \subset\subset V_1 \bigcap V_2$ be three open sets containing x. If $f_1 \in \mathcal{O}(V_1)$, $f_2 \in \mathcal{O}(V_2)$ and $(f_1)_x = (f_2)_x$, then $f_1 \circ \gamma_t|_U = f_2 \circ \gamma_t|_U$ for all sufficiently small t. Therefore the right hand side in (1) depends only on f_x. The mapping thus obtained is obviously a derivation of the stalk algebra $\mathcal{O}_x$. Since x in (1) can be an arbitrary point of U and the subset $\{f_x \mid x \in U\} \subset \mathcal{O}$ is open by definition of the sheaf topology, the sheaf mapping $\mathcal{O}_X \to \mathcal{O}_X$ defined by (1) is continuous. □

The holomorphic vector field arising from (1) is called the *infinitesimal transformation of* $\{\gamma_t\}$.We shall also consider *holomorphic* local one-parameter transformation groups $\{\gamma_t\}$, i.e., holomorphic local actions of $G = \mathbb{C}$. The infinitesimal transformation in this case is defined along the same lines. Namely, t in (1) is a complex variable and the derivative exists because the action is holomorphic.

Theorem *Let $A \in \mathcal{T}_X(X)$ be a holomorphic vector field. There exists a holomorphic local one-parameter transformation group $\{\gamma_t\}$ of X, for which A is the infinitesimal transformation. Given two open subsets $U \subset\subset V$ in X there exists $\epsilon = \epsilon(U, V) > 0$ such that*

$$f \circ \gamma_t|_U = (\exp tA_U)(f|_U) \tag{2}$$

for all $f \in \mathcal{O}(V)$ and $t \in \mathbb{C}$, $|t| < \epsilon$. If $\{\delta_t\}$ is a local one-parameter transformation group of X (a local $\mathbb{R}$-action) with the same infinitesimal transformation, then $\{\delta_t\}$ is equivalent to the restriction of $\{\gamma_t\}$ to $\mathbb{R}$.

One says that $A \in \mathcal{T}_X(X)$ *generates* the holomorphic local one-parameter transformation group $\{\gamma_t\}$. The same terminology is used in the real case.

Corollary *Any local one-parameter transformation group $\{\delta_t\}$ of a complex space X is real analytic.*

Proof of the corollary Let A be the infinitesimal transformation of $\{\delta_t\}$ and let $\{\gamma_t\}$ be the holomorphic local one-parameter transformation group of X generated by A. Since $\{\delta_t\}$ is equivalent to the restriction of $\{\gamma_t\}$ to $\mathbb{R}$, our statement follows from the remark at the end of §1.2. □

Lemma *Let $\{\gamma_t\}$ be a local one-parameter transformation group of X with infinitesimal transformation A. Then, for any pair of open subsets $U \subset\subset V$ of X and $f \in \mathcal{O}(V)$ we have*

$$\frac{d^n}{dt^n}\bigg|_{t=0} (f \circ \gamma_t|_U) = A_U^n(f|_U) \tag{3}$$

and

$$A_U^n(f \circ \gamma_t|_U) = A_V^n(f) \circ \gamma_t|_U, \tag{4}$$

where $t \in \mathbb{R}$, $|t| < \epsilon = \epsilon(U, V)$.

Proof We begin with (3). The case $n = 0$ is trivial. For $n = 1$ the equality (3) follows from the definition of the infinitesimal transformation. Let V' be an open subset in X such that $U \subset\subset V' \subset\subset V$. Then $\gamma_s(V') \subset V$ and $f \circ \gamma_s|_{V'} \in \mathcal{O}(V')$ for all sufficiently small s. Assume that (3) is true for some n and replace the open set V by its subset V' and the function f by the transformed function $f \circ \gamma_s|_{V'}$. We obtain:

$$\left.\frac{d^n}{dt^n}\right|_{t=s} (f \circ \gamma_t|_U) = \left.\frac{d^n}{dt^n}\right|_{t=0} (f \circ \gamma_{t+s}|_U) =$$

$$= \left.\frac{d^n}{dt^n}\right|_{t=0} ((f \circ \gamma_s|_{V'}) \circ \gamma_t|_U) = A_U^n\, (f \circ \gamma_s|_U).$$

Therefore

$$\left.\frac{d^{n+1}}{dt^{n+1}}\right|_{t=o} (f \circ \gamma_t|_U) = \left.\frac{d}{ds}\right|_{s=0} \left.\frac{d^n}{dt^n}\right|_{t=s} (f \circ \gamma_t|_U) = \left.\frac{d}{ds}\right|_{s=0} A_U^n(f \circ \gamma_s|_U) =$$

$$= A_U^n \left.\frac{d}{ds}\right|_{s=0} (f \circ \gamma_s|_U) = A_U^n A_U(f|_U) = A_U^{n+1}(f|_U).$$

This proves (3) for all n.

We now apply (3) to prove (4). Namely, for s, t small enough we have

$$A_U^n(f \circ \gamma_t|_U) = \left.\frac{d^n}{ds^n}\right|_{s=0} ((f \circ \gamma_t|_{V'}) \circ \gamma_s|_U) = \left.\frac{d^n}{ds^n}\right|_{s=0} ((f \circ \gamma_s|_{V'}) \circ \gamma_t|_U) =$$

$$= \left\{ \left.\frac{d^n}{ds^n}\right|_{s=0} (f \circ \gamma_s|_{V'}) \right\} \circ \gamma_t|_U = A_{V'}^n(f|_{V'}) \circ \gamma_t|_U = A_V^n(f) \circ \gamma_t|_U. \qquad \square$$

Proof of the theorem Let $\{\gamma_t\}$ be a holomorphic local one-parameter transformation group of X with infinitesimal transformation $A \in \mathcal{T}_X(X)$. For $U \subset\subset V$ define $\epsilon = \epsilon(U, V)$ so that $\gamma_t(\overline{U}) \subset V$ for all $t \in \mathbb{C}$, $|t| < \epsilon$. Given $f \in \mathcal{O}(V)$ there exist holomorphic functions $a_n \in \mathcal{O}(U)$ such that

$$f \circ \gamma_t|_U \;=\; \sum_0^\infty \frac{a_n t^n}{n!} \;\in \mathcal{O}(U)$$

and the series converges for all $t \in \mathbb{C}$, $|t| < \epsilon$, since the mapping $t \mapsto f \circ \gamma_t|_U$ is holomorphic in the open disk $|t| < \epsilon$. By (3) we have

$$a_n \;=\; A_U^n(f|_U), \quad n = 0, 1, 2, \ldots\ .$$

This implies (2).

We now prove the existence of $\{\gamma_t\}$. Assume at first that X is a complex model space in a Stein domain $D \subset \mathbb{C}^n$, defined by a coherent ideal sheaf $\mathcal{J} \subset \mathcal{O}_D$. The Extension Lemma of §1.4 tells us that any holomorphic vector field $A \in \mathcal{T}_X(X)$ is a restriction of a holomorphic vector field $A' \in \mathcal{T}_D(D)$ which preserves $\mathcal{J}$. Denote by

z_i holomorphic coordinates in D and let $a_i(z) := A'_D(z_i)$. The system of differential equations

$$\dot{z}_i = a_i(z), \qquad i = 1, 2, \ldots, n,$$

has the unique solution

$$z_i = \varphi_i(t, z_0), \qquad i = 1, 2, \ldots, n,$$

satisfying the initial conditions

$$\varphi_i(0, z_0) = (z_0)_i,$$

where $z_0 \in D$. The mapping

$$z \mapsto \gamma_t(z) := (\varphi_1(t, z), \ldots, \varphi_n(t, z))$$

is holomorphic for all $z \in D_1 \subset\subset D$ and $|t| < \epsilon = \epsilon(D_1)$. Clearly, $\{\gamma_t\}$ is a holomorphic local one-parameter transformation group of D with infinitesimal transformation A'. Since A' preserves $\mathfrak{J}$, it follows that $\{\gamma_t\}$ acts on X.

Now, in the general case, there exists an open covering $\{V_i\}_{i \in J}$ of X such that each $\overline{V}_i$ is compact and each V_i is isomorphic to a complex model space of the above kind. Therefore, for each $i \in J$ our construction yields a holomorphic local one-parameter transformation group $\{\gamma_{i,t}\}$ on V_i with infinitesimal transformation $A|_{V_i}$.

Recall that X is always assumed to have countable topology. Consequently X is a paracompact topological space so that one can apply the well-known shrinking procedure (see e.g. [Ke], p.171). As a result, we obtain a covering $\{U_i\}_{i \in J}$ of X with the same index set J such that $U_i \subset\subset V_i$ for all $i \in J$.

Denote by Δ_ϵ the disk of radius ϵ with center 0 in $\mathbb{C}$. For each $i \in J$ we have a positive number $\epsilon_i > 0$ and a mapping

$$\Phi_i : \Delta_{\epsilon_i} \to \mathrm{Hol}(U_i, V_i), \qquad \Phi_i(t) = \gamma_{i,t}.$$

By (2) a holomorphic local one-parameter transformation group is uniquely determined by its infinitesimal transformation. Therefore Φ_i, $i \in J$, are compatible, i.e.,

$$\Phi_i(t)\Big|_{U_i \cap U_j} = \Phi_j(t)\Big|_{U_i \cap U_j}, \qquad |\, t \,| < \min\,(\epsilon_i, \epsilon_j),$$

for all $i, j \in J$. This shows that the system $\{\Phi_i\}_{i \in J}$ defines a holomorphic local one-parameter transformation group of X.

We still have to prove the last statement of the theorem. In order to do this choose V' as in the above lemma. Applying (2) to $\{\gamma_t\}$ and (4) to $\{\delta_t\}$, we obtain

$$(f \circ \gamma_s|_{V'}) \circ \delta_t|_U = \{(\exp sA_{V'})\,(f|_{V'})\} \circ \delta_t|_U = \sum_0^\infty \frac{s^n}{n!} A^n_{V'}(f|_{V'}) \circ \delta_t|_U =$$

$$= \sum_0^\infty \frac{s^n}{n!} A^n_U(f \circ \delta_t|U) = (\exp sA_U)\,(f \circ \delta_t|_U) = (f \circ \delta_t|_{V'}) \circ \gamma_s|_U\,,$$

where s and t are real and small enough. Since γ_s and δ_t commute, the mappings

$$\nu_t = \delta_t \circ \gamma_{-t} \in \mathrm{Hol}(U, V), \quad |t| < \epsilon = \epsilon(U, V),$$

make up a local one-parameter transformation group. Clearly, the infinitesimal transformation of $\{\nu_t\}$ is 0. Therefore, applying the same argument as in the above lemma, we obtain

$$\frac{d}{dt}(f \circ \nu_t|_U) = 0 \qquad \text{or, equivalently,} \qquad f \circ \nu_t|_U = f|_U$$

for all sufficiently small t. It follows that $\{\gamma_t\}$ and $\{\delta_t\}$ are equivalent. □

Remark The proof shows that a local one-parameter transformation group with real parameter can be extended to a holomorphic local one-parameter transformation group. However, for a global one-parameter transformation group $\{\gamma_t\}_{t\in\mathbb{R}}$ a global holomorphic extension may not exist (for example, $X = \{z \in \mathbb{C} \mid \operatorname{Im} z > 0\}$, $\gamma_t(z) = z + t$).

The following useful fact is a consequence of the proof.

Proposition *If X is compact then each holomorphic vector field $A \in \mathcal{T}_X(X)$ generates a global one-parameter transformation group $\{\gamma_t\}$ of X.*

Proof We may assume that J is a finite set. Letting $\epsilon = \min_{i\in J}\epsilon_i$ we obtain a mapping

$$\Phi: \Delta_\epsilon \to \operatorname{Aut}(X),$$

such that

$$\Phi(t)|_{U_i} = \Phi_i(t)|_{U_i}$$

for all $i \in J$. It is clear that Φ is a local homomorphism from $\mathbb{C}$ to $\operatorname{Aut}(X)$, i.e., $\Phi(t+s) = \Phi(t)\circ\Phi(s)$, when $|t|, |s|, |t+s| < \epsilon$. Extend Φ to the whole group $\mathbb{C}$ by

$$\Phi(t) = \underbrace{\Phi(\frac{t}{N})\circ\ldots\circ\Phi(\frac{t}{N})}_{N \text{ times}}, \qquad \text{where } |t| < N\epsilon.$$

A straightforward verification shows that this extension is a well-defined homomorphism.

1.6 Analyticity of Lie group actions

Let G be a Lie group, $\mathfrak{g}$ the Lie algebra of G, and $\exp : \mathfrak{g} \to G$ the exponential mapping. Suppose we have a local G-action on a complex space X. Then, for any $\mathrm{A} \in \mathfrak{g}$, the homomorphism $\mathbb{R} \ni t \mapsto \exp(t\mathrm{A}) \in G$ gives rise to a local one-parameter transformation group of X, which will be denoted by $\{\gamma_{\mathrm{A},t}\}$. We have seen in the preceding section that each $\{\gamma_{\mathrm{A},t}\}$ extends to a holomorphic local one-parameter transformation group. We use for the latter the same notation $\{\gamma_{\mathrm{A},t}\}$, where $t \in \mathbb{C}$.

Theorem *A local action of a Lie group G on a complex space X is real analytic.*

Proof Let

$$\Delta_r^k := \{s = (s_1,\ldots,s_k) \in \mathbb{C}^k \mid |s_j| < r,\ j = 1,2,\ldots,k\}$$

be the polydisk in $\mathbb{C}^k$. Given two open subsets $U \subset\subset V \subset X$ choose $\epsilon > 0$ so that

$$\gamma_{A,t}(U) \subset\subset V \qquad \text{for all} \qquad t \in \Delta^1_\epsilon .$$

In this situation we claim:

(*) *if a series*

$$f(s) := \sum f_{i_1,\dots,i_k}\, s_1^{i_1} \cdot \ldots \cdot s_k^{i_k}, \qquad \textit{where } f_{i_1,\dots,i_k} \in \mathcal{O}(V),$$

converges absolutely with respect to any continuous seminorm on $\mathcal{O}(V)$ *for all* $s \in \Delta^k_r$, *then the mapping*

$$\Delta^k_r \times \Delta^1_\epsilon \to \mathcal{O}(U), \qquad (s,t) \mapsto f(s) \circ \gamma_{A,t}|_U,$$

is holomorphic.

In fact, by the Banach-Steinhaus theorem for any continuous seminorm p on $\mathcal{O}(U)$ there exists a continuous seminorm q on $\mathcal{O}(V)$, such that

$$p(f \circ \gamma_{A,t}|_U) \le Cq(f) \qquad \text{for all} \qquad f \in \mathcal{O}(V) \qquad \text{and} \qquad |\,t\,| \le \epsilon_1 < \epsilon .$$

Therefore, the series

$$f(s) \circ \gamma_{A,t}|_U = \sum (f_{i_1,\dots,i_k} \circ \gamma_{A,t}|_U)\, s_1^{i_1} \cdot \ldots \cdot s_k^{i_k}$$

converges absolutely with respect to p and the convergence is uniform if $(s,t) \in \overline{\Delta^k_{r_1}} \times \overline{\Delta^1_{\epsilon_1}}$, where $r_1 < r$, $\epsilon_1 < \epsilon$. Since the Weierstrass uniform convergence theorem is valid for holomorphic functions with values in Fréchet spaces, this implies (*).

Let $A_1, \dots, A_m \in \mathfrak{g}$. Applying (*) several times we obtain that for any $f \in \mathcal{O}(V)$ the mapping

$$\Delta^m_r \to \mathcal{O}(U), \qquad (t_1,\dots,t_m) \mapsto f \circ (\gamma_{A_1,t_1} \circ \ldots \circ \gamma_{A_m,t_m})\big|_U ,$$

is holomorphic for some $r > 0$. Choose $A_1, \dots, A_m$ to be a basis of $\mathfrak{g}$. Then

$$(t_1, \dots, t_m) \mapsto \exp(t_1 A_1) \cdot \ldots \cdot \exp(t_m A_m)$$

is a real analytic diffeomorphism of a neighborhood of $0 \in \mathbb{R}^d$ onto a neighborhood of $e \in G$. It follows that the mapping

$$g \mapsto f \circ g|_U \in \mathcal{O}(U),$$

defined in a neighborhood of $e \in G$, is real analytic at $e \in G$. The observation made at the end of §1.2 completes the proof. □

1.7 Lie homomorphism

We retain the notation of the preceding section. The multiplication in G and the bracket in $\mathfrak{g}$ are related by

$$\exp \mathrm{A} \cdot \exp \mathrm{B} = \exp\,(\mathrm{A} + \mathrm{B} + \frac{1}{2}[\mathrm{A}, \mathrm{B}] + \ldots),$$

where A, B are in a neighborhood of $0 \in \mathfrak{g}$ and the dots denote the terms of order ≥ 3 in the coordinates of A, B. For any element $\mathrm{A} \in \mathfrak{g}$ let $\lambda(\mathrm{A})$ be the infinitesimal transformation of $\{\gamma_{-\mathrm{A},t}\}$.

Theorem *Suppose we have a local action of a Lie group G on a complex space X. Then the mapping*

$$\lambda : \mathfrak{g} \to \mathcal{T}_X(X)$$

is a Lie algebra homomorphism. If G is a complex Lie group and the local G-action is holomorphic, then λ is a homomorphism of complex Lie algebras.

Proof The definition of the infinitesimal transformation (see (1) of §1.5) shows that $\lambda(c\mathrm{A}) = c\lambda(\mathrm{A})$, where $c \in \mathbb{R}$. Moreover, if the action is holomorphic the same is true for $c \in \mathbb{C}$. We have to show that λ is additive and that λ sends $[\mathrm{A}, \mathrm{B}] \in \mathfrak{g}$ to the bracket of the vector fields $[\lambda(\mathrm{A}),\ \lambda(\mathrm{B})]$.

Given $\mathrm{A}, \mathrm{B} \in \mathfrak{g}$ define $\mathrm{C}(t), \mathrm{D}(t) \in \mathfrak{g}$ for small $t \in \mathbb{R}$ by

$$\exp(t\mathrm{A}) \exp(t\mathrm{B}) \exp \mathrm{C}(t) = \exp\big(t(\mathrm{A} + \mathrm{B})\big)$$

and

$$\exp(t\mathrm{A}) \exp(t\mathrm{B}) \exp(-t\mathrm{A}) \exp(-t\mathrm{B}) \exp \mathrm{D}(t) = \exp\big(t^2[\mathrm{A}, \mathrm{B}]\big).$$

Since

$$\exp \mathrm{P}(t) \exp \mathrm{Q}(t) = \exp\,\big(\mathrm{P}(t) + \mathrm{Q}(t) + \frac{1}{2}[\mathrm{P}(t), \mathrm{Q}(t)] + O(t^3)\big),$$

where $t \mapsto \mathrm{P}(t) \in \mathfrak{g}$ and $t \mapsto \mathrm{Q}(t) \in \mathfrak{g}$ are two analytic curves with $\mathrm{P}(0) = \mathrm{Q}(0) = 0$, it follows that $\mathrm{C}(t) = O(t^2)$ and $\mathrm{D}(t) = O(t^3)$ as t tends to 0. Let $U \subset\subset V$ be two open subsets in X and $f \in \mathcal{O}(V)$. Suppose we have an analytic curve $t \mapsto \mathrm{P}(t) \in \mathfrak{g}$ defined for small $t \in \mathbb{R}$, such that $\mathrm{P}(t) = O(t^k)$ as t tends to 0. We have seen in §1.6 that a local action of a Lie group on a complex space is real analytic. Therefore

$$f \circ \exp \mathrm{P}(t)\big|_U = f\big|_U + O(t^k)$$

in $\mathcal{O}(U)$ as $t \to 0$. Applying this to the curves $\mathrm{C}(t)$ and $\mathrm{D}(t)$, we arrive at following equalities in $\mathcal{O}(U)$:

$$f \circ \big(\gamma_{\mathrm{A},t} \circ \gamma_{\mathrm{B},t}\big)|_U = f \circ \gamma_{(\mathrm{A}+\mathrm{B}),t}|_U + O(t^2)$$

and

$$f \circ \big(\gamma_{\mathrm{A},t} \circ \gamma_{\mathrm{B},t} \circ \gamma_{-\mathrm{A},t} \circ \gamma_{-\mathrm{B},t}\big)|_U = f \circ \gamma_{[\mathrm{A},\mathrm{B}],t^2}|_U + O(t^3)$$

as $t \to 0$. By using (2) of §1.5 several times, we obtain

$$\lambda(\mathrm{A}+\mathrm{B}) = \lambda(\mathrm{A}) + \lambda(\mathrm{B}) \qquad \text{and} \qquad \lambda([\mathrm{A},\mathrm{B}]) = [\lambda(\mathrm{A}), \lambda(\mathrm{B})].$$

□

The homomorphism λ is called the *Lie homomorphism* associated with a local action of G on X. This homomorphism gives a precise description of a local action in the following sense.

If two local actions $\{\Phi_\pi\}$ and $\{\Phi'_\pi\}$ of G have the same Lie homomorphism, then they are equivalent.

Proof Let $\pi = (U_\pi, V_\pi) \in \Pi_X$ be a pair of open subsets such that $U_\pi \subset\subset V_\pi$. From the definition of a local action we obtain two neighborhoods G_π, G'_π of $e \in G$ and two mappings $\Phi_\pi : G_\pi \to \mathrm{Hol}(U_\pi, V_\pi)$, $\Phi'_\pi : G'_\pi \to \mathrm{Hol}(U_\pi, V_\pi)$. We may assume that $G_\pi = \exp(B)$ and $G'_\pi = \exp(B')$, where B and B' are convex neighborhoods of $0 \in \mathfrak{g}$ on which the exponential mapping is one-to-one. Take any $\mathrm{A} \in \mathfrak{g}$. By the theorem of §1.5 the local one-parameter transformation groups, induced by the homomorphism $t \mapsto \exp t\mathrm{A}$ from $\{\Phi_\pi\}$ and $\{\Phi'_\pi\}$, are equivalent. Therefore,

$$\Phi_\pi(\exp t\mathrm{A}) = \Phi'_\pi(\exp t\mathrm{A}) \qquad \text{if } \exp t\mathrm{A} \in G_\pi \cap G'_\pi,$$

showing that $\{\Phi_\pi\}$ and $\{\Phi'_\pi\}$ are equivalent. □

We are now in a position to prove the main theorem of this chapter.

Theorem (second fundamental theorem of S.Lie) *Let X be a complex space, $\mathfrak{g}$ a Lie algebra, and $\lambda : \mathfrak{g} \to \mathcal{T}_X(X)$ a Lie algebra homomorphism. There exists a local action of a Lie group G on X, such that the Lie algebra of G is $\mathfrak{g}$ and the associated Lie homomorphism coincides with λ. If $\mathfrak{g}$ is a complex Lie algebra and λ is $\mathbb{C}$-linear, then G is a complex Lie group and the action is holomorphic.*

Proof If $\mathfrak{g}$ is a real Lie algebra, then λ extends to a homomorphism of complex Lie algebras $\mathfrak{g}_\mathbb{C} := \mathfrak{g} \otimes \mathbb{C} \xrightarrow{\lambda_\mathbb{C}} \mathcal{T}_X(X)$. Suppose $G_\mathbb{C}$ is a complex Lie group with Lie algebra $\mathfrak{g}_\mathbb{C}$, acting holomorphically on X, so that the associated Lie homomorphism is $\lambda_\mathbb{C}$. Take a Lie subgroup $G \subset G_\mathbb{C}$ which corresponds to the subalgebra $\mathfrak{g} \subset \mathfrak{g}_\mathbb{C}$. Then the restriction of the action from $G_\mathbb{C}$ to G is a required action of G. In what follows we assume that $\mathfrak{g}$ is a complex Lie algebra and $\lambda : \mathfrak{g} \to \mathcal{T}_X(X)$ is a homomorphism of complex Lie algebras.

It suffices to prove the theorem for complex model spaces. In fact, one can take a covering $\{V_i\}_{i \in J}$, find a desired local action for each V_i and then glue them together exactly as in §1.5. (The compatibility on intersections follows from the fact that a local action is uniquely defined up to equivalence by its Lie homomorphism.) From now on we assume that X is a complex model space in a Stein domain $D \subset \mathbb{C}^n$, defined by a coherent ideal sheaf $\mathcal{J} \subset \mathcal{O}_D$.

The Extension Lemma of §1.4 says that the natural mapping $\rho : \mathcal{T}_{D,X}(D) \to \mathcal{T}_X(X)$ is onto (recall that $\mathcal{T}_{D,X}$ is the subsheaf of derivations of $\mathcal{O}_D$ preserving $\mathcal{J}$). Choose a linear mapping $\lambda' : \mathfrak{g} \to \mathcal{T}_{D,X}(D)$, so that $\rho \circ \lambda' = \lambda$.

We now repeat the integration procedure of §1.5. The only difference is that, instead of one vector field, we have a finite dimensional vector space consisting of vector fields, namely, $\lambda'(\mathfrak{g})$.

Denote by z_i holomorphic coordinates in D and let

$$a_i(z, \mathrm{A}) := -\lambda'(\mathrm{A})_D\,(z_i), \qquad \text{where } z \in D,\ \mathrm{A} \in \mathfrak{g}.$$

The system of differential equations

$$\dot{z}_i = a_i(z, \mathrm{A}), \quad i = 1, 2, \ldots, n,$$

has the unique solution

$$z_i = \varphi_i(t, \mathrm{A}, z_0), \quad i = 1, 2, \ldots, n,$$

satisfying the initial conditions

$$\varphi_i(0, \mathrm{A}, z_0) = (z_0)_i,$$

where $\mathrm{A} \in \mathfrak{g}$, $z_0 \in D$. Since $a_i(z, s\mathrm{A}) = s a_i(z, \mathrm{A})$, $s \in \mathbb{C}$, we have

$$\varphi_i(st, \mathrm{A}, z_0) = \varphi_i(t, s\mathrm{A}, z_0)$$

Let X_1 be a complex model space, defined by $\mathfrak{J}|D_1$ in D_1, where $D_1 \subset\subset D$ is an open subset. Then there exists a neighborhood Δ of $0 \in \mathfrak{g}$, such that the functions $\varphi_i(\mathrm{A}, z) := \varphi_i(1, \mathrm{A}, z)$ are well-defined and holomorphic in $\Delta \times D_1$. We may assume that Δ is a polydisk (in some coordinates) and $\varphi(\mathrm{A}, z) := (\varphi_1(\mathrm{A}, z), \ldots, \varphi_n(\mathrm{A}, z)) \in D$, if $\mathrm{A} \in \Delta$, $z \in D_1$. Since $\lambda'(\mathfrak{g}) \subset \mathcal{T}_{D,X}(D)$, the mapping $z \mapsto \varphi(\mathrm{A}, z)$ restricts to a holomorphic mapping $X_1 \to X$, again denoted by $\varphi(\mathrm{A}, .)$.

Let G be any (connected) complex Lie group with Lie algebra $\mathfrak{g}$. Suppose that the polydisk Δ is small enough, so that $\exp|_\Delta : \Delta \to \Omega := \exp(\Delta)$ is biholomorphic. Then we have a mapping $\Phi : \Omega \to \mathrm{Hol}(X_1, X)$,

$$\Phi(\exp \mathrm{A}) \cdot z := \varphi(\mathrm{A}, z), \qquad \mathrm{A} \in \Delta, z \in X_1.$$

Now, the group law in G is locally defined by the *Campbell-Hausdorff formula* (see e.g. [Se3])

$$(\mathrm{A}, \mathrm{B}) \mapsto \mathrm{A} \odot \mathrm{B} = \mathrm{A} + \mathrm{B} + \frac{1}{2}[\mathrm{A}, \mathrm{B}] + \frac{1}{12}[\mathrm{A}, [\mathrm{A}, \mathrm{B}]] + \frac{1}{12}[\mathrm{B}, [\mathrm{B}, \mathrm{A}]] + \ldots\ .$$

The right hand side can be written as a power series in the coordinates of A and B. There exists a polydisk $\Delta_1 \subset\subset \Delta$ centered at $0 \in \mathfrak{g}$ such that for $\mathrm{A}, \mathrm{B} \in \Delta_1$ this series converges and the sum is in Δ. The equality in G

$$\exp \mathrm{A} \cdot \exp \mathrm{B} = \exp \mathrm{C}, \qquad \mathrm{A}, \mathrm{B} \in \Delta_1,\ \mathrm{C} \in \Delta,$$

implies that $\mathrm{C} = \mathrm{A} \odot \mathrm{B}$. By our construction

$$f \circ \Phi(\exp \mathrm{A}) = \exp(-\lambda(\mathrm{A})_{X_1})\,(f|_{X_1})$$

for any $f \in \mathcal{O}(X)$ and $\mathrm{A} \in \Delta$. (The symbol 'exp' on the right hand side denotes the exponent of an operator in $\mathcal{O}(X_1)$.)

Let $\pi = (U_\pi, V_\pi) \in \Pi_X$, i.e., $U_\pi \subset\subset V_\pi$. Fix an open subset U' in X such that $U_\pi \subset\subset U' \subset\subset V_\pi$. Choose Δ and Δ_1 so that, instead of the pair (X, X_1), they serve both for (V_π, U') and (U', U_π). If $g \in \Omega$ then, depending on the context, we use the same notation $\Phi(g)$ for holomorphic mappings $U_\pi \to U'$, $U' \to V_\pi$, and $U_\pi \to V_\pi$.

Let $g = \exp \mathrm{A}$, $h = \exp \mathrm{B}$, where $\mathrm{A}, \mathrm{B} \in \Delta_1$, and $k = gh$. Then $k = \exp \mathrm{C}$, where $\mathrm{C} \in \Delta$. For any $f \in \mathcal{O}(V_\pi)$ we have

$$f' := f \circ \Phi(g) = \exp(-\lambda(\mathrm{A})_{U'})\,(f|_{U'}) \in \mathcal{O}(U')$$

and

$$f' \circ \Phi(h) = \exp(-\lambda(\mathrm{B})_{U_\pi})\,(f'|_{U_\pi}) \in \mathcal{O}(U_\pi).$$

Furthermore,

$$f \circ \Phi(g) \circ \Phi(h) = \exp(-\lambda(\mathrm{B})_{U_\pi})\,\exp(-\lambda(\mathrm{A})_{U_\pi})\,(f|_{U_\pi}) \in \mathcal{O}(U_\pi)$$

is the sum of a power series in the coordinates of A and B. Similarly,

$$f \circ \Phi(k) = \exp(-\lambda(\mathrm{C})_{U_\pi})\,(f|_{U_\pi}) \in \mathcal{O}(U_\pi)$$

is the sum of a power series in the coordinates of C. Recall that $\lambda : \mathfrak{g} \to \mathcal{T}_X(X)$ is a Lie algebra homomorphism. Therefore the formal substitution $\mathrm{C} = \mathrm{A} \odot \mathrm{B} = -(-\mathrm{B}) \odot (-\mathrm{A})$ in the second series gives the first one. Since both series are convergent, it follows that

$$f \circ \Phi(g) \circ \Phi(h) = f \circ \Phi(k) \in \mathcal{O}(U_\pi),$$

and so we obtain a local action of G on X. It is clear that the associated Lie homomorphism coincides with λ. □

The above results can be summarized as follows.

Corollary *Let G be a Lie group, $\mathfrak{g}$ the Lie algebra of G, and X a complex space. The construction of the Lie homomorphism gives rise to a bijection:*

$$\left\{\begin{matrix}\textit{equivalence classes of}\\ \textit{local } G-\textit{actions on } X\end{matrix}\right\} \longleftrightarrow \left\{\begin{matrix}\textit{Lie algebra homomorphisms}\\ \mathfrak{g} \to \mathcal{T}_X(X)\end{matrix}\right\}.$$

For a complex Lie group G we have a bijection:

$$\left\{\begin{matrix}\textit{equivalence classes of}\\ \textit{holomorphic local}\\ G-\textit{actions on } X\end{matrix}\right\} \longleftrightarrow \left\{\begin{matrix}\textit{complex Lie algebra}\\ \textit{homomorphisms } \mathfrak{g} \to \mathcal{T}_X(X)\end{matrix}\right\}$$

□

The construction of the Lie homomorphism has many applications. Here are some examples.

Proposition 1 *Let $I \subset G$ be the ineffectivity kernel of a global continuous Lie group action on a complex space X. Then the Lie algebra of I is* Ker λ. *In particular, if G is a complex Lie group acting holomorphically on X, then I is a complex Lie subgroup of G.*

Proof Take $\mathrm{A} \in \mathfrak{g}$. Then the one-parameter subgroup $\exp(t\mathrm{A})$ $(t \in \mathbb{R})$ is contained in I if and only if

$$f \circ \gamma_{\mathrm{A},t}|_U = f$$

for all $U \subset\subset V$ and all $f \in \mathcal{O}(V)$. By (2) of §1.5 this is equivalent to $\lambda(\mathrm{A}) = 0$. Since in the holomorphic case λ is $\mathbb{C}$-linear, the second assertion follows. □

Let Y be a closed complex subspace of X defined by a coherent ideal sheaf $\mathcal{J} \subset \mathcal{O}_X$. Given a transformation group G of X, let $G_Y \subset G$ denote the subset of all elements $g \in G$ such that for any open set $V \subset X$ and for any $f \in \mathcal{J}(V)$ one has $f \circ g \in \mathcal{J}(g^{-1}V)$. Then G_Y is a subgroup of G, which is called the subgroup preserving the closed complex subspace Y.

Proposition 2 *If G is a topological transformation group of X then G_Y is a closed subgroup of G. If G is a (complex) Lie transformation group of X then G_Y is a closed (complex) Lie subgroup of G.*

Proof Let $U \subset\subset X$, $V \subset X$ be open subsets, $C = \overline{U}$, and $g \cdot C \subset V$ for some $g \in G$. Assume that $g = \lim g_n$, where $g_n \in G_Y$. Take $f \in \mathcal{J}(V)$. Then $g_n \cdot C \subset V$ and $f \circ g_n|_U \in \mathcal{J}(U)$ for all sufficiently large n. By the definition of a topological transformation group we have $f \circ g|_U = \lim (f \circ g_n|_U)$ in $\mathcal{O}(U)$. But $\mathcal{J}(U)$ is closed in $\mathcal{O}(U)$, see [GR2], Ch.5, §6. Therefore $f \circ g|_U \in \mathcal{J}(U)$. Thus $g \in G_Y$ showing that G_Y is a closed subgroup of G.

By a theorem of É. Cartan a closed subgroup of a Lie group is a Lie subgroup, see e.g. [He], Ch.2, Theorem 2.3. In particular, G_Y is a Lie subgroup of G. Denote by $\mathfrak{g}_Y$ the corresponding Lie subalgebra of $\mathfrak{g}$. We claim that

$$\mathfrak{g}_Y = \{\mathrm{A} \in \mathfrak{g} \mid \lambda(\mathrm{A})_x \cdot \mathcal{J}_x \subset \mathcal{J}_x \quad \text{for all } x \in X\}.$$

Indeed, if a one-parameter subgroup $\exp(t\mathrm{A})$ $(t \in \mathbb{R})$ is contained in G_Y then $\lambda(\mathrm{A})|_U$ preserves the closed subspace $\mathcal{J}(U) \subset \mathcal{O}(U)$. Since U is an arbitrary open subset of X, it follows that $\lambda(\mathrm{A})_x \cdot \mathcal{J}_x \subset \mathcal{J}_x$ for all $x \in X$. Conversely, if this condition is fulfilled then $(\exp t\lambda(\mathrm{A})|_U) \cdot \mathcal{J}(U) \subset \mathcal{J}(U)$ for arbitrary open $U \subset X$. By (2) of §1.5 this implies that $\exp(t\mathrm{A}) \in G_Y$ for all $t \in \mathbb{R}$.

Finally, if G is a complex Lie transformation group then λ is $\mathbb{C}$-linear. Therefore $\mathfrak{g}_Y$ is a complex Lie subalgebra and so G_Y is a complex Lie subgroup. □

For a vector field A on a complex space X let $A(x)$ denote the value of A in the tangent space $T_x(X)$, $x \in X$. More precisely, $T_x(X) = \mathrm{Der}(\mathcal{O}_x, \mathbb{C})$ and $A(x)$ is the composition

$$\mathcal{O}_x \xrightarrow{A_x} \mathcal{O}_x \xrightarrow{\nu_x} \mathbb{C},$$

where A_x is the germ of A at x and $\nu_x : \mathcal{O}_x \to \mathbb{C}$ is the canonical epimorphism.

Proposition 3 *Let G be a connected complex Lie group acting holomorphically on a reduced complex space X. Let $\mathrm{A}_1, \ldots, \mathrm{A}_m$ be a basis of the Lie algebra $\mathfrak{g}$ and put*

$A_j := \lambda(\mathrm{A}_j)$, $j = 1, \ldots, m$. *Then the tangent space to the orbit Gx at x is spanned by $A_1(x), \ldots, A_m(x)$. The subsets*

$$E_k := \{x \in X \mid \dim Gx < k\}$$

are analytic.

Proof The tangent space to Gx at x is, by definition, the subspace $\mathrm{Im}(d\theta_x)_o \subset T_x(X)$, where $o := e \cdot G_x \in G/G_x$ and $\theta_x : G/G_x \to X$ is defined in §1.2. Identify $T_o(G/G_x)$ with $\mathfrak{g}/\mathfrak{g}_x$, where $\mathfrak{g}_x$ is the Lie subalgebra of G_x, and let $\overline{\mathrm{A}}$ denote the coset of $\mathrm{A} \in \mathfrak{g}$ modulo $\mathfrak{g}_x$. Then $A_j(x) = -(d\theta_x)_o(\overline{\mathrm{A}}_j)$ by the definition of λ, and the first assertion follows. Thus

$$E_k = \{x \in X \mid \mathrm{rank}\{A_1(x), \ldots, A_m(x)\} < k\}.$$

In order to prove the second assertion, we may assume that X is an analytic set in a Stein domain $D \subset \mathbb{C}^n$. By the Extension Lemma $A_j = A'_j|_X$, where $A'_j \in \mathcal{T}_{D,X}(D)$, $j = 1, \ldots, m$, are holomorphic vector fields in D tangent to X (see §1.4). Obviously,

$$E'_k := \{x \in D \mid \mathrm{rank}\{A'_1(x), \ldots, A'_m(x)\} < k\}$$

is an analytic subset in D. Therefore $E_k = E'_k \cap X$ is analytic in X. □

Proposition 4 *Let X be an irreducible complex space almost homogeneous under the holomorphic action of a complex Lie group G. Then an open G-orbit in X is unique and connected. Its complement is an analytic set.*

Proof As usual, denote by G° the connected component of G containing e. Observe that any G-orbit is the disjoint union of G°-orbits of equal dimension. In particular, an open G-orbit is the disjoint union of open G°-orbits. Let $n := \dim X$. Then $E := E_n$ is a proper analytic subset and, since X is irreducible, the complement $\Omega := X - E$ is also irreducible, see [GR3], Ch.9, § 1. On the other hand, Proposition 3 tells us that Ω is the union of all open G°-orbits. In particular, Ω is non-singular and, consequently, connected. Therefore Ω is one (open) G°-orbit. This orbit is G-invariant and G acts on it transitively. □

1.8 Global actions

Trying to extend a local G-action in order to obtain a global one on the same complex space X, we encounter two basic difficulties. Firstly, it can happen that X has not enough points (see the remark in §1.5). Secondly, far from $e \in G$ the extension may become ambiguous. In this section we give some sufficient conditions for the existence of such an extension.

Proposition *Let G be a connected Lie group, X a complex space, and $\lambda : \mathfrak{g} \to \mathcal{T}_X(X)$ a Lie algebra homomorphism. Suppose that there exists a basis $\mathrm{A}_1, \ldots, \mathrm{A}_m$ of $\mathfrak{g}$ such that each vector field $\lambda(\mathrm{A}_j)$ generates a global one-parameter transformation group $\{\gamma_{j,t}\}_{t\in\mathbb{R}}$. Choose $\sigma > 0$ so that the mapping*

$$(t_1, \ldots, t_m) \mapsto \exp(t_1\mathrm{A}_1) \cdot \ldots \cdot \exp(t_m\mathrm{A}_m)$$

is a diffeomorphism of the cube $|t_j| < \sigma$ *onto a neighborhood* Ω *of* $e \in G$. *For* $g = \exp(t_1 A_1) \cdot \ldots \cdot \exp(t_m A_m) \in \Omega$ *define* $\Phi(g) \in \mathrm{Aut}(X)$ *by*

$$\Phi(g) := \gamma_{1,t_1} \circ \ldots \circ \gamma_{m,t_m}.$$

Then Φ *is a local homomorphism, i.e.,* $\Phi(gh) = \Phi(g) \circ \Phi(h)$ *in some neighborhood of* $e \in G$.

Proof The set Ω is called a *cubical neighborhood* of $e \in G$ with parameter σ. By the second Lie theorem there exists a local G-action with Lie homomorphism λ. As in §1.2, this action is given by a system of mappings $\Phi_\pi : G_\pi \to \mathrm{Hol}(U_\pi, V_\pi)$, where G_π is a neighborhood of $e \in G$ and $\pi \in \Pi_X$. We claim:

(*) *for each pair* $\pi = (U_\pi, V_\pi) \in \Pi_X$ *and for any open set* $U \subset\subset U_\pi$ *there exists a cubical neighborhood* Ω' *of* $e \in G$ *such that* $\Omega' \subset G_\pi \cap \Omega$ *and* $\Phi_\pi(g)|_U = \Phi(g)|_U$ *for* $g \in \Omega'$.

In order to show (*) fix a sequence of open sets

$$U = U^{(1)} \subset\subset U^{(2)} \subset\subset \ldots \subset\subset U^{(m-1)} \subset\subset U^{(m)} = U_\pi.$$

One can find a neighborhood $O \subset G_\pi$ of $e \in G$ with the following properties:

$$\Phi_\pi(g)\ (U_\pi^{(j)}) \subset U_\pi^{(j+1)} \qquad \text{for all } \ g \in O, \ \ j = 1, \ldots, m-1,$$

and

$$\underbrace{O \cdot \ldots \cdot O}_{m \text{ times}} \subset G_\pi.$$

Let $\sigma' < \sigma$ be a positive number such that $\exp(tA_j) \in O$ and $\Phi_\pi(\exp(tA_j)) = \gamma_{j,t}|_{U_\pi}$ for all j, $j = 1, \ldots, m$, if $|t| < \sigma'$. (Here we use the fact that the Lie homomorphism of the local action is λ.) Define Ω' to be the cubical neighborhood of e with parameter σ'. Then $\Omega' \subset G_\pi \cap \Omega$. Furthermore, if $g = g_1 \cdot \ldots \cdot g_m$, where $g_j = \exp(t_j A_j)$, $|t_j| < \sigma'$, then

$$\Phi_\pi(g)|_U = \Phi_\pi(g_1) \circ \ldots \circ \Phi_\pi(g_m)|_U = \gamma_{1,t_1} \circ \ldots \circ \gamma_{m,t_m}|_U = \Phi(g)|_U,$$

and so we obtain (*).

Shrinking Ω' if necessary, we may assume that

$$\Phi_\pi(g) \subset U_\pi \qquad \text{for all } \ g \in \Omega'.$$

Then condition (a) in the definition of a local G-action (see §1.2) shows that

$$\Phi_\pi(gh)|_U = \Phi_\pi(g) \circ \Phi_\pi(h)|_U,$$

when g, h, and gh are in Ω'. Applying (*) we obtain the same equality for Φ. As a result, we have the following:

(**) *for every* $x \in X$ *there exists a neighborhood* U_x *of* x *and a cubical neighborhood* $\Omega_x \subset \Omega$ *of* $e \in G$ *such that if* $g, h, gh \in \Omega_x$ *then*

$$\Phi(gh)|_{U_x} = \Phi(g) \circ \Phi(h)|_{U_x}.$$

We now want to show that one can replace Ω_x in (**) by a neighborhood of $e \in G$ which does not depend on $x \in X$. For this we possibly have to shrink U_x. Namely, we shall assume that $U_x \subset\subset V_x$, where V_x is a complex model space in a domain in $\mathbb{C}^n$. In particular, there is a finite number of holomorphic functions $f_\alpha \in \mathcal{O}(V_x)$, $\alpha = 1, 2, \ldots, k$, generating $\mathcal{O}_y$ as an analytic local algebra for all $y \in V_x$.

Define Ω^* (resp. Ω^*_x) to be a cubical neighborhood of $e \in G$ such that $\Omega^* \cdot \Omega^* \subset \Omega$ (resp. $\Omega^*_x \cdot \Omega^*_x \subset \Omega_x$). Let

$$\Psi(g,h) := \Phi(h)^{-1} \circ \Phi(g)^{-1} \circ \Phi(gh), \qquad \text{where} \quad (g,h) \in \Omega^* \times \Omega^*,$$

and

$$\Xi := \{(g,h) \in \Omega^* \times \Omega^* \mid \Psi(g,h)|_{U_x} = \mathrm{id}\}.$$

Then Ξ is closed in $\Omega^* \times \Omega^*$ and $\Xi \supset \Omega^*_x \times \Omega^*_x$.

Assume that $\Xi \neq \Omega^* \times \Omega^*$. Let $(g_0, h_0) \in \partial\Xi$ be an accumulation point of the interior of Ξ. Each mapping

$$(g,h) \;\mapsto\; f_\alpha \circ \Psi(g,h)|_{U_x} \;\in\; \mathcal{O}(U_x), \quad \alpha = 1, \ldots, k,$$

is well-defined and real analytic in a neighborhood of (g_0, h_0). By definition, its restriction to Ξ is a constant mapping. It follows that $f_\alpha \circ \Psi(g,h)|_{U_x} = f_\alpha|_{U_x}$ for all α, $\alpha = 1, \ldots, k$, and for all (g,h) in some neighborhood of (g_0, h_0). In other words, for (g,h) in this neighborhood we have $\Psi(g,h)|_{U_x} = \mathrm{id}$. Thus (g_0, h_0) is an interior point of Ξ. This contradiction shows that our assumption is false and $\Xi = \Omega^* \times \Omega^*$.

Now we can replace in (**) Ω_x by Ω^*. Since Ω^* does not depend on x, it follows that Φ is a local homomorphism. (In fact, $\Phi(gh) = \Phi(g) \circ \Phi(h)$ for $g, h \in \Omega^*$.) □

Let G be a simply connected topological group. A local homomorphism from G to an arbitrary group, which is defined in a connected neighborhood of e, extends to a (global) homomorphism of G. Therefore the above result has as a consequence the following theorem first proved by R.Palais for differentiable manifolds, see [Pa], Ch. 4, Theorem III.

Theorem *Let G be a simply connected Lie group, X a complex space, and $\lambda : \mathfrak{g} \to \mathcal{T}_X(X)$ a Lie algebra homomorphism. Suppose that there exists a basis $\mathrm{A}_1, \ldots, \mathrm{A}_m$ of $\mathfrak{g}$ such that each vector field $\lambda(\mathrm{A}_j)$ generates a global one-parameter transformation group $\{\gamma_{j,t}\}_{t \in \mathbb{R}}$ of X. Then there exists a global real analytic G-action on X with Lie homomorphism λ.* □

Corollary 1 *If G is simply connected and X is compact then for any Lie algebra homomorphism $\lambda : \mathfrak{g} \to \mathcal{T}_X(X)$ there exists a global real analytic G-action on X with Lie homomorphism λ. If G is a complex Lie group and λ a homomorphism of complex Lie algebras then the action is holomorphic.*

Proof By the proposition in §1.5 any holomorphic vector field on X generates a global one-parameter transformation group. Therefore the first statement follows from the preceding theorem and the second one results from §1.7. □

Remark Another proof of this corollary can be obtained as a consequence of the theorem in §2.3.

A closed subgroup G of a connected complex Lie group $G_\mathbb{C}$ is called a *real form* of $G_\mathbb{C}$ if the inclusion $\mathfrak{g} \hookrightarrow \mathfrak{g}_\mathbb{C}$ gives rise to an isomorphism of complex Lie algebras $\mathfrak{g} \bigotimes \mathbb{C} \simeq \mathfrak{g}_\mathbb{C}$. In other words, G is a real form of $G_\mathbb{C}$ if and only if G is closed in $G_\mathbb{C}$ and $\mathfrak{g}_\mathbb{C} = \mathfrak{g} \bigoplus i\mathfrak{g}$.

Corollary 2 *Let G be a connected real form of a simply connected complex Lie group $G_\mathbb{C}$. Then any continuous G-action on a complex compact space X extends to a holomorphic $G_\mathbb{C}$-action on X.*

Proof Let $\lambda : \mathfrak{g} \to \mathcal{T}_X(X)$ be the Lie homomorphism of the given G-action. Since $\mathcal{T}_X(X)$ is a complex Lie algebra, one has an obvious $\mathbb{C}$-linear extension $\lambda_\mathbb{C} : \mathfrak{g}_\mathbb{C} \to \mathcal{T}_X(X)$. By Corollary 1 the homomorphism $\lambda_\mathbb{C}$ is a Lie homomorphism of some holomorphic $G_\mathbb{C}$-action, whose restriction to G is certainly equivalent to the given G-action. Since G is connected and both actions are globally defined, they coincide everywhere. □

Example Assume G is a connected real form of a simply connected complex Lie group $G_\mathbb{C}$. Suppose $Z \neq \{e\}$ a finite normal subgroup of $G_\mathbb{C}$, such that $G \cap Z = \{e\}$. Then $G_\mathbb{C}/Z$ is another complex Lie group locally isomorphic to $G_\mathbb{C}$ and having G as a real form. On the other hand, if $H \subset G_\mathbb{C}$ is a closed complex Lie subgroup satisfying $H \cap Z = \{e\}$, then the manifold $X := G_\mathbb{C}/H$ is acted on by G, but not by $G_\mathbb{C}/Z$. Moreover, by an appropriate choice of G, $G_\mathbb{C}$, Z, and H one can achieve that X is compact. This shows that the conclusion in Corollary 2 is false if $G_\mathbb{C}$ is not simply connected. One can take $G = \mathrm{SL}(3,\mathbb{R})$, $G_\mathbb{C} = \mathrm{SL}(3,\mathbb{C})$,

$$Z = \left\{ \begin{pmatrix} \epsilon & 0 & 0 \\ 0 & \epsilon & 0 \\ 0 & 0 & \epsilon \end{pmatrix} \,\middle|\, \epsilon^3 = 1 \right\} \simeq \mathbb{Z}_3, \quad H = H_d = \left\{ \begin{pmatrix} d^n & * & * \\ 0 & * & * \\ 0 & * & * \end{pmatrix} \,\middle|\, n \in \mathbb{Z} \right\},$$

where $|d| \neq 1$ and $*$ denotes arbitrary complex numbers such that the determinant of the matrix is 1. Then $X = G_\mathbb{C}/H_d$ is a homogeneous Hopf 3-fold, see §3.6, Example 1.

2 Automorphism Groups

The automorphism group $\mathrm{Aut}(X)$ of a complex space X, equipped with a natural topology, is a topological group. Our goal in this chapter is to show that there are two important classes of complex spaces, for which $\mathrm{Aut}(X)$ has a Lie group structure compatible with its topology. The first class consists of all (not necessarily reduced) compact spaces, the second one is the class of all bounded domains in $\mathbb{C}^n$.

The result about compact spaces goes back to S.Bochner and D.Montgomery. The proof makes use of powerful theorems from the general theory of topological groups. They are stated here without proof, but the reduction to the topological setting is complete. This reduction requires, among other things, the identity theorem for compact groups of holomorphic transformations with a fixed point. The identity theorem, in its turn, is deduced from the local linearization theorem for such groups. In addition to general results, we consider several examples of compact complex spaces, for which $\mathrm{Aut}(X)$ is explicitly determined. In particular, we prove that if compact spaces X and Y are reduced then $\mathrm{Aut}^\circ(X \times Y) \simeq \mathrm{Aut}^\circ(X) \times \mathrm{Aut}^\circ(Y)$.

The proof for bounded domains is based on the classical compactness principle due to C.Carathéodory and H.Cartan. This principle implies also that the action of the automorphism group on a bounded domain is proper. Therefore a preliminary discussion of proper actions on complex spaces is helpful. We show that the dimension of an automorphism group acting properly on an irreducible complex space X is bounded from above by $n^2 + 2n$, where $n = \dim_{\mathbb{C}} X$, and we list all possibilities, when the equality holds. One example from this list is the full automorphism group of the unit ball $\mathbb{B}_n \subset \mathbb{C}^n$. At the end we return to this example again. Namely, we prove that $\mathbb{B}_n$ can be characterized as a bounded domain $D \subset \mathbb{C}^n$ with C^2 boundary, such that: (a) D is strictly pseudoconvex and $\mathrm{Aut}(D)$ is non-compact; or (b) $D/\mathrm{Aut}(D)$ is compact.

2.1 Topology in $\mathbf{Hol}(X, Y)$

Let $X = (X, \mathcal{O}_X)$ and $Y = (Y, \mathcal{O}_Y)$ be two complex spaces. If X and Y are reduced, then the set $\mathrm{Hol}(X,Y)$ can be naturally endowed with the *compact-open topology.* More generally, it is possible to introduce a topology into $\mathrm{Hol}(X,Y)$ for arbitrary X and Y. Let U be an open subset in X and $V = (V, \mathcal{O}_Y|V)$ an open subspace of Y with a fixed isomorphism onto a complex model space in a domain $E \subset \mathbb{C}^n$. Each holomorphic mapping $\phi \in \mathrm{Hol}(U,V)$ can be regarded as a mapping from U to $E \subset \mathbb{C}^n$ and thus identified with an n-tuple of sections $s_i \in \mathcal{O}(U)$, $s_i = z_i \circ \phi$, where z_i are coordinate functions on $\mathbb{C}^n$. The topology of the Fréchet space $\mathcal{O}(U)$ induces a topology on $\mathrm{Hol}(U,V) \subset \mathcal{O}(U)^n$. Assume that $C = \bar{U}$ is compact and let

$$W_{C,V} := \{\varphi \in \mathrm{Hol}(X,Y) \mid \varphi(C) \subset V\}.$$

The topology in $\mathrm{Hol}(X,Y)$ is defined as the coarsest topology such that:

(i) all subsets $W_{C,V}$ are open;

(ii) all mappings $W_{C,V} \to \mathrm{Hol}(U,V)$, $\varphi \mapsto \varphi|_U$, are continuous.

Dropping (ii) we obtain the definition of the compact-open topology. In the reduced case the Fréchet topology in $\mathcal{O}(U)$ is the topology of compact convergence, i.e., the compact-open topology in the space of mappings $\mathrm{Hol}(U,\mathbb{C}) \simeq \mathcal{O}(U)$. Therefore (ii) is automatically fulfilled and the topology in $\mathrm{Hol}(X,Y)$ is nothing else than the compact-open topology. However, generally speaking, the topology defined by (i) and (ii) is essentially finer than the latter one.

Example Let A and B be two local analytic algebras of finite dimension, $X = (\{p\}, A)$, and $Y = (\{q\}, B)$. Then $\mathrm{Hol}(X,Y) = \{$algebra homomorphisms $B \to A\}$. The sequence topology on a finite dimensional local analytic algebra coincides with its Euclidean topology, see [GR1], Ch.2, §4.2. It follows that the topology in $\mathrm{Hol}(X,Y)$, defined by (i) and (ii), is induced by the Euclidean topology of the finite dimensional vector space $\mathrm{Hom}_{\mathbb{C}}(B,A)$. On the other hand, the compact-open topology in $\mathrm{Hol}(X,Y)$ is trivial.

By assumption, all complex spaces in this book have countable topology. A straightforward verification shows that $\mathrm{Hol}(X,Y)$ *is Hausdorff and has countable topology.* We want to describe convergent sequences in $\mathrm{Hol}(X,Y)$. Note that the convergence of sections in the canonical topology is a local property with respect to the topology of the base. More precisely, let $\mathcal{F}$ be a coherent analytic sheaf on X and let $\{s_\nu\}_{\nu=1,2,\ldots}$ be a sequence in $\mathcal{F}(X)$. Then $\{s_\nu\}_{\nu=1,2,\ldots}$ converges to $s \in \mathcal{F}(X)$ if and only each point $x \in X$ has a neighborhood U_x such that $\{s_\nu|_{U_x}\}_{\nu=1,2,\ldots}$ converges to $s|_{U_x}$, see [GR2], Ch. 5, §6.

Proposition 1 *A sequence* $\{\varphi_\nu\}_{\nu=1,2,\ldots}$ *in* $\mathrm{Hol}(X,Y)$ *converges to* $\varphi \in \mathrm{Hol}(X,Y)$ *if and only if for each point* $x \in X$ *and for each open neighborhood* $V_{\varphi(x)}$ *of* $\varphi(x) \in Y$*, isomorphic to a complex model space in a domain* $E \subset \mathbb{C}^q$*, there exists an open neighborhood* U_x *of* $x \in X$*, isomorphic to a complex model space in a domain* $D \subset \mathbb{C}^p$*, such that:*

$\varphi(U_x) \subset V_{\varphi(x)}$ *and* $\varphi|_{U_x}$ *extends to a holomorphic mapping* $\Phi = (\Phi^1, \ldots, \Phi^q) : D \to E$*;*

for all sufficiently large ν *we have* $\varphi_\nu(U_x) \subset V_{\varphi(x)}$ *and* $\varphi_\nu|_{U_x}$ *extends to a holomorphic mapping* $\Phi_\nu = (\Phi^1_\nu, \ldots, \Phi^q_\nu) : D \to E$*;*

Φ^k_ν *converges to* Φ^k *for each* k, $k = 1, 2, \ldots, q$*, uniformly on compact subsets in* D.

Proof We first deduce from these conditions that φ_ν converges to φ in $\mathrm{Hol}(X,Y)$. Let $U \subset X$ be an open subset with compact closure $C := \bar{U}$ and let $V \subset Y$ be an open subset with a fixed isomorphism of $(V, \mathcal{O}_Y|_V)$ onto a complex model space in a domain in $\mathbb{C}^n$. We may assume that $\varphi \in W_{C,V}$. Then we have to show the following:

(*) $\varphi_\nu \in W_{C,V}$ for all $\nu \geq \nu_0$;

(**) $\varphi_\nu|_U \to \varphi|_U$ in $\mathrm{Hol}(U,V) \subset \mathcal{O}(U)^n$.

For any $x \in C$ and $V_{\varphi(x)} := V$ fix a neighborhood U_x as above. There exists a finite set $\{x_1, \ldots, x_m\} \subset C$ such that $C \subset U_{x_1} \cup \ldots \cup U_{x_m}$. Since $\varphi_\nu(U_{x_i}) \subset V$ for all $\nu \geq \nu_i$, $i = 1, \ldots, m$, we obtain (*) with $\nu_0 := \max(\nu_1, \ldots, \nu_m)$. Furthermore,

$\varphi_\nu|_{U_{x_i}} \to \varphi|_{U_{x_i}}$ in Hol$(U_{x_i}, V) \subset \mathcal{O}(U_{x_i})^n$. As we have noted above, this implies (**).

Conversely, suppose that $\varphi_\nu \to \varphi$ in Hol(X, Y). Given $x \in X$ and $V_{\varphi(x)} \subset Y$ take any open neighborhood U'_x of x, isomorphic to a complex model space in a Stein domain $D' \subset\subset \mathbb{C}^p$ and such that $\varphi(U'_x) \subset V_{\varphi(x)}$. We identify U'_x as a set with its image in D'. Consider $\varphi|_{U'_x}$ as a holomorphic mapping $U'_x \to \mathbb{C}^q$. Since D' is Stein, $\varphi|_{U'_x}$ extends to a holomorphic mapping $\Phi : D' \to \mathbb{C}^q$. Let $D'' \subset\subset D'$ be another Stein domain, such that $U''_x := D'' \cap U'_x$ contains x and $\Phi(\overline{D''}) \subset E$. Then $\varphi_\nu(\overline{U''_x}) \subset V_{\varphi(x)}$ for all sufficiently large ν and $\varphi_\nu|_{U''_x} \to \varphi|_{U''_x}$ in $\mathcal{O}_X(U''_x)^q$. By the Selection Theorem (see §1.3) there exists a sequence of holomorphic mappings $\Phi_\nu : D'' \to \mathbb{C}^q$, extending $\varphi_\nu|_{U''_x}$ and converging to Φ uniformly on compact subsets in D''. Finally, let $D \subset\subset D''$ be any domain such that $U_x := D \cap U'_x$ contains x. Then $\Phi_\nu(D) \subset E$ for all sufficiently large ν. Therefore U_x has all the required properties. □

Proposition 2 *Let $(X, \mathcal{O}_X), (Y, \mathcal{O}_Y)$, and $(Z, \mathcal{O}_Z)$ be three complex spaces. Then the mapping*

$$\mathrm{Hol}(X, Y) \times \mathrm{Hol}(Y, Z) \to \mathrm{Hol}(X, Z), \quad (\varphi, \psi) \mapsto \psi \circ \varphi,$$

is continuous.

Proof By Proposition 1 it suffices to prove the statement for three domains in complex vector spaces. But then it is an elementary property of the compact-open topology. □

Lemma *Let D be a domain in $\mathbb{R}^q$ and*

$$\Phi_\nu : D \to \mathbb{R}^q, \qquad \nu = 1, 2, \ldots,$$

a sequence of C^1 mappings. Assume that $\{\Phi_\nu\}$ converges to id_D in the C^1 topology uniformly on compact subsets in D. Let D_1 be a relatively compact subdomain in D. Then, for all sufficiently large ν we have:

(1) *$\Phi_\nu|_{D_1} : D_1 \to \Phi_\nu(D_1)$ is a C^1 diffeomorphism;*

(2) *$D_1 \subset \Phi_\nu(D)$;*

(3) *there exists a C^1 mapping $\Psi_\nu : D_1 \to D$ such that $\Phi_\nu \circ \Psi_\nu = \mathrm{id}_{D_1}$;*

Furthermore,

(4) *$\{\Psi_\nu\}$ converges to id_{D_1} in the C^1 topology uniformly on compact subsets of D_1.*

Proof A proof of (1) can be found in any differential topology textbook, see e.g. [H], Ch. 2, Lemma 1.3.

In order to prove (2) fix a subdomain $D_2 \subset\subset D$ such that $D_1 \subset\subset D_2$. For all sufficiently large ν we have

$$D_1 \cap \Phi_\nu(\partial D_2) = \emptyset\ ,$$

so that either $D_1 \subset \Phi_\nu(D_2)$ or $D_1 \cap \Phi_\nu(D_2) = \emptyset$. Since $\Phi_\nu(x) \to x$ for $x \in D_1$, the second case is possible only for a finite number of ν and (2) follows.

If ν is large enough then (1) is valid for D_2 instead of D_1 and (2) is valid for D_2 instead of D. Defining Ψ_ν as the composition

$$D_1 \hookrightarrow \Phi_\nu(D_2) \overset{\left(\Phi_\nu|_{D_2}\right)^{-1}}{\longrightarrow} D_2 \hookrightarrow D,$$

we obtain (3).

Finally, in order to prove (4), consider a compact set $C \subset D_1$. Applying (2) to $\Phi_\nu|_{D_1}$ we see that $C \subset \Phi_\nu(D_1)$ for all sufficiently large ν. Thus, for $x \in C$ and for any norm in $\mathbb{R}^n$ we have

$$|\Psi_\nu(x) - x| \leq \sup_{y\in D_1} |\Psi_\nu(\Phi_\nu(y)) - \Phi_\nu(y)| = \sup_{y\in D_1} |\Phi_\nu(y) - y|,$$

so that $\Psi_\nu(x) \to x$ uniformly on C. The estimate shows also that if ν is large enough then $\Psi_\nu(x) \in D_1$ for all $x \in C$. It follows that the sequence of matrices $\Psi'_\nu(x) = \left(\Phi'_\nu(\Psi_\nu(x))\right)^{-1}$ converges to the unit matrix uniformly on C. □

The group $\mathrm{Aut}(X)$ as a subset of $\mathrm{Hol}(X, X)$ is endowed with the induced topology.

Proposition 3 *The mapping*

$$\mathrm{Aut}(X) \to \mathrm{Aut}(X), \quad \varphi \mapsto \varphi^{-1},$$

is continuous.

Proof In view of Proposition 2 it is enough to prove the continuity of this mapping at the neutral point $\mathrm{id}_X \in \mathrm{Aut}(X)$. Let $\{\varphi_\nu\}_{\nu=1,2,\ldots}$ be a sequence in $\mathrm{Aut}(X)$, converging to id_X. We have to show that $\{\varphi_\nu^{-1}\}$ converges to id_X. Fix a point $x \in X$ and a neighborhood V_x of x, isomorphic to a complex model space in a domain $E \subset \mathbb{C}^q$. We identify V_x with this model space. Choose a neighborhood U_x of $x \in X$ as in Proposition 1. The proof of this proposition shows that one can take U_x to be the restriction of the model space to a subdomain $D \subset\subset E$. By construction, $\varphi_\nu(U_x) \subset V_x$ and $\varphi_\nu|_{U_x}$ extends to a holomorphic mapping $\Phi_\nu : D \to E$ so that the sequence $\{\Phi_\nu\}$ converges to id_D uniformly on compact subsets in D. (Note that in Proposition 1 the extension Φ of the limit mapping $\varphi|_{U_x}$ is arbitrary. The only condition is that Φ should be defined on a larger Stein domain. Since in our case $\varphi = \mathrm{id}_X$, we can take $\Phi := \mathrm{id}_D$.)

The compact convergence of holomorphic functions implies the compact convergence of their derivatives. Thus, applying the lemma to $\{\Phi_\nu\}$, we obtain a sequence of holomorphic mappings $\Psi_\nu : D_1 \to D$, converging to id_{D_1} and such that $\Phi_\nu \circ \Psi_\nu = \mathrm{id}_{D_1}$, where D_1 is any fixed relatively compact subdomain of D. In particular, suppose that D_1 contains x and denote by W_x the restriction of our model space to D_1. Then Ψ_ν determines a holomorphic mapping of W_x, namely, $\varphi_\nu^{-1}|_{W_x}$. Since x and U_x are arbitrary, $\{\varphi_\nu^{-1}\}$ converges to id_X by Proposition 1. □

Corollary $\mathrm{Aut}(X)$ *is a topological group.* □

Remark Let G be a topological group. Using the topology in $\mathrm{Aut}(X)$ one can rephrase the definition of a topological G-action on X (see §1.2). Namely, a homomorphism $G \to \mathrm{Aut}(X)$ defines a continuous G-action if and only if this homomorphism is continuous.

Theorem *For a compact complex space X the group* $\mathrm{Aut}(X)$ *is locally compact.*

Proof Let $T_x \subset\subset U_x \subset\subset V_x \subset W_x$ be four open neighborhoods of $x \in X$ chosen in the following way. The neighborhood W_x is isomorphic to a complex model space in a domain $D \subset \mathbb{C}^n$, so that x corresponds to $0 \in D$, the polydisk

$$\Delta_c^n = \{z = (z_1, \dots, z_n) \in \mathbb{C}^n \mid |z_j| < c,\ j = 1, 2, \dots, n\}$$

is contained in D, and the neighborhoods T_x, U_x, and V_x are obtained by restriction of the model space in D to Δ_a^n, Δ_b^n, and Δ_c^n respectively, where $0 < a < b < c$. We identify T_x, U_x, V_x, and W_x with the model spaces in $\Delta_a^n, \Delta_b^n, \Delta_c^n$, and D respectively. The restriction mapping

$$\mathcal{O}(\Delta_c^n)^n \simeq \mathrm{Hol}(\Delta_c^n, \mathbb{C}^n) \xrightarrow{\rho} \mathrm{Hol}(V_x, \mathbb{C}^n) \simeq \mathcal{O}(V_x)^n$$

is an epimorphism of Fréchet spaces. Let

$$\Upsilon_{a,b,c} := \{\Phi \in \mathrm{Hol}(\Delta_c^n, \mathbb{C}^n) \mid \Phi(\overline{\Delta_b^n}) \subset \Delta_c^n, \Phi(\overline{\Delta_a^n}) \subset \Delta_b^n,\ \Phi(0) \in \Delta_a^n\}.$$

Since $\Upsilon_{a,b,c}$ is an open subset in $\mathrm{Hol}(\Delta_c^n, \mathbb{C}^n)$, the image $\rho(\Upsilon_{a,b,c})$ is open in $\mathrm{Hol}(V_x, \mathbb{C}^n)$ by the Banach open mapping theorem. Therefore, by the definition of the topology in $\mathrm{Aut}(X)$, the subset

$$\Omega := \{\varphi \in \mathrm{Aut}(X) \mid \varphi(\overline{V_x}) \subset W_x \ \text{ and } \ \varphi|_{V_x} \in \rho(\Upsilon_{a,b,c})\}$$

is an open neighborhood of id_X. (We identify $\varphi|_{V_x}$ with the composition $V_x \to W_x \hookrightarrow D \hookrightarrow \mathbb{C}^n$.)

Consider a sequence $\{\varphi_\nu\}_{\nu=1,2,\dots}$ in $\Omega \cap \Omega^{-1}$. Each restriction $\varphi_\nu|V_x$ (resp. $\varphi_\nu^{-1}|V_x$) is induced by $\Phi_\nu \in \mathrm{Hol}(\Delta_c^n, \mathbb{C}^n)$ (resp. by $\Psi_\nu \in \mathrm{Hol}(\Delta_c^n, \mathbb{C}^n)$) such that $\Phi_\nu(\overline{\Delta_b^n}) \subset \Delta_c^n$, $\Phi_\nu(\overline{\Delta_a^n}) \subset \Delta_b^n$, and $\Phi_\nu(0) \in \Delta_a^n$ (resp. $\Psi_\nu(\overline{\Delta_b^n}) \subset \Delta_c^n$, $\Psi_\nu(\overline{\Delta_a^n}) \subset \Delta_b^n$, and $\Psi_\nu(0) \in \Delta_a^n$). By Montel's theorem one can find subsequences $\{\Phi_{\nu_k}\}$ and $\{\Psi_{\nu_k}\}$ converging in $\mathrm{Hol}(\Delta_b^n, \mathbb{C}^n)$ to Φ and Ψ respectively.

It is clear that $\Phi(0) \in \overline{\Delta_a^n}$ and $\Phi(\Delta_b^n) \subset \overline{\Delta_c^n}$. This implies that $\Phi(\Delta_b^n) \subset \Delta_c^n$, for otherwise, according to the Maximum Principle, $\Phi(\Delta_b^n) \subset \partial\Delta_c^n$, which is obviously impossible. The same argument shows that $\Psi(\Delta_b^n) \subset \Delta_c^n$, $\Phi(\Delta_a^n) \subset \Delta_b^n$, and $\Psi(\Delta_a^n) \subset \Delta_b^n$. Therefore the compositions $\Psi \circ (\Phi|_{\Delta_a^n})$ and $\Phi \circ (\Psi|_{\Delta_a^n})$ are well-defined. By construction, they induce the identity mappings on T_x.

Now, X is covered by a finite number of $T_{x_i}, i = 1, 2, \dots, m$. For each x_i our construction yields a neighborhood $\Omega = \Omega_i$ of id_X in $\mathrm{Aut}(X)$. The above argument along with Proposition 1 shows that the intersection $\Omega_1 \cap \Omega_1^{-1} \cap \dots \cap \Omega_m \cap \Omega_m^{-1}$ has compact closure. □

2.2 Local linearization of a compact group with a fixed point

Let X be an arbitrary complex space, G a group acting on X, and $x \in X$ a fixed point of G. Then G acts naturally on the local algebra $\mathcal{O}_x$ by algebra automorphisms. Namely, for $g \in G$ we have an automorphism $f_x \mapsto (f \circ g^{-1})_x$, where f is a holomorphic function defined in a neighborhood U of x. It should be noted that $(f \circ g^{-1})_x$ depends only on the germ f_x, but not on its representative $f \in \mathcal{O}(U)$.

This representation of G on $\mathcal{O}_x$ gives rise to a series of finite-dimensional representations

$$G \to \mathrm{GL}(\mathfrak{m}_x/\mathfrak{m}_x^{d+1}), \quad d = 1, 2, \dots .$$

Let

$$\tau_x^d : G \to \mathrm{GL}((\mathfrak{m}_x/\mathfrak{m}_x^{d+1})^*), \quad d = 1, 2, \dots ,$$

be the dual representations. In particular, $\tau_x := \tau_x^1$ is a representation on the tangent space $T_x(X) = (\mathfrak{m}_x/\mathfrak{m}_x^2)^*$, which is called the *isotropy representation* of G at x.

Linearization Theorem *Let K be a compact group acting continuously on a complex space X. Suppose that K has a fixed point $x \in X$. Then there exists a K-invariant neighborhood U of x in X, a $\tau_x(K)$-invariant neighborhood V of 0 in $T_x(X)$, and a closed K-equivariant embedding $\varphi : U \to V$.*

Remark Assume that X is a complex manifold. Then the theorem says that one can choose local coordinates at x, in which the action of K is linear. This classical result goes back to H.Cartan and S.Bochner. The generalization to arbitrary complex spaces is due to W.Kaup, see [Ka1].

Lemma 1 *Let A, B be analytic local $\mathbb{C}$-algebras, $\mathfrak{m}(A), \mathfrak{m}(B)$ their maximal ideals, and $\alpha : A \to B$ any homomorphism. Denote by $\dot{\alpha}$ the induced map $\mathfrak{m}(A)/\mathfrak{m}(A)^2 \to \mathfrak{m}(B)/\mathfrak{m}(B)^2$. If $\dot{\alpha}$ is surjective then α is also surjective.*

Proof This is a well-known consequence of the Weierstrass Division Theorem, see [GR1], Ch. 2, §3, Theorem 1. □

Lemma 2 *Let K be a compact group acting continuously on a Hausdorff topological space. Then, for any open subset $U \subset X$ the intersection $U' = \bigcap_{k \in K} k \cdot U$ is also open.*

Proof Assume the contrary and suppose $p \in U'$ is not an interior point. Then there exists a sequence $\{p_\nu\}_{\nu=1,2,\dots}$ such that $p_\nu \notin U'$ for all ν and $\lim p_\nu = p$. For each ν one can find $k_\nu \in K$ satisfying $p_\nu \notin k_\nu \cdot U$ or, equivalently, $q_\nu := k_\nu^{-1} p_\nu \notin U$. Since K is compact, we may assume that $\{k_\nu\}$ has a limit, say, $k \in K$. Then $q := \lim q_\nu = k^{-1}p$. It is clear that $q \notin U$, so that $p = kq \notin k \cdot U$. This is a contradiction. □

Proof of the theorem Let $f_{1,x}, \dots, f_{n,x}$ be any basis of $\mathfrak{m}_x$ modulo $\mathfrak{m}_x^2$. There exists a neighborhood U of x and holomorphic functions $f_i \in \mathcal{O}(U)$ having the germs $f_{i,x}$ at x. By Lemma 2 we may assume that U is K-invariant. Using the canonical isomorphism

$$\mathfrak{m}_x/\mathfrak{m}_x^2 \simeq \mathrm{Hom}_{\mathbb{C}}\left((\mathfrak{m}_x/\mathfrak{m}_x^2)^*, \mathbb{C}\right),$$

we obtain from $f_{1,x}, \dots, f_{n,x}$ a coordinate system $z_1, \dots, z_n$ on $T_x(X)$.

Define a holomorphic mapping $\phi : U \to T_x(X)$ by

$$z_i \circ \phi = f_i, \quad i = 1, \dots, n.$$

For $k \in K$ denote by ϕ^k the composition

$$U \xrightarrow{k} U \xrightarrow{\phi} T_x(X) \xrightarrow{\tau_x(k)^{-1}} T_x(X).$$

Then $k \mapsto \phi^k$ is a continuous function on K with values in the Fréchet space $\mathrm{Hol}(U, T_x(X))$. Averaging ϕ^k we obtain a holomorphic mapping

$$\varphi := \int_K \phi^k \cdot d\mu(k) \in \mathrm{Hol}(U, T_x(X)),$$

where μ is the normalized Haar measure on K. The invariance of μ implies that

$$\varphi \circ l = \tau_x(l) \circ \varphi \qquad \text{for all } \ l \in K,$$

i.e., φ is an equivariant mapping. Now, let $(a_{ij}(k))$ be a matrix defined by

$$(f_j \circ k^{-1})_x \equiv \sum_i a_{ij}(k) f_{i,x} \mod \mathfrak{m}_x^2,$$

where $k \in K$. Then

$$z_j \circ \tau_x(k)^{-1} = \sum_i a_{ij}(k) z_i.$$

Therefore

$$z_j \circ \varphi = z_j \circ \int_K \phi^k \cdot d\mu(k) = \int_K z_j \circ \phi^k \cdot d\mu(k) =$$

$$= \int_K [z_j \circ \tau_x(k)^{-1} \circ \phi \circ k] \cdot d\mu(k) = \int_K \Big[\sum_p a_{pj}(k) z_p \circ \phi \circ k \Big] \cdot d\mu(k) =$$

$$= \int_K \Big[\sum_p a_{pj}(k) f_p \circ k \Big] \cdot d\mu(k) = \int_K \Big[\sum_{p,q} a_{pj}(k) a_{qp}(k^{-1}) f_q \Big] \cdot d\mu(k) = f_j \ + \ \tilde{f}_j,$$

where $\tilde{f}_{j,x} \in \mathfrak{m}_x^2$ for all j, $j = 1, 2, \ldots, n$. By construction

$$\varphi(x) = 0 \in T_x(X) = \mathbb{C}^n,$$

and so we have the homomorphism of local algebras

$$\tilde{\varphi}_x : \mathcal{O}_{\mathbb{C}^n,0} \to \mathcal{O}_{X,x}.$$

The above calculation shows that the induced map

$$\mathfrak{m}_0/\mathfrak{m}_0^2 \to \mathfrak{m}_x/\mathfrak{m}_x^2$$

is surjective. By Lemma 1 the homomorphism $\tilde{\varphi}_x$ is also surjective. Geometrically speaking, this means that one can shrink the neighborhood U and find a neighborhood V of $0 \in T_x(X)$ so that $\varphi(U) \subset V$ and $\varphi|_U : U \to V$ is a closed embedding. By Lemma 2 we can choose U and V to be invariant. Then $\varphi|_U$ is automatically equivariant. □

As a corollary we obtain the following important result.

Identity Theorem *Let K be a compact group acting continuously on a complex space X and $Kx = \{x\}$. Then the following conditions are equivalent:*

(i) *the representation of K on $\mathcal{O}_x$ is faithful;*

(ii) *there exists a base for the topology at x, consisting of K-invariant neighborhoods with effective action;*

(iii) *the representation τ_x is faithful.*

If X is reduced and X' is the union of all irreducible components of X containing x, then (i) - (iii) *are equivalent to*

(ii') *the K-action on X' is effective.* □

Remark It follows from (iii) that K is a Lie group. By the result of §1.6 the K-action on X is analytic.

Example For $a = (a_1, \ldots, a_n) \in \mathbb{C}^n$ let φ_a be the automorphism of $\mathbb{P}_n$ given by $\varphi_a(z_0 : z_1 : \ldots : z_n) = (z_0 + a_1 z_1 + \ldots + a_n z_n : z_1 : \ldots : z_n)$. Since $\varphi_a \circ \varphi_b = \varphi_{a+b}$, we obtain an action of $G = \mathbb{C}^n$ on $\mathbb{P}_n$, which is obviously effective. Note that $x = (1 : 0 : \ldots : 0)$ is a fixed point. In the local coordinates $\zeta_i = z_i/z_0$ at x the mapping φ_a has the form

$$(\zeta_1, \ldots, \zeta_n) \mapsto \Big(\frac{\zeta_1}{1 + \sum a_i \zeta_i}, \ldots, \frac{\zeta_n}{1 + \sum a_i \zeta_i} \Big).$$

It follows that $\tau_x(a)$ is the unit matrix for all $a \in \mathbb{C}^n$.

Thus, the compactness assumption in the Identity Theorem is essential. On the other hand, the following weak version of the Identity Theorem does not require this assumption.

Proposition *Let G be a Lie group acting analytically on a complex space X. Let $x \in X$ be a fixed point of G and assume that the representation of G on $\mathcal{O}_x$ is faithful. (For example, X is reduced and G is effective on X'.) Then there exists a number d such that* Ker τ_x^d *is discrete. Moreover, if G is connected then for some d the representation τ_x^d is faithful.*

Proof Denote by $\mathfrak{h}_d$ the Lie algebra of $H_d :=$ Ker τ_x^d. Since $\bigcap_{d=1}^{\infty} \mathfrak{m}_x^d = \{0\}$, it follows that

$$\bigcap_{d=1}^{\infty} H_d = \{e\} \qquad \text{and} \qquad \bigcap_{d=1}^{\infty} \mathfrak{h}_d = \{0\}.$$

Obviously, $\mathfrak{h}_d \supset \mathfrak{h}_{d+1}$ and $H_d \supset H_{d+1}$ for all d. Since each $\mathfrak{h}_d$ is a finite-dimensional vector space, we obtain $\mathfrak{h}_d = 0$ for some d.

Fix d with this property. Then H_d is a discrete normal subgroup in G. Since G is connected, H_d is abelian (in fact, central in G) and finitely generated. We claim that

$$(*) \quad \text{card } H_d/H_{d+1} < \infty \implies H_d = H_{d+1}.$$

Assume this is proved and consider the descending sequence of groups $\{H_j\}_{j \geq d}$. For all sufficiently large j the rank of H_j is independent of j. Thus H_{j+1}/H_j is finite and, consequently, $H_j = H_{j+1}$. Since the intersection of all H_j is trivial, each of these groups equals $\{e\}$.

In order to prove (∗) take an $h \in H_d$ and write

$$(f_i \circ h^{-1})_x \equiv f_{i,x} + q_{i,x} \qquad \bmod \quad \mathfrak{m}_x^{d+2},$$

where f_i are the same as in the proof of the theorem and $q_{i,x} \in \mathfrak{m}_x^{d+1}$. More precisely,

$$q_{i,x} = \sum c_{i,\alpha} f_{1,x}^{\alpha_1} \cdot \ldots \cdot f_{n,x}^{\alpha_n},$$

where $\alpha = (\alpha_1, \ldots, \alpha_n) \in \mathbb{Z}^n$, $\alpha_i \geq 0$, $\sum \alpha_i = d+1$, and $c_{i,\alpha} \in \mathbb{C}$. By iterating we obtain

$$(f_i \circ h^{-m})_x \equiv f_{i,x} + m q_{i,x} \qquad \bmod \quad \mathfrak{m}_x^{d+2}.$$

Since $h^m \in H_{d+1}$ for some $m \geq 1$, it follows that $q_{i,x} \in \mathfrak{m}_x^{d+2}$. Hence $h \in H_{d+1}$. □

2.3 The automorphism group of a compact complex space

Let X be a compact complex space. In this section we prove that $\mathrm{Aut}(X)$ is a complex Lie group and that the action of $\mathrm{Aut}(X)$ on X is holomorphic. This theorem in the non-singular case is due to S.Bochner and D.Montgomery, see [BM3]. R.Gunning [Gu1] and H.Kerner [Ker] extended the result to reduced complex spaces. The final version is due to W.Kaup, see [Ka1]. The proof is based on the general theory of topological groups. For the convenience of the reader we state all necessary results and give the corresponding references. All proofs can be also found in [MZ].

A topological group G is called a *group without small subgroups*, if there exists a neighborhood of $e \in G$ which contains no subgroups of G except $\{e\}$.

Theorem I (see [Gl],[Ya]) *A locally compact group without small subgroups is a Lie group (i.e., has a structure of a Lie group compatible with the topology).*

Theorem II (see [MZ], p.172) *Let G be a locally compact group and let Ω be a neighborhood of $e \in G$. There exists an open subgroup $H \subset G$ and a compact subgroup $K \subset \Omega \cap H$, such that K is normal in H and H/K has no small subgroups.*

Theorem III (see [BM2]) *Let G be a locally compact effective transformation group of a connected manifold M of class C^k, $k \geq 1$. If each transformation of G is of class C^1, then G is a Lie group and the mapping $G \times M \to M$ is of class C^k.*

Remark Theorem III is used below for disconnected manifolds with a finite number of connected components. Since the subgroup of G preserving each component is open and of finite index, the generalization is immediate.

We shall also need some preparation, which is mostly necessary to control nilpotent elements of the structure sheaf.

Lemma *Let X be a complex space, $\mathcal{F}$ a coherent sheaf of $\mathcal{O}_X$-modules, and $A :=$* supp $\mathcal{F}$. *For each $s \in \mathcal{F}(X)$ the set $A_s := \{x \in X | s_x \neq 0\}$ is the union of some irreducible components of the analytic set A.*

Proof We have

$$A = \mathrm{supp}\ \mathcal{F}/\mathcal{O}s \bigcup \mathrm{supp}\ \mathcal{O}s.$$

Both sets on the right hand side are analytic and the second one coincides with A_s. □

Let $\mathcal{F} := \mathcal{N}$, $A := \operatorname{supp} \mathcal{N}$, and suppose that X is compact. There exists a finite covering $\{U_i\}_{i=1,\dots,r}$ of X with the following properties: (i) $A \cap U_i$ has finitely many irreducible components; (ii) there is a finite number of sections $f_{i1}, \dots, f_{ik_i} \in \mathcal{O}(U_i)$ generating $\mathcal{O}_x$ as an analytic local algebra for each $x \in U_i$ $(i = 1, \dots, r)$. Denote by Γ_i, $i = 1, \dots, r$, a finite subset of $A \cap U_i$ containing at least one point of each irreducible component of $A \cap U_i$.

Proposition *Let φ be an automorphism of X such that $\varphi(x) = x$ for all $x \in X$. Denote by φ_x the induced automorphism of the stalk algebra $\mathcal{O}_x$. Assume that $\varphi_x = \mathrm{id}$ for all $x \in \Gamma := \Gamma_1 \cup \dots \cup \Gamma_r$. Then $\varphi = \mathrm{id}_X$.*

Proof Since φ acts on red X trivially, it follows that the corresponding automorphism of $\mathcal{O}$ induces the identity mapping on $\mathcal{O}/\mathcal{N}$. Therefore

$$n_{ij} := f_{ij} \circ \varphi - f_{ij} \in \mathcal{N}(U_i), \qquad j = 1, \dots, k_i.$$

It is clear that

$$\{x \in U_i \mid \varphi_x \neq \mathrm{id}\} = \bigcup_{j=1}^{k_i} \{x \in U_i \mid (n_{ij})_x \neq 0\}.$$

Our assumption implies that this set does not contain an irreducible component of $A \cap U_i$. Thus by the above lemma it must be empty. □

Theorem *Let X be a compact complex space. The automorphism group* Aut(X) *can be endowed with a structure of a complex Lie group so that the action of* Aut(X) *on X is holomorphic. The Lie algebra of* Aut(X) *is isomorphic to the Lie algebra $\mathcal{T}_X(X)$ of all holomorphic vector fields on X.*

Proof Since each automorphism of X determines in a natural way an automorphism of red X, we have a canonical mapping

$$\chi : \mathrm{Aut}(X) \to \mathrm{Aut}(\mathrm{red}\, X).$$

It is clear that χ is a continuous homomorphism of topological groups. We know from §2.1 that both groups are locally compact. The second one acts effectively on the complex manifold red $X - S(\mathrm{red}\, X)$. Thus, in view of Theorem III, Aut(red X) is a Lie group and, in particular, a group without small subgroups. We want to show that Aut(X) is also a group without small subgroups.

Assume the contrary. Then Ker χ must be a group with small subgroups. Furthermore, using Theorem II one can find a compact subgroup $K \subset \operatorname{Ker} \chi$, which also has small subgroups. By the proposition Ker χ acts effectively on $\bigoplus_{x \in \Gamma} \mathcal{O}_x$. Consider the direct sum of the isotropy representations

$$\tau_\Gamma : \operatorname{Ker} \chi \ \to \ \mathrm{GL}\,\Big(\bigoplus_{x \in \Gamma} T_x(X)\Big).$$

By the Identity Theorem the restriction of τ_Γ to K is a faithful representation, so that K is a Lie group. Thus we obtain a contradiction showing that $\mathrm{Aut}(X)$ is a group without small subgroups. Applying Theorem I we see that $\mathrm{Aut}(X)$ is a Lie group.

To simplify the notation write G instead of $\mathrm{Aut}(X)$. Since the G-action on X is effective, the resulting Lie homomorphism $\lambda : \mathfrak{g} \to \mathcal{T}_X(X)$ is an embedding (see §1.7). By the proposition of §1.5 any element of $\mathcal{T}_X(X)$ generates a global one-parameter transformation group of X. It follows that λ is an isomorphism of real Lie algebras. But $\mathcal{T}_X(X)$ is a complex Lie algebra. With the help of λ the complex structure from $\mathcal{T}_X(X)$ can be carried over to $\mathfrak{g}$. Thus G is a complex Lie group and the G-action on X is holomorphic. □

Corollary *For a compact complex space X the Lie algebra $\mathcal{T}_X(X)$ has finite dimension.* □

Remark This is, of course, a special case of the Finiteness Theorem due to H.Cartan and J.-P. Serre.

We now consider the automorphism groups of several well-known compact complex spaces.

Example 1: The complex projective space $\mathbb{P}_n$ Each invertible $(n+1)\times(n+1)$ complex matrix $A = (a_{ij})$ determines an automorphism φ_A of the complex projective space $\mathbb{P}_n$. In homogeneous coordinates φ_A is given by

$$(z_0 : z_1 : \ldots : z_n) \mapsto (z'_0 : z'_1 : \ldots : z'_n), \quad \text{where } z'_i = \sum a_{ij} z_j .$$

We claim that each $\varphi \in \mathrm{Aut}(\mathbb{P}_n)$ is of the form $\varphi = \varphi_A$ for some $A \in \mathrm{GL}(n+1, \mathbb{C})$. This follows from the theorem of Hurwitz-Weierstrass, according to which each meromorphic function on $\mathbb{P}_n$ is the ratio of two homogeneous polynomials of the same degree (see e.g. [GR3], Ch. 9, §5). Applying this theorem to the meromorphic functions $(z_j/z_0) \circ \varphi$ we see that

$$\varphi((z_0 : z_1 : \ldots : z_n)) = (p_0 : p_1 : \ldots : p_n),$$

where $p_j = p_j(z_0, z_1, \ldots, z_n)$ are homogeneous polynomials of the same degree. Since this is also true for φ^{-1}, all p_j must be linear. Clearly, $\varphi_A = \mathrm{id}$ if and only if $a_{ij} = \lambda\delta_{ij}$. In other words, we have an epimorphism

$$\alpha : \mathrm{GL}(n+1, \mathbb{C}) \to \mathrm{Aut}(\mathbb{P}_n), \quad \alpha(A) = \varphi_A,$$

and Ker α coincides with the center of $\mathrm{GL}(n+1, \mathbb{C})$. This yields an isomorphism

$$\mathrm{Aut}(\mathbb{P}_n) \simeq \mathrm{PGL}(n+1, \mathbb{C}).$$

Remark It is easily seen that for any point $o \in \mathbb{P}_n$ the isotropy subgroup of o in $\mathrm{Aut}(\mathbb{P}_n)$ is transitive on $\mathbb{P}_n - \{o\}$. Thus, $\mathbb{P}_n$ is an almost homogeneous manifold of a connected complex Lie group and the complement to the open orbit has an isolated point. This property characterizes $\mathbb{P}_n$ in the class of compact complex manifolds, see [Oe1].

Example 2: Complex tori A complex torus $T = T_\Gamma$ is the quotient group of $\mathbb{C}^n$ by a lattice $\Gamma \subset \mathbb{C}^n$ of rank $2n$. A torus is a connected compact complex Lie group. Conversely, a connected compact complex Lie group is abelian and, therefore, isomorphic to some T_Γ. We write the group operation in T additively. The neutral element is denoted by 0.

An automorphism $\varphi : T \to T$ fixing the origin $0 \in T$ is a group automorphism.

Proof Define a holomorphic mapping $\psi : T \times T \to T$ by

$$\psi(x,y) \;=\; \varphi(x+y) \;-\; \varphi(x) \;-\; \varphi(y)$$

and put $\psi_x(y) := \psi(x,y)$. Choose a Stein neighborhood of 0, say, U. Since $\psi(\{0\} \times T) = \{0\}$ and T is compact, there exists an open neighborhood V_1 of 0 such that $\psi(V_1 \times T) \subset U$. For any $x \in V_1$ we have $\psi_x(T) \subset U$. Since U is Stein, the image $\psi_x(T)$ is a point. Thus, $\psi|(V_1 \times T)$ does not depend on y. Similarly, there exists a neighborhood V_2 of 0 such that $\psi|(T \times V_2)$ does not depend on x. Therefore, the restriction $\psi|(V_1 \times V_2)$ is a constant mapping. It follows that $\psi(x,y) = 0$ for all $x, y \in T$. □

Holomorphic group automorphisms of T can be identified, after lifting to $\mathbb{C}^n$, with invertible linear transformations of $\mathbb{C}^n$ preserving Γ. For any $x \in T$ denote by t_x the corresponding group translation

$$y \mapsto x + y, \qquad y \in T.$$

The mapping $x \mapsto t_x$ is a group monomorphism of T into $\mathrm{Aut}(T)$, and we identify T with its image in $\mathrm{Aut}(T)$. Take any $\varphi \in \mathrm{Aut}(T)$ and let $x := \varphi(0)$. By the above result $t_{-x} \circ \varphi$ is a group automorphism of T. Thus $\mathrm{Aut}(T)$ is generated by group automorphisms and group translations. Moreover, T is normal in $\mathrm{Aut}(T)$ and

$$\mathrm{Aut}(T) \;=\; \{A \in \mathrm{GL}(n,\mathbb{C}) \mid A(\Gamma) = \Gamma\} \ltimes T,$$

where $\ltimes$ denotes a semidirect product.

Example 3: The n-fold point This is the complex model space defined in $\mathbb{C}$ by the ideal sheaf $z^n\mathcal{O}$. The underlying topological space consists of one point. The stalk of the structure sheaf over this point is the local $\mathbb{C}$-algebra $\mathbb{C}+\mathbb{C}\epsilon+\ldots+\mathbb{C}\epsilon^{n-1}$, where $\epsilon^n = 0$. Automorphisms of the n-fold point can be identified with automorphisms of this algebra. Denote by G_n the automorphism group of the n-fold point. Then

G_n is a solvable group for all $n \geq 1$, and $\mathrm{G}_1 = \{e\}$, $\mathrm{G}_2 \simeq \mathbb{C}^*$, $\mathrm{G}_3 \simeq \mathrm{Aut}(\mathbb{C})$.

Proof For all $n \geq 2$ we have an epimorphism $\alpha_n : \mathrm{G}_n \to \mathrm{G}_{n-1}$. It is clear that $\mathrm{G}_1 = \{e\}$, $\mathrm{G}_2 \simeq \mathbb{C}^*$, and $\operatorname{Ker} \alpha_n \simeq \mathbb{C}$ for $n \geq 3$. Thus G_n is solvable for all n.

We now consider the case $n = 3$. Any automorphism of $\mathbb{C}$

$$z \mapsto \varphi_{a,b}(z) := az + b, \qquad a \in \mathbb{C}^*,\ b \in \mathbb{C},$$

extends to an automorphism of $\mathbb{P}_1$, which fixes the point $p = \infty$ and induces an automorphism $\tilde{\varphi}_{a,b} : \mathcal{O}_{\mathbb{P}_1,p} \to \mathcal{O}_{\mathbb{P}_1,p}$. Let $\zeta := 1/z$ and $\epsilon := \zeta_p \in \mathcal{O}_{\mathbb{P}_1,p}$. Then

$$\tilde{\varphi}_{a,b}(\epsilon) \equiv \sum_{k=1}^{n-1} \frac{(-1)^{k-1} b^{k-1}}{a^k} \epsilon^k \qquad \bmod \mathfrak{m}_p^n.$$

Each automorphism of $\mathcal{O}_{\mathbb{P}_1,p}$ induces an automorphism of $\mathcal{O}_{\mathbb{P}_1,p}/\mathfrak{m}_p^n$, i.e., an element of G_n. Let $\tilde{\varphi}_{a,b}^{(n)} \in \mathrm{G}_n$ be the automorphism obtained in this way from $\tilde{\varphi}_{a,b}$. Denote the coset of ϵ modulo $\mathfrak{m}_p^n$ again by ϵ. Then for $n = 3$ we have

$$\tilde{\varphi}_{a,b}^{(3)}(\epsilon) = \frac{1}{a}\epsilon - \frac{b}{a^2}\epsilon^2, \qquad \epsilon^3 = 0.$$

The mapping $\varphi_{a,b} \mapsto \tilde{\varphi}_{a,b}^{(3)}$ is the desired group isomorphism. □

For X non-compact it happens quite often that $\mathrm{Aut}(\mathrm{red}\, X)$ is locally compact, but $\mathrm{Aut}(X)$ is not.

Example 4 Consider the *double z-axis* in $\mathbb{C}^2$, i.e., the closed complex subspace defined by the ideal sheaf $\mathcal{J} = w^2\mathcal{O}_{\mathbb{C}^2} \subset \mathcal{O}_{\mathbb{C}^2}$. Denote by ζ and η the images of z and w in $\mathcal{O}_X(X)$. The function $[\zeta]$ determines an isomorphism $\mathrm{red}\, X \simeq \mathbb{C}$. The multiplication by η_x in each stalk algebra $\mathcal{O}_{X,x}$ gives rise to a sheaf epimorphism $\mathcal{O} \to \mathcal{N}$, whose kernel is also equal to $\mathcal{N}$. Since X is a Stein space, the associated sequence of modules of global sections

$$0 \to \mathcal{N}(X) \to \mathcal{O}(X) \xrightarrow{\cdot\eta} \mathcal{N}(X) \to 0$$

is exact. Hence $\mathcal{N}(X) = \mathcal{O}(X)\cdot\eta$ and for $f_1, f_2 \in \mathcal{O}(X)$ one has $f_1\cdot\eta = f_2\cdot\eta$ if and only if $[f_1] = [f_2]$. Using the isomorphism $\mathrm{red}\, X \simeq \mathbb{C}$ we shall write any element of $\mathcal{N}(X)$ as $f\cdot\eta$, where $f \in \mathcal{O}(\mathbb{C})$.

Let φ be an automorphism of X. Then $[\zeta]\circ\varphi = a[\zeta]+b$ for some $a \in \mathbb{C}^*, b \in \mathbb{C}$. Thus $\zeta\circ\varphi - (a\zeta + b) \in \mathcal{N}(X)$ and, of course, $\eta\circ\varphi \in \mathcal{N}(X)$. Consequently

$$\zeta\circ\varphi = a\zeta + b + g\cdot\eta \qquad \text{and} \qquad \eta\circ\varphi = f\cdot\eta,$$

where $f \in \mathcal{O}^*(\mathbb{C}), g \in \mathcal{O}(\mathbb{C})$. Conversely, given a, b, f, g as above one can construct φ using the above data in the following way. Consider the holomorphic mapping

$$\Phi : \mathbb{C}^2 \to \mathbb{C}^2, \quad (z, w) \mapsto (az + b + g(z)w, f(z)w).$$

Observe that Φ preserves $\mathcal{J}$ and define φ to be the restriction of Φ to X. In order to see that φ is an automorphism of X take the mapping

$$\Psi : \mathbb{C}^2 \to \mathbb{C}^2, \quad (z, w) \mapsto \left(z^* - \frac{g(z^*)w}{af(z^*)}, \frac{w}{f(z^*)} \right), \qquad \text{where} \quad z^* = \frac{z-b}{a}.$$

One checks immediately that Ψ restricts to X and that the restriction is φ^{-1}. Thus $\mathrm{Aut}(X)$ is homeomorphic to $\mathbb{C}^* \times \mathbb{C} \times \mathcal{O}^*(\mathbb{C}) \times \mathcal{O}(\mathbb{C})$ and, in particular, $\mathrm{Aut}(X)$ is not locally compact. Note that the group operation in $\mathrm{Aut}(X)$ is given by

$$(a, b|f, g)\cdot(c, d|h, k) = (\ ac\ ,\ ad + b\ |\ f^{c,d}h\ ,\ ak + g^{c,d}h\ ),$$

where $a, c \in \mathbb{C}^*, b, d \in \mathbb{C}, f, h \in \mathcal{O}^*(\mathbb{C}), g, k \in \mathcal{O}(\mathbb{C})$, and $f^{a,b}(t) = f(at + b)$.

2.4 Automorphisms of fiber bundles

The results of this section in the non-singular case are due to A.Blanchard, see [Bl].

Lemma 1 *Let $\pi : X \to Y$ be a proper holomorphic mapping between reduced complex spaces X and Y. Suppose we have a (real analytic) action of a connected (real) Lie group G on X. For each $g \in G$ denote by the same letter the corresponding holomorphic transformation of X. Let $y_0 \in Y$ and let A be a connected analytic subset of the fiber $\pi^{-1}(y_0)$. Then*

$$\pi(gA) = \{one\ point\} \tag{1}$$

for all $g \in G$.

Proof. Fix a Stein neighborhood $U = U(y_0) \subset Y$. Since $\pi^{-1}(y_0)$ is compact, one has

$$g \cdot \pi^{-1}(y_0) \subset \pi^{-1}(U),$$

if g is in some neighborhood of $e \in G$. In particular, $\pi \circ g$ maps A into U. But A is a connected compact analytic set, and so a holomorphic mapping from A to a Stein space is constant. Therefore $\pi(gA)$ is a point.

Denote by Ω the set of all $g \in G$ satisfying (1). The above argument shows that Ω is open. On the other hand, let $g_k \in \Omega$ and suppose that $g_k \to g \in G$ as $k \to \infty$. For $x_1, x_2 \in A$ one has $\pi(g_k x_1) = \pi(g_k x_2)$ for all k. Passing to the limit we obtain

$$\pi(gx_1) = \pi(gx_2).$$

Thus $\pi(gA)$ is a one-point set. Hence $g \in \Omega$, and Ω is closed. Since G is connected, it follows that $G = \Omega$. □

Assume that π has connected fibers. Then, according to Lemma 1, each $g \in G$ interchanges the fibers of π. Therefore g induces a homeomorphism of Y (denoted by the same letter) so that the diagram

$$\begin{array}{ccc} X & \xrightarrow{g} & X \\ \pi\downarrow & & \downarrow\pi \\ Y & \xrightarrow{g} & Y \end{array}$$

commutes.

Example 1 The induced transformation $g : Y \to Y$ is not necessarily holomorphic. Let $X := \mathbb{C}$, $Y := \{(z,w) \in \mathbb{C}^2 \mid z^2 = w^3\}$, and $G \simeq \mathbb{C}$ the translation group of $\mathbb{C}$. The mapping $\pi : X \to Y$ defined by $\pi(t) := (t^3, t^2)$ is one-to-one, but the transformation $g : Y \to Y$ is holomorphic only if $g = e$. In all other cases the image $g(o)$ of the singular point $o = (0,0) \in Y$ is a non-singular point.

Lemma 2 *Assume that $\pi_* \mathcal{O}_X = \mathcal{O}_Y$. Then $g : Y \to Y$ is holomorphic for all $g \in G$. The induced G-action on Y is real analytic. If G is a complex Lie group* *and if the given G-action on X is holomorphic, then the induced G-action on Y is also holomorphic. The diagram*

$$\begin{array}{ccc} G \times X & \rightarrow & X \\ \mathrm{id} \times \pi \downarrow & & \downarrow \pi \\ G \times Y & \rightarrow & Y \end{array}$$

commutes.

Proof Let U be a neighborhood of $y_0 \in Y$, $f \in \mathcal{O}_Y(U)$, and g a fixed element of G. The holomorphic function $f \circ (g \circ \pi) = f \circ (\pi \circ g)$ in $\pi^{-1}(g^{-1}U)$ defines an element of $\pi_* \mathcal{O}_X(g^{-1}U)$. Since $\pi_* \mathcal{O}_X = \mathcal{O}_Y$, one can shrink U so that this section is defined by a holomorphic function in $g^{-1}U$. This means that the continuous function $z \mapsto f(gz)$ is holomorphic in some neighborhood of $g^{-1}(y_0)$.

The induced G-action on Y is continuous and therefore, as we know from §1.6, real analytic. The fact that the diagram commutes is obvious. If G is a complex Lie group acting holomorphically on X, then

$$(\mathrm{id} \times \pi)_* \mathcal{O}_{G \times X} = \mathcal{O}_{G \times Y}$$

and the same argument as above shows that the induced action on Y is holomorphic. □

Remark The assumption $\pi_* \mathcal{O}_X = \mathcal{O}_Y$ implies that π has connected fibers.

Proposition 1 *Let (X, Y, π) be a locally trivial holomorphic fiber bundle, where the total space X and the base Y are reduced compact complex spaces and the bundle projection $\pi : X \rightarrow Y$ has connected fibers. Then each automorphism from* $\mathrm{Aut}^\circ(X)$ *interchanges the fibers of π and induces an automorphism of Y. One has a canonical homomorphism of complex Lie groups*

$$\mathrm{Aut}^\circ(X) \rightarrow \mathrm{Aut}^\circ(Y). \tag{2}$$

Proof Since the condition $\pi_* \mathcal{O}_X = \mathcal{O}_Y$ holds for locally trivial bundles, the proposition follows from Lemma 1 and Lemma 2. □

Example 2 We keep the notation of Proposition 1. Let X be a compact non-singular surface, $Y = \mathbb{P}_1$, and assume that the typical fiber of π is isomorphic to $\mathbb{P}_1$. Then X is a projective algebraic surface (see [K], Theorem 8), which is rational and geometrically ruled. For each $n \in \mathbb{Z}^+$ there is exactly one, up to an isomorphism, fibre bundle (X, Y, π) with $Y = \mathbb{P}_1$ and with fiber $\mathbb{P}_1$. It is characterized by the following property: there exists a unique section $\sigma : Y \rightarrow X$ such that the image $C := \sigma(\mathbb{P}_1) \subset X$ is a curve with self-intersection index $-n$. The surface X with this property is denoted by F_n, see [Sh], Ch. 5, §1. We determine here the group $\mathrm{Aut}^\circ(F_n)$. It should be mentioned that for $n > 0$ the group $\mathrm{Aut}(F_n)$ is in fact connected, see [HuOe2], Ch. 2, §2, Prop. 7.

In terms of Lie transformation groups F_n admits the following description. Let B be the subgroup of upper triangular matrices in $\mathrm{SL}(2, \mathbb{C})$ and let

$$\varphi_n(b) := \alpha^n, \qquad \text{where} \quad b = \begin{pmatrix} \alpha & \beta \\ 0 & \alpha^{-1} \end{pmatrix} \in B.$$

Then F_n is obtained from the product $SL(2,\mathbb{C}) \times \mathbb{P}_1$ as the quotient space with respect to the free B-action

$$(g,\zeta) \overset{b}{\mapsto} (gb^{-1}, \varphi_n(b)^{-1}\zeta), \qquad \text{where} \quad g \in SL(2,\mathbb{C}),\ b \in B,\ \zeta \in \mathbb{P}_1 = \mathbb{C} \cup \{\infty\}.$$

(The action is given by a morphism of algebraic varieties. Since all orbits have the same dimension, they are closed.) Denote the equivalence class of (g,ζ) in F_n by $[g,\zeta]$. Then the projection mapping $\pi : F_n \to \mathbb{P}_1 = SL(2,\mathbb{C})/B$ and the section $\sigma : \mathbb{P}_1 \to F_n$ are given by

$$\pi([g,\zeta]) = gB, \quad \sigma(gB) = [g,\infty].$$

Let us show that the self-intersection index of $C = \sigma(\mathbb{P}_1)$ is equal to $-n$. Take a non-zero regular function $q(g)$, $g \in SL(2,\mathbb{C})$, which depends only on the first column of the matrix g and is a homogeneous polynomial of degree n in the entries of this column. Then, as one can easily check, the mapping

$$\sigma_t : \mathbb{P}_1 \to F_n, \quad \sigma_t(gB) := [g, tq(g)],$$

is well-defined for every $t \in \mathbb{C}$. It is also clear that σ_t is a section of the bundle $(F_n, \mathbb{P}_1, \pi)$. Let $C_t := \sigma_t(\mathbb{P}_1)$. Then C_0 and C_t $(t \neq 0)$ have n intersection points counting multiplicities, so that $C_0^2 = (C_0, C_t) = n$. On the other hand, $f([g,\zeta]) := \zeta/q(g)$ is a well-defined rational function on F_n with the divisor $(f) \sim C_0 - C - nF$, where F is a fiber of $\pi : F_n \to \mathbb{P}_1$. Therefore, $C \sim C_0 - nF$ and $C^2 = C_0^2 - 2n(C_0, F) + n^2F^2 = n - 2n + 0 = -n$.

The group $SL(2,\mathbb{C})$ acts on F_n in a natural way, namely,

$$[g,\zeta] \overset{s}{\mapsto} [sg,\zeta], \qquad \text{where} \quad s, g \in SL(2,\mathbb{C}).$$

It follows that the homomorphism (2) is surjective. Let J be its kernel and assume that $n > 0$. (Since $F_0 = \mathbb{P}_1 \times \mathbb{P}_1$ the group $\mathrm{Aut}(F_0)$ is easily understood.) In order to determine J, observe that C is invariant under all automorphisms of F_n. Thus each point of C is fixed by J. This means that an element $j \in J$ acts on F_n in the following way:

$$j \cdot [g,\zeta] = [g, p(g)\zeta + q(g)],$$

where p and q are regular functions on $SL(2,\mathbb{C})$, satisfying

$$p(gb) = p(g)$$

and

$$q(gb) = \varphi_n(b)q(g)$$

for all $g \in SL(2,\mathbb{C})$, $b \in B$. Since $SL(2,\mathbb{C})/B$ is compact, the first equation shows that $p(g)$ is constant. From the second equation one can easily deduce that $q(g)$ is a homogeneous polynomial of degree n in the entries of the first column of $g \in SL(2,\mathbb{C})$.

Therefore J is a solvable group of dimension $n+2$. More precisely, one has the exact sequence

$$\{e\} \to J' \to J \to \mathbb{C}^* \to \{e\},$$

where the commutator subgroup $J' \subset J$ is given by $p = 1$ so that $J' \simeq \mathbb{C}^{n+1}$. (The extension is in fact trivial, i.e., $J \simeq \mathbb{C}^* \ltimes \mathbb{C}^{n+1}$.)

Since the kernel of the homomorphism (2) is solvable and since the image is isomorphic to $\mathrm{PSL}(2, \mathbb{C})$, it follows that J is the radical of $\mathrm{Aut}^\circ(F_n)$. The Levi decomposition of $\mathrm{Aut}^\circ(F_n)$ is of the form $\mathrm{Aut}^\circ(F_n) = S \cdot J$, where S is locally isomorphic to $\mathrm{SL}(2, \mathbb{C})$. The adjoint action of S on J' is given by an irreducible linear representation (of dimension $n + 1$). In particular,

$$\dim \mathrm{Aut}(F_n) = n + 5.$$

Proposition 2 *Let X_1 and X_2 be reduced compact complex spaces. Then the complex Lie groups* $\mathrm{Aut}^\circ(X_1 \times X_2)$ *and* $\mathrm{Aut}^\circ(X_1) \times \mathrm{Aut}^\circ(X_2)$ *are canonically isomorphic.*

Proof Since a connected group of automorphisms preserves each connected component of a complex space, one may assume that X_1 and X_2 are connected. By Proposition 1 we have canonical homomorphisms

$$\phi_i : \mathrm{Aut}^\circ(X_1 \times X_2) \to \mathrm{Aut}^\circ(X_i), \quad i = 1, 2.$$

On the other hand, there is a canonical embedding

$$\iota : \mathrm{Aut}^\circ(X_1) \times \mathrm{Aut}^\circ(X_2) \to \mathrm{Aut}^\circ(X_1 \times X_2),$$

such that $\phi_i \circ \iota = \mathrm{id}_{X_i}$, $i = 1, 2$. Namely, $\iota(g_1, g_2)(x_1, x_2) = (g_1 x_1, g_2 x_2)$, where $x_i \in X_i$, $g_i \in \mathrm{Aut}(X_i)$, $i = 1, 2$. It follows that the homomorphism $\phi := \phi_1 \times \phi_2$ is surjective. Since X_1 and X_2 are reduced, ϕ is also injective. □

Example 3 If X_1 and X_2 are not reduced, then Proposition 2 is in general false. Let X_1 (resp. X_2) be the double point with the local algebra $A_1 = \mathbb{C} + \mathbb{C}\epsilon_1$ (resp. $A_2 = \mathbb{C} + \mathbb{C}\epsilon_2$), where $\epsilon_1^2 = \epsilon_2^2 = 0$. Then $X = X_1 \times X_2$ is a one-point complex space equipped with the local algebra $A = A_1 \otimes A_2 = \mathbb{C} + \mathbb{C}\epsilon_1 + \mathbb{C}\epsilon_2 + \mathbb{C}\epsilon_1\epsilon_2$. The automorphism groups of X_1, X_2, and X are identified with the automorphism groups of the local algebras A_1, A_2, and A respectively. In particular,

$$\mathrm{Aut}(X_i) \simeq \mathbb{C}^* \quad (i = 1, 2)$$

(see §2.3, Example 3). On the other hand, in the basis $1, \epsilon_1, \epsilon_2, \epsilon_1\epsilon_2$ an automorphism of A is given by a matrix of one of two types:

$$\begin{pmatrix} 1 & 0 & 0 & 0 \\ 0 & a & 0 & 0 \\ 0 & 0 & b & 0 \\ 0 & u & v & ab \end{pmatrix} \quad \text{or} \quad \begin{pmatrix} 1 & 0 & 0 & 0 \\ 0 & 0 & b & 0 \\ 0 & a & 0 & 0 \\ 0 & u & v & ab \end{pmatrix},$$

where $a, b \in \mathbb{C}^*$, $u, v \in \mathbb{C}$. Thus $\mathrm{Aut}(X)$ has dimension 4 and consists of two connected components.

2.5 Proper actions

Let G be a topological group acting continuously on a locally compact topological space X. The action is called *proper* if the mapping

$$G \times X \to X \times X, \quad (g, x) \mapsto (gx, x),$$

is proper, i.e., the inverse image of a compact set is compact. This means that the subset

$$\{g \in G \mid gC_1 \cap C_2 \neq \emptyset\}$$

is compact for any two compact sets $C_1, C_2 \subset X$.

Lemma *Assume that a G-action on X is proper. Then:*
(i) *for any compact set $C \subset X$ the set GC is closed;*
(ii) *for any point $x \in X$ the isotropy subgroup G_x is compact.*
(iii) *G is locally compact.*
(iv) *the quotient space X/G is Hausdorff.*

Proof (i) Suppose $x_\nu \in C, g_\nu \in G$ and $y_\nu := g_\nu x_\nu$ tends to $y \in X$. Since C is compact, we may assume that $\{x_\nu\}$ converges to $x \in C$. By the definition of a proper action one can find a subsequence of $\{g_\nu\}$ converging to some $g \in G$. Apparently, $y = gx \in GC$.

(ii) This is clear from the following description of G_x:

$$G_x = \{g \in G \mid g\{x\} \cap \{x\} \neq \emptyset\}.$$

(iii) Fix $x \in X$ and let U be a neighborhood of x with compact closure. Then

$$\Omega := \{g \in G \mid gx \in U\}$$

is a neighborhood of $e \in G$ and

$$\Omega \subset \{g \in G \mid g\{x\} \cap \overline{U} \neq \emptyset\}.$$

Thus Ω is contained in a compact set.

(iv) By (i) all orbits are closed. Suppose that $Gx \neq Gy$. Then there exists a neighborhood U of x such that $\overline{U}$ is compact and $\overline{U} \cap Gy = \emptyset$. Since $y \notin G\overline{U}$ and since, again by (ii), $G\overline{U}$ is closed, there exists a neighborhood V of y such that $V \cap G\overline{U} = \emptyset$. (For example, one can take $V := X - G\overline{U}$.) It follows that $GU \cap GV = \emptyset$. In other words, any two distinct points of X/G have mutually disjoint open neighborhoods. □

Let M be a connected, locally compact metric space. Denote by Iso(M) the full group of isometries of M, endowed with the compact-open topology. The classical theorem due to van Dantzig and van der Waerden says that the action of Iso (M) on M is proper and that Iso(M) is locally compact (see [KoNo], vol. I, Ch. 1, Theorem 4.7).

We now recall a geometric fact which will be used in the proof of the subsequent theorem. One says that M is *two point homogeneous* if Iso(M) is transitive on equidistant pairs of points. Two-point homogeneous, connected, locally compact metric spaces were classified by H.-C. Wang [Wa1] and J.Tits [Ti1]. We shall use this classification only for Riemannian manifolds. (In fact, the list in the general case is the same, but the proof for Riemannian manifolds is considerably easier.) A connected Riemannian manifold M is called *isotropic* if Iso(M) is transitive on M and the isotropy subgroup of Iso(M) is transitive on the unit sphere in the tangent space. We refer the reader to [W1], § 8.12, for the proof of the following result.

Let M be a connected Riemannian manifold. Then M is two point homogeneous if and only if M is isotropic. The only isotropic Riemannian manifolds are Euclidean spaces and symmetric spaces of rank 1.

In what follows $\mathbb{B}_n$ denotes the unit ball in $\mathbb{C}^n$, $\mathrm{U}(p,q) \subset \mathrm{GL}(n,\mathbb{C})$ (resp. $\mathrm{PU}(p,q) \subset \mathrm{PGL}(n,\mathbb{C})$) the *pseudo-unitary group* of signature $(p,q), p+q=n$ (resp. the corresponding projective group), and dim G the real dimension of G. (As a rule, G has no complex structure.)

Theorem (W.Kaup [Ka2]) *Let X be an irreducible (reduced) complex space of complex dimension n and let $G \subset \mathrm{Aut}(X)$ be a subgroup acting properly on X. Then G has a structure of a Lie group, compatible with the compact-open topology, and*

$$\dim G \leq n^2 + 2n. \tag{$\dagger$}$$

The G-action on X is real analytic. The equality in ($\dagger$) holds exactly in three cases:

(a) $X = \mathbb{B}_n,\ G = \mathrm{PU}(n,1)$;

(b) $X = \mathbb{C}^n,\ \ G = \mathrm{U}(n) \ltimes \mathbb{C}^n$;

(c) $X = \mathbb{P}_n,\ \ G = \mathrm{PU}(n+1)$.

Remark The group G in (a) is in fact the full automorphism group of $\mathbb{B}_n$, see Proposition 3 of §2.6.

Proof By (iii) of the lemma G is locally compact. Consider the G-action on the manifold $X - S(X)$. It follows from Theorem III of §2.3 that G is a Lie group. The analyticity of the action is then a general fact proved in §1.6.

In order to obtain the estimate ($\dagger$) take any regular point $x \in X$. By (ii) of the lemma the isotropy subgroup G_x is compact. The Identity Theorem (see §2.2) says that the isotropy representation $\tau_x : G_x \to \mathrm{GL}\,(T_x(D)) \simeq \mathrm{GL}\,(n,\mathbb{C})$ is faithful. Thus

$$\dim G_x \leq \dim \mathrm{U}(n) = n^2$$

and, consequently,

$$\dim G = \dim G(x) + \dim G_x \leq 2n + n^2.$$

Finally, suppose that in ($\dagger$) we have an equality. Then $G_x \simeq \mathrm{U}(n)$ and $G(x)$ has (real) dimension $2n$. Thus $G(x)$ is open in X. On the other hand, by (i) of the lemma $G(x)$ is closed. Since X is irreducible and, in particular, connected, we have $G(x) = X$. Therefore, X is a homogeneous manifold of G. Since G_x is compact we can choose a G-invariant Hermitian metric on X. Then G_x acts transitively on the unit sphere in $T_x(X)$ so that X is isotropic. According to the classification of isotropic Riemannian manifolds, those which have an invariant complex structure are either complex Euclidean spaces or Hermitian symmetric spaces of rank 1. Thus X is isomorphic to $\mathbb{B}_n$, $\mathbb{C}^n$, or $\mathbb{P}_n$, see [He], Ch. 10, §6. The group G in each case coincides with the full group of holomorphic isometries. □

Remarks 1) Suppose that an irreducible complex space X admits a continuous metric, which is invariant under all automorphisms. Then $\mathrm{Aut}(X)$ is a closed subgroup of $\mathrm{Iso}(X)$ and the theorem of van Dantzig and van der Waerden implies that $\mathrm{Aut}(X)$ acts properly on X. By the above theorem $\mathrm{Aut}(X)$ is a Lie group of dimension $\leq \frac{n^2+2n}{2}$. This method of proof along with its application to hyperbolic complex spaces is due to S.Kobayashi, see [Ko], Ch. 5, § 2.

2) Assume that X in the above theorem is a Stein manifold and $H_k(X, \mathbb{Z}) \neq 0$. Using Morse theory, E.Bedford proved that

$$\dim \mathrm{Aut}(X) \leq (n-k)^2 + \frac{k(k+1)}{2} + 2n$$

and this estimate is sharp, see [Be].

2.6 The automorphism group of a bounded domain

We start with a simple topological lemma. By a homeomorphic mapping we mean a homeomorphism onto an open subspace of the target space.

Lemma 1 *Let W be a domain in $\mathbb{R}^n$. Suppose that a sequence of homeomorphic mappings*

$$\phi_\nu : W \to \phi_\nu(W) \subset \mathbb{R}^n, \quad \nu = 1, 2, \ldots,$$

converges uniformly on compact subsets to a homeomorphic mapping $\phi : W \to \phi(W) \subset \mathbb{R}^n$. Then for any two domains U, V, $U \subset\subset V \subset\subset W$, there exists a number ν_0 such that

$$\phi(U) \subset \phi_\nu(V) \qquad \textit{for all } \nu \geq \nu_0.$$

Proof Since $\partial(\phi(V))$ and $\phi(\overline{U})$ are mutually disjoint, there exists a neighborhood Ω of $\partial(\phi(V))$ such that $\Omega \cap \phi(\overline{U}) = \emptyset$. Take any point $x \in U$ and choose ν_0 so that

(a) $\phi_\nu(\partial V) \subset \Omega$

and

(b) $\phi_\nu(x) \in \phi(U)$

for all $\nu \geq \nu_0$. We claim that $\phi(U) \subset \phi_\nu(V)$ if $\nu \geq \nu_0$. Assume the contrary. Let $\phi(y) \notin \phi_\nu(V)$ for some $y \in U$, $\nu \geq \nu_0$. It follows from (a) that

$$\phi(U) = \{\phi(U) \cap \phi_\nu(V)\} \cup \{\phi(U) \cap (\mathbb{R}^n - \overline{\phi_\nu(V)})\}.$$

According to (b), the point $\phi_\nu(x)$ is contained in the first set. By assumption $\phi(y)$ is contained in the second one. Since $\phi(U)$ is connected, we obtain a contradiction. □

The next lemma goes back to A. Hurwitz.

Lemma 2 *Let D be a domain in $\mathbb{C}^n$ and $\{f_\nu\}_{\nu=1,2,\dots}$ a sequence in $\mathcal{O}(D)$. Suppose that each f_ν has no zeros and $f_\nu \to f$ uniformly on compact subsets in D. Then either $f = 0$ or f does not vanish.*

Proof Assume that $A := \{f = 0\}$ is a hypersurface in D. Choose a non-singular point $a \in A$, draw a complex line L through a transversally to A and take a small disk in L about a. Restricting all functions to this disk, we reduce our problem to the one-dimensional case. Namely, after a coordinate change, we obtain a sequence of non-vanishing holomorphic functions f_ν in the unit disk, such that $f = \lim f_\nu$ has only one zero $z = 0$ of some order p. Clearly, $z^{p-1} f_\nu(z)^{-1}$ converges to $z^{p-1} f(z)^{-1}$ uniformly on the circle $|z| = r < 1$. Hence

$$\operatorname{Res}_{z=0} \frac{z^{p-1}}{f(z)} = \lim_{\nu\to\infty} \operatorname{Res}_{z=0} \frac{z^{p-1}}{f_\nu(z)} = 0.$$

Since $z = 0$ is a simple pole of $z^{p-1} f(z)^{-1}$, this yields a contradiction. □

The following classical result is due to C.Carathéodory and H.Cartan (see [C], Satz 10, and [Ca2], Thèoréme 2; see also [N]).

Compactness Principle *Let D be a bounded domain in $\mathbb{C}^n$ and $\{\varphi_\nu\}_{\nu=1,2,\dots}$ a sequence in $\operatorname{Aut}(D)$. Consider each φ_ν as an element of $\operatorname{Hol}(D, \mathbb{C}^n)$. Then $\{\varphi_\nu\}$ has a convergent subsequence and the limit mapping, φ, maps D into $\overline{D}$. Furthermore, either $\varphi \in \operatorname{Aut}(D)$ or $\varphi(D) \subset \partial D$.*

Proof The existence of a convergent subsequence follows from Montel's theorem. Without loss of generality we may assume that $\varphi_\nu \to \varphi$ in $\operatorname{Hol}(D, \mathbb{C}^n)$. Since the topology in $\operatorname{Hol}(D, \mathbb{C}^n)$ is the topology of compact convergence, the inclusion $\varphi(D) \subset \overline{D}$ is clear. Applying Montel's theorem to the sequence $\{\psi_\nu\}$, where $\psi_\nu := \varphi_\nu^{-1}$, we may assume that $\psi_\nu \to \psi \in \operatorname{Hol}(D, \mathbb{C}^n)$. Suppose that there exists a point $p \in D$ such that $q := \varphi(p) \in D$. Then we have to prove that φ is an automorphism of D. We shall do it in three steps.

Step 1 The mapping φ is locally homeomorphic. Denote by j_φ the Jacobian of φ. We claim that $j_\varphi(p) \neq 0$. In order to prove this, let $q_\nu := \varphi_\nu(p)$. Since $\varphi_\nu \to \varphi$ and $\psi_\nu \to \psi$ in $\operatorname{Hol}(D, \mathbb{C}^n)$, it follows that $j_{\varphi_\nu} \to j_\varphi$ and, respectively, $j_{\psi_\nu} \to j_\psi$ uniformly on compact subsets in D. Note that $q_\nu \to q$ so that all q_ν are contained in a compact subset of D. Thus the inequality

$$|j_{\psi_\nu}(q_\nu) - j_\psi(q)| \leq |j_{\psi_\nu}(q_\nu) - j_\psi(q_\nu)| + |j_\psi(q_\nu) - j_\psi(q)|$$

shows that

$$\lim_{\nu\to\infty} j_{\psi_\nu}(q_\nu) = j_\psi(q).$$

Therefore

$$j_\psi(q) j_\varphi(p) = \lim_{\nu\to\infty} j_{\psi_\nu}(q_\nu)\, j_{\varphi_\nu}(p) = 1.$$

In particular, $j_\varphi(p) \neq 0$. By Lemma 2 the Jacobian j_φ does not vanish in D so that φ is locally homeomorphic.

Step 2 One has $\varphi(D) \subset D$. Let $D^* := \{x \in D | \varphi(x) \in D\}$. Then D^* is an open subset of D and it is enough to show that D^* is also closed in D. Let $\{x_k\}_{k=1,2,\dots}$

be a sequence in D^* converging to some point $x \in D$. Fix three subdomains $U \subset\subset V \subset\subset W$ in D so that $x \in U$ and $\varphi|_W : W \to \varphi(W)$ is a homeomorphism. By Lemma 1 there exists a number ν with the property that $\varphi(U) \subset \varphi_\nu(V)$. Since $x_k \in U$ for all sufficiently large k, we have

$$\varphi(x) = \lim_{k\to\infty} \varphi(x_k) \in \overline{\varphi_\nu(V)} = \varphi_\nu(\overline{V}) \subset D,$$

so that $x \in D^*$.

Step 3 The mapping φ is an automorphism of D. For any $x \in D$ and for any norm function in $\mathbb{C}^n$ we can write

$$|\psi(\varphi(x)) - x| = |\psi(\varphi(x)) - \psi_\nu(\varphi_\nu(x))| \leq$$

$$\leq |\psi(\varphi(x)) - \psi(\varphi_\nu(x))| + |\psi(\varphi_\nu(x)) - \psi_\nu(\varphi_\nu(x))|.$$

The first summand tends to zero as $\nu \to \infty$. Since $\varphi_\nu(x)$, $\nu = 1, 2, \ldots$, remain in a compact subset of D, the same is true for the second summand. Consequently $\psi \circ \varphi = \text{id}$. In particular, $\psi(q) = p$ so that all preceding considerations apply to ψ instead of φ. Therefore $\psi(D) \subset D$ and $\varphi \circ \psi = \text{id}$. □

Theorem (H.Cartan [Ca3]) *Let D be a bounded domain in $\mathbb{C}^n$. Then the action of* Aut(D) *on D is proper and* Aut(D) *is locally compact. Furthermore,* Aut(D) *has a structure of a Lie group, compatible with the compact-open topology, and*

$$\dim \operatorname{Aut}(D) \leq n^2 + 2n.$$

The action of Aut(D) *on D is real analytic.*

Proof Let $C_1, C_2 \subset D$ be two compact subsets. Suppose we have a sequence $\{\varphi_\nu\}$ in Aut(D) such that $\varphi_\nu(C_1) \cap C_2 \neq \emptyset$ for all ν. We claim that a subsequence of $\{\varphi_\nu\}$ converges in Aut(D). Indeed, we have $\varphi_\nu(x_\nu) = y_\nu$ for some $x_\nu \in C_1$ and $y_\nu \in C_2$. Since C_1 and C_2 are compact, we may assume that $x_\nu \to x \in D$ and $y_\nu \to y \in D$. By Montel's theorem we may also assume that $\varphi_\nu \to \varphi$ in Hol($D, \mathbb{C}^n$). But then $\varphi(x) = y$ and the Compactness Principle implies that $\varphi \in$ Aut(D).

This argument shows that the action of Aut(D) on D is proper. All other statements follow from the theorem of §2.5. □

Since the isotropy subgroup is compact, the Identity Theorem shows that an automorphism of D with a fixed point is uniquely determined by its differential at this point. H.Cartan gave a direct proof of the Identity Theorem for this case. The method of proof does not use the compactness of the isotropy group and yields the following more general result.

Proposition 1 *Let D be a bounded domain in $\mathbb{C}^n$, $\varphi \in$* Hol(D, D)*, and $\varphi(x_0) = x_0$ for some $x_0 \in D$. If $d\varphi_{x_0}$ is the identity map then $\varphi = \text{id}_D$.*

Proof We may assume that $x_0 = 0$. Then in a neighborhood of 0 we have an expansion

$$\varphi(z) = \sum_{k=1}^{\infty} a_k(z), \qquad (*)$$

where $a_k(z) \in \mathbb{C}^n$ and the coordinates $a_{kj}(z), j = 1, \dots, n$, of $a_k(z)$ are homogeneous polynomials of degree k. The differential $d\varphi_0$ is the linear map $z \mapsto a_1(z)$.

Assume that $a_1(z) = z$, but $\varphi(z) \neq z$. Then

$$\varphi(z) = z + a_p(z) + \sum_{k>p} a_k(z),$$

where $a_p \neq 0,\ p > 1$. Denote by φ^m the m-th iterate of φ. A straightforward calculation shows that in a neighborhood of 0, possibly depending on m,

$$\varphi^m(z) \ = \ z + ma_p(z) + \sum_{k>p} \tilde{a}_k(z),$$

or, in the coordinate form,

$$\varphi_j^m(z) = z_j + ma_{pj}(z) + \sum_{k>p} \tilde{a}_{kj}(z), \qquad j = 1, \dots, n,$$

where the $\tilde{a}_{kj}(z)$ are homogeneous polynomials of degree k. Fix j with $a_{pj}(z) \neq 0$ and choose $\alpha_1, \dots, \alpha_n,\ \ \alpha_1 + \dots + \alpha_n = p$, so that

$$\frac{\partial^p a_{pj}\ (0)}{\partial z_1^{\alpha_1}\ \dots\ \partial z_n^{\alpha_n}} = c \neq 0.$$

Since φ^m maps D into D, we have $|\varphi_j^m(z)| < M_j := \sup_D |z_j|$ and, by the Cauchy inequality for the corresponding derivative,

$$m \cdot |c| = \Big| \frac{\partial^p \varphi_j^m (0)}{\partial z_1^{\alpha_1}\ \dots\ \partial z_n^{\alpha_n}} \Big| \leq M_j\ \alpha_1!\ \dots\ \alpha_n!\ \rho^{-p}$$

for all m, where $\rho > 0$ is a fixed number such that the polydisk $\Delta_\rho^n = \{z \in \mathbb{C}^n \mid |z_k| < \rho,\ k = 1, \dots, n\}$ is contained in D. Since m is arbitrary and $c \neq 0$, we get a contradiction. □

As usual, denote by $\mathfrak{g}$ the Lie algebra of $G = \mathrm{Aut}(D)$. The Lie homomorphism of the G-action on D is injective. We identify $\mathfrak{g}$ with a subalgebra of $\mathcal{T}_D(D)$. In contrast with the case of compact complex spaces we have the following proposition.

Proposition 2 (i) *A holomorphic action of a connected complex Lie group G on a bounded domain D is trivial.*

(ii) $\mathfrak{g} \cap i\mathfrak{g} = \{0\}$.

Proof (i) Since any connected Lie group is generated by its one-parameter subgroups, it is enough to show that a holomorphic $\mathbb{C}$-action $(t, x) \mapsto \gamma_t(x),\ t \in \mathbb{C},\ x \in D$, is trivial. Let $f = z_k,\ k = 1, 2, \dots, n$, be one of the coordinate functions on D. Since f is bounded on D, the entire function $t \mapsto f(\gamma_t(x))$ is constant by Liouville's theorem. Thus $\gamma_t(x) = x$ for all $t \in \mathbb{C},\ x \in D$.

(ii) A vector field from $\mathfrak{g} \cap i\mathfrak{g}$ generates a global holomorphic one-parameter transformation group of D. By (i) this vector field is the zero one. □

Remarks 1) Theorem III appeared more than 10 years after [Ca3]. The proof of the above theorem in [Ca3] is substantially different. Another proof is based on the existence of an invariant metric (see §2.5, Remark 1). For bounded domains one can use the Bergman, the Carathéodory, or the Kobayashi metric as well.

2) H.Fujimoto proved that $\mathrm{Aut}(X)$ is a Lie group for a large class of complex spaces including both compact spaces and bounded domains in $\mathbb{C}^n$, see [Fu].

2.7 The automorphism groups of the polydisk and the ball

A bounded domain $D \subset \mathbb{C}^n$ containing 0 is called *circular* if D is invariant under the one-parameter transformation group $z \mapsto \gamma_t(z) := e^{it}z$, $t \in \mathbb{R}$. The following proposition is due to H.Cartan, see [Ca1].

Proposition 1 *Let $\varphi : D \to D'$ be an isomorphism between two circular domains and $\varphi(0) = 0$. Then φ is a linear map.*

Proof Put $G := \mathrm{Aut}(D)$ and let $G_0 \subset G$ be the isotropy subgroup of $0 \in D$. Consider the isotropy representation $\tau = \tau_0 : G_0 \to \mathrm{GL}(n, \mathbb{C})$ and denote by E the identity $n \times n$ matrix. Observe that $\varphi^{-1} \circ \gamma_{-t} \circ \varphi \circ \gamma_t \in G_0$. We have

$$\tau(\varphi^{-1} \circ \gamma_{-t} \circ \varphi \circ \gamma_t) = d\varphi_0^{-1} \cdot e^{-it}E \cdot d\varphi_0 \cdot e^{it}E = E.$$

Thus

$$\varphi^{-1} \circ \gamma_{-t} \circ \varphi \circ \gamma_t = \mathrm{id}_D$$

by the Identity Theorem. In other words, φ commutes with γ_t. In a neighborhood of $0 \in D$ we have the expansion $(*)$ of §2.6. By Lemma 2 of §2.2 we may assume that this neighborhood is circular. Then we can write

$$\varphi(e^{it}z) = \sum_{k=1}^{\infty} e^{ikt}a_k(z).$$

This is equal to $e^{it}\varphi(z)$ if and only if $a_k = 0$ for all $k \geq 2$. □

We now apply Proposition 1 in order to determine the automorphism groups of two circular domains: the unit polydisk

$$\Delta^n = \{z = (z_1, \ldots, z_n) \in \mathbb{C}^n \mid |z_k| < 1,\ k = 1, \ldots, n\}$$

and the unit ball

$$\mathbb{B}_n = \{z = (z_1, \ldots, z_n) \in \mathbb{C}^n \mid |z_1|^2 + \ldots + |z_n|^2 < 1\}.$$

Lemma 1 *Let G_2 be a group acting on a set X and G_1 a subgroup of G_2. Suppose there exists an $x \in X$ such that $G_1(x) = G_2(x)$ and $G_{1,x} = G_{2,x}$. Then $G_1 = G_2$.*

Proof For any $g \in G_2$ we have $gx = hx$, where $h \in G_1$. Hence $k := h^{-1}g \in G_{2,x} = G_{1,x}$ and, consequently, $g = kh \in G_1$. □

Lemma 2 *Let $\mu, \mu' : \mathbb{R}^n \to \mathbb{R}^+$ be two homogeneous functions of the same degree λ. Let $D := \{v \in \mathbb{R}^n \mid \mu(v) < 1\}$, $D' := \{v \in \mathbb{R}^n \mid \mu'(v) < 1\}$. If $A \cdot D = D'$ for some $A \in \mathrm{GL}(n, \mathbb{R})$, then $\mu'(Av) = \mu(v)$ for all $v \in \mathbb{R}^n$.*

Proof Let $r = \mu(v)$. For any $\epsilon > 0$ we have

$$\frac{1}{(r+\epsilon)^{\frac{1}{\lambda}}} \cdot v \in D,$$

hence

$$\frac{1}{(r+\epsilon)^{\frac{1}{\lambda}}} \cdot Av \in D'$$

or, equivalently, $\mu'(Av) < r + \epsilon$. Since $\epsilon > 0$ is arbitrary, we obtain that $\mu'(Av) \leq r = \mu(v)$. By the same reason $\mu(v) = \mu(A^{-1}Av) \leq \mu'(Av)$. □

Let S_n denote the permutation group, acting on Δ^n by

$$(z_1, \ldots, z_n) \overset{\sigma}{\longmapsto} (z_{\sigma(1)}, \ldots, z_{\sigma(n)}), \quad \sigma \in S_n.$$

Then S_n normalizes the subgroup consisting of all product automorphisms

$$(z_1, \ldots, z_n) \mapsto (\varphi_1(z_1), \ldots, \varphi_n(z_n)),$$

where $\varphi_k \in \mathrm{Aut}(\Delta^1)$, $k = 1, \ldots, n$.

Proposition 2 *The automorphism group of Δ^n is the semidirect product*

$$\mathrm{Aut}(\Delta^n) = S_n \ltimes \{\underbrace{\mathrm{Aut}(\Delta^1) \times \ldots \times \mathrm{Aut}(\Delta^1)}_{n \text{ times}}\} \simeq \qquad (\dagger)$$

$$\simeq S_n \ltimes \{\underbrace{\mathrm{PGL}(2, \mathbb{R}) \times \ldots \times \mathrm{PGL}(2, \mathbb{R})}_{n \text{ times}}\}.$$

Proof Let $G_2 := \mathrm{Aut}(\Delta^n)$ and let G_1 be the group on the right hand side of (†). Then G_1 is a subgroup of G_2 and G_1 acts transitively on Δ^n. By Lemma 1 it is enough to prove that the isotropy subgroups of G_1 and G_2 at the origin coincide. The first of them consists of automorphisms

$$z = (z_1, \ldots, z_n) \mapsto \varphi(z) = (e^{i\alpha_1} z_{\sigma(1)}, \ldots, e^{i\alpha_n} z_{\sigma(n)}), \qquad \sigma \in S_n.$$

Thus, we have to show that any automorphism of Δ^n fixing 0 is of this form. We use the same notation z for the column with entries z_k. From Proposition 1 it follows that $\varphi(z) = Az$, where $A = (a_{kl})$ is a non-degenerate complex $n \times n$ matrix. By Lemma 2 we have $\mu(Az) = \mu(z)$, where $\mu(z) := \max |z_k|$. In particular,

$$\max_k |a_{kl}| = 1$$

for all l, $l = 1, 2, \ldots, n$. Fix k, $1 \leq k \leq n$, and take

$$z_l := \begin{cases} \frac{\overline{a_{kl}}}{|a_{kl}|}, & \text{if } a_{kl} \neq 0 \\ 0, & \text{if } a_{kl} = 0. \end{cases}$$

Then $\mu(Az) = 1$, hence

$$\sum_l |a_{kl}| \leq 1$$

for all k, $k = 1, 2, \ldots, n$. Therefore, if $|a_{pq}| = 1$ for some p, q, then all other entries of the p-th row of A are zeros. It follows that in each column and in each row of A there is exactly one non-zero element and this element has absolute value 1. □

Denote by $\mathrm{U}(n,1)$ the subgroup of $\mathrm{GL}(n+1,\mathbb{C})$ preserving the Hermitian form $h(z) = |z_0|^2 - |z_1|^2 - \ldots - |z_n|^2$. The unit ball $\mathbb{B}_n \subset \mathbb{C}^n$ can be realized as a domain in $\mathbb{P}_n$:

$$\mathbb{B}_n = \{z = (z_0 : z_1 : \ldots : z_n) \in \mathbb{P}_n \mid h(z) > 0\}.$$

For any $A \in \mathrm{U}(n,1)$ the linear automorphism of $\mathbb{P}_n$ with matrix A preserves this domain. Therefore we obtain a Lie group homomorphism $\mathrm{U}(n,1) \to \mathrm{Aut}(\mathbb{B}_n)$, whose kernel coincides with the (one-dimensional) center of $\mathrm{U}(n,1)$ and whose image is, by definition, the corresponding projective group $\mathrm{PU}(n,1)$.

Proposition 3 $\mathrm{Aut}(\mathbb{B}_n) = \mathrm{PU}(n,1)$.

Proof Let $G_1 := \mathrm{PU}(n,1)$, $G_2 := \mathrm{Aut}(\mathbb{B}_n)$. Given two vectors $z, w \in \mathbb{C}^{n+1}$ such that $h(z) > 0$ and $h(w) > 0$, there exists a matrix $A \in \mathrm{U}(n,1)$ with the property that $Az = \lambda w$, $\lambda \in \mathbb{C}^*$. Thus G_1 is transitive on $\mathbb{B}_n$ and we only have to prove that the isotropy subgroups of G_1 and G_2 coincide. It is convenient to take the point $x = (1 : 0 : \ldots : 0)$. Then $G_{1,x}$ can be easily calculated. Namely, an element $g \in G_1$ is contained in $G_{1,x}$ if and only if g is represented by a matrix

$$\left(\begin{array}{c|ccc} e^{i\alpha} & 0 & \ldots & 0 \\ \hline 0 & & & \\ \vdots & & U & \\ 0 & & & \end{array}\right),$$

where $U \in \mathrm{U}(n)$. Note that g determines such a matrix uniquely up to a multiplication by $e^{i\theta}$. In particular, the product $U_g := e^{-i\alpha} \cdot U$ depends only on g. We identify $T_x(\mathbb{B}_n)$ with $\mathbb{C}^n$ making use of the coordinate system $\zeta_k = z_k/z_0$, $k = 1, 2, \ldots, n$. Consider the isotropy representation $\tau_x : G_{2,x} \to \mathrm{GL}(n,\mathbb{C})$. Obviously, any $g \in G_{1,x}$ is linear in the coordinates ζ_k and $\tau_x(g) = U_g$. Thus $\tau_x(G_{1,x}) = \mathrm{U}(n)$. Now, $\tau_x(G_{2,x})$ is a compact subgroup of $\mathrm{GL}(n,\mathbb{C})$ containing $\tau_x(G_{1,x})$. Since $\mathrm{U}(n)$ is a maximal compact subgroup of $\mathrm{GL}(n,\mathbb{C})$, we obtain that $\tau_x(G_{1,x}) = \tau_x(G_{2,x})$. By the Identity Theorem $G_{1,x} = G_{2,x}$. □

Corollary *If $n > 1$ then Δ^n and $\mathbb{B}_n$ are not isomorphic.*

Proof The automorphism groups have different dimension:

$$\dim \mathrm{Aut}(\Delta^n) = 3n, \qquad \dim \mathrm{Aut}(\mathbb{B}_n) = n^2 + 2n.$$ □

Remarks 1) The fact that Δ^n and $\mathbb{B}_n$ are not isomorphic was discovered by Poincaré (for $n = 2$) in 1907. Here is a short proof. Assume that $\varphi : \Delta^n \to \mathbb{B}_n$ is a biholomorphic mapping. By homogeneity of Δ^n we may assume that $\varphi(0) = 0$. Applying Proposition 1, we see that φ is a linear mapping, i.e., $\varphi(z) = Az$. By

Lemma 2 we have $|w_1|^2 + \ldots + |w_n|^2 = \max_k |z_k|^2$, where $w = Az$. This is, however, impossible, since the left hand side is smooth and the right hand side is not.

2) The polydisk and the ball are examples of bounded symmetric domains, which were listed by É. Cartan. A bounded symmetric domain D is homogeneous, the group $G := \mathrm{Aut}(D)$ is semisimple, and D is a Hermitian symmetric space with connected isometry group G°. Conversely, any Hermitian symmetric space of non-compact type can be realized as a bounded domain in $\mathbb{C}^n$, see [He]. É.Cartan proved that all homogeneous bounded domains in $\mathbb{C}^2$ and $\mathbb{C}^3$ are symmetric and posed a question whether it is so in higher dimension, see [Ca]. The answer was given by I.I.Piatetsky-Shapiro in 1959. He constructed first examples of non-symmetric homogeneous bounded domains in $\mathbb{C}^4$ and $\mathbb{C}^5$. He also introduced the notion of a Siegel domain playing a crucial role both for the construction and for the future development of this area, see [P]. As it was supposed by É.Cartan, the discovery of non-symmetric homogeneous bounded domains led to a classification of all homogeneous bounded domains, see [GPV1].

2.8 A characterization of the ball

A real C^2 function ρ, defined in a domain $\Omega \subset \mathbb{C}^n$, is called *strictly plurisubharmonic* if the complex Hessian

$$H_\rho(x;\xi) := \sum_{k,l=1}^{n} \frac{\partial^2 \rho}{\partial z_k \partial \bar{z}_l}(x)\ \xi_k \bar{\xi}_l$$

is positive definite for all $x \in \Omega$. Let D be a bounded domain in $\mathbb{C}^n$ and let $q \in \partial D$. The boundary ∂D (and the domain D itself) is said to be *strictly pseudoconvex at* q if there exist a neighborhood Ω of q and a strictly plurisubharmonic function (of class C^2) ρ in Ω, such that $d\rho(x) \neq 0$ for all $x \in \Omega$ and $D \cap \Omega = \{x \in \Omega \mid \rho(x) < 0\}$. The domain D is called *strictly pseudoconvex* if D is strictly pseudoconvex at each boundary point q. Note that, by our definition, strictly pseudoconvex domains always have C^2 boundary. We refer the reader to [GuRo], Ch. 9, for a detailed exposition of the pseudoconvexity theory including the solution of the Levi problem.

As we have seen in the preceding section, the automorphism group of $\mathbb{B}_n$ is isomorphic to $\mathrm{PU}(n,1)$. In particular, $\mathrm{Aut}(\mathbb{B}_n)$ is non-compact. It turns out that this property characterizes $\mathbb{B}_n$ among all strictly pseudoconvex bounded domains. The following results were first obtained by B.Wong [Wo] and J.-P.Rosay [Ro]. The method of proof given here is due S.I.Pinchuk [Pi2].

Theorem 1 *Let $D \subset \mathbb{C}^n$ be a bounded domain, $p \in D$, and $q \in \partial D$. Suppose that the boundary ∂D is pseudoconvex at q. Assume that there exists a sequence $\{\varphi_\nu\}$ in $\mathrm{Aut}(D)$ such that $\varphi_\nu(p) \to q$ as $\nu \to \infty$. Then D is isomorphic to $\mathbb{B}_n$.*

Theorem 2 *Let $D \subset \mathbb{C}^n$ be a strictly pseudoconvex bounded domain. If $\mathrm{Aut}(D)$ is non-compact then D is isomorphic to $\mathbb{B}_n$.*

Proof of Theorem 2 Let $\{\varphi_\nu\}_{\nu=1,2,\ldots}$ be a discrete sequence of automorphisms in $\mathrm{Aut}(D)$. By the Compactness Principle (see §2.6) we may assume that $\{\varphi_\nu\}$ converges to $\varphi \in \mathrm{Hol}(D, \mathbb{C}^n)$ and $\varphi(D) \subset \partial D$. For any point $p \in D$ we have $\varphi_\nu(p) \to \varphi(p) =: q \in \partial D$. Therefore D is isomorphic to $\mathbb{B}_n$ by Theorem 1. □

For the proof of Theorem 1 we need several lemmas.

Lemma 1 *$\mathbb{B}_n$ is isomorphic to the unbounded domain*

$$\mathbb{H}_n := \{w = (w_1, \ldots, w_n) \in \mathbb{C}^n \mid \mathrm{Re}\, w_n + |w_1|^2 + \ldots + |w_{n-1}|^2 < 0\}.$$

Proof Consider a holomorphic mapping $\phi : \mathbb{B}_n \to \mathbb{C}^n$ defined by

$$w_k = \frac{iz_k}{z_n - 1} \quad (k = 1, \ldots, n-1), \quad w_n = \frac{z_n + 1}{z_n - 1}.$$

The identity

$$\mathrm{Re}\, w_n + \sum_{k=1}^{n-1} |w_k|^2 = \frac{1}{|z_n - 1|^2} \cdot \{\sum_{k=1}^{n} |z_k|^2 - 1\}$$

shows that $\phi(\mathbb{B}_n) \subset \mathbb{H}_n$. One checks immediately that $\phi : \mathbb{B}_n \to \mathbb{H}_n$ is biholomorphic. Namely, the inverse mapping is given by

$$z_k = \frac{2iw_k}{1 - w_n} \quad (k = 1, \ldots, n-1), \quad z_n = \frac{w_n + 1}{w_n - 1}.$$

□

Lemma 2 *Let ρ be a C^2 strictly plurisubharmonic function in a neighborhood of $q \in \mathbb{C}^n$. Assume that $d\rho(q) \neq 0$. Then there exists a local coordinate system $\zeta = (\zeta_1, \ldots, \zeta_n)$ with center q, such that*

$$\rho(x) = \rho(q) + 2\mathrm{Re} \sum_k a_k \zeta_k + |\zeta|^2 + o(|\zeta|^2),$$

as $|\zeta| := \sqrt{|\zeta_1|^2 + \ldots + |\zeta_n|^2} \to 0$, where $\zeta_k = \zeta_k(x)$.

Proof We may assume that $\frac{\partial \rho}{\partial z_1}(q) \neq 0$. Let

$$\zeta_1 := \sum_k \frac{\partial \rho}{\partial z_k}(q)\ (z_k - z_k(q)) + \frac{1}{2} \sum_{k,l} \frac{\partial^2 \rho}{\partial z_k \partial z_l}(q)\ (z_k - z_k(q))(z_l - z_l(q)),$$

$$\zeta_k := z_k - z_k(q) \qquad \text{for } k > 1.$$

Then

$$\rho(x) = \rho(q) + 2\mathrm{Re}\, \zeta_1 + \sum_{k,l} c_{kl} \zeta_k \overline{\zeta}_l + o(|\zeta|^2),$$

where $\overline{c}_{kl} = c_{lk}$. Since ρ is strictly plurisubharmonic and since this property does not depend on the coordinate system, the Hermitian form $\sum_{k,l} c_{kl} \zeta_k \overline{\zeta}_l$ is positive definite. By a linear change of coordinates one can bring this form to $|\zeta|^2$. □

Proof of Theorem 1 Let Ω be a neighborhood of $q \in \partial D$, in which D is defined by the condition $\rho < 0$, where ρ is a C^2 strictly plurisubharmonic function and $d\rho \neq 0$ everywhere in Ω. By the Compactness Principle we may assume that $\varphi_\nu \to \varphi \in \mathrm{Hol}(D, \mathbb{C}^n)$ uniformly on compact subsets in D. We claim that $\varphi(D) = \{q\}$. In fact, it is clear that $\varphi(p) = q$. Therefore there exists a neighborhood $V \subset D$ of p

such that $\varphi(V) \subset \Omega$. Since $\varphi(D) \subset \overline{D}$, it follows that $\rho \circ \varphi|_V \leq 0$. The function $\rho \circ \varphi|_V$ is plurisubharmonic and attains its maximum value 0 at p. By the Maximum Principle for plurisubharmonic functions $\rho \circ \varphi|_V = 0$. Consequently

$$H_\rho(\varphi(x);\ d\varphi(x) \cdot \xi) = \frac{\partial^2}{\partial t\, \partial \bar{t}}\Big|_{t=0} \rho(\varphi(x + t\xi)) = 0$$

for all $x \in V$ and $\xi \in \mathbb{C}^n$. The positivity of H_ρ implies that $d\varphi(x) = 0$. Hence φ is a constant mapping.

Shrinking Ω if necessary, choose a coordinate system $\zeta_1, \ldots, \zeta_n$ in Ω as in Lemma 2. We may assume that $\zeta = (\zeta_1, \ldots, \zeta_n) : \Omega \to \mathbb{C}^n$ is a biholomorphic mapping onto a ball with center 0 in $\mathbb{C}^n$. From now on we identify Ω with this ball and the point q with its center $0 \in \mathbb{C}^n$. Let $S := \partial D \cap \Omega = \{x \in \Omega \mid \rho(x) = 0\}$ and

$$\mu(x) := \sqrt{\sum_k |\frac{\partial \rho}{\partial \zeta_k}(x)|^2}, \qquad \text{where} \quad x \in \Omega.$$

It is convenient to normalize ρ so that $\mu(0) = \frac{1}{2}$. Define $U(x)$ to be any unitary $n \times n$ matrix satisfying

$$U(x) \cdot \frac{\partial \rho}{\partial \bar{\zeta}}(x) = \mu(x) \cdot (0, \ldots, 0, 1)^t.$$

Suppose that $x \in S$ and the length of $w \in \mathbb{C}^n$ is small enough. Then, by the choice of local coordinates,

$$\rho(U(x)^{-1} w + x) = 2\mu(x) \text{Re}\ w_n + |w|^2 (1 + \sigma(x, w)), \tag{1}$$

where

$$|\sigma(x, w)| \leq \alpha(|x|) + \beta(|w|) \qquad \text{and} \qquad \alpha(\tau), \beta(\tau) \to 0 \quad \text{as } \tau \to +0.$$

Recall that $p_\nu := \varphi_\nu(p) \to q = 0$. Without loss of generality we may assume that $p_\nu \in \Omega$ for all ν and that there exists a point $q_\nu \in S$ such that $|p_\nu - q_\nu| = \min_{x \in S} |p_\nu - x|$. Note that the segment joining p_ν and q_ν is orthogonal to S. Therefore

$$p_\nu - q_\nu = \lambda_\nu \cdot \frac{\partial \rho}{\partial \bar{\zeta}}\ (q_\nu),$$

where $\lambda_\nu \in \mathbb{R}$. Since ρ is negative in $D \cap \Omega$, we have $\lambda_\nu < 0$. Letting

$$\mu_\nu := \mu(q_\nu), \ \ U_\nu := U(q_\nu), \ \ \delta_\nu := -\lambda_\nu\ \mu_\nu,$$

we obtain from (1) that

$$\rho(U_\nu^{-1} w + q_\nu) = 2\mu_\nu \text{Re}\ w_n + |w|^2 (1 + \sigma_\nu(w)),$$

where $|\sigma_\nu(w)| \leq \alpha_\nu + \beta(|w|)$, $\alpha_\nu \to 0$ as $\nu \to \infty$ and $\beta(\tau) \to 0$ as $\tau \to +0$. It follows that

$$\frac{1}{\delta_\nu}\ \rho(U_\nu^{-1} A_\nu^{-1} w + q_\nu) = 2\mu_\nu \text{Re}\ w_n + \{\sum_{k=1}^{n-1} |w_k|^2\ +\ \delta_\nu\ |w_n|^2\} \cdot (1 + \sigma_\nu(A_\nu^{-1} w)), \tag{2}$$

where A_ν^{-1} is the diagonal $n \times n$ matrix with eigenvalues $\sqrt{\delta_\nu}, \ldots, \sqrt{\delta_\nu}, \delta_\nu$.

Suppose that $x \in D$ remains in a compact subset of D. Since $\varphi_\nu(x)$ tends uniformly to $\varphi(x) = q$, the mapping

$$x \mapsto \psi_\nu(x) := A_\nu U_\nu(\varphi_\nu(x) - q_\nu) \in \mathbb{C}^n$$

is well-defined for all sufficiently large ν. Substituting

$$w = \psi_\nu(x) = (\psi_\nu^1(x), \ldots, \psi_\nu^n(x))$$

in (2) we obtain

$$\frac{1}{\delta_\nu}\rho(\varphi_\nu(x)) = 2\mu_\nu \mathrm{Re}\, \psi_\nu^n(x) + \{\sum_{k=1}^{n-1} |\psi_\nu^k(x)|^2 + \delta_\nu |\psi_\nu^n(x)|^2\} \cdot (1 + \sigma_\nu(A_\nu^{-1}\psi_\nu(x))). \quad (3)$$

The estimate

$$|\sigma_\nu(A_\nu^{-1}\psi_\nu(x))| = |\sigma_\nu(U_\nu(\varphi_\nu(x) - q_\nu))| \le \alpha_\nu + \beta(|\varphi_\nu(x) - q_\nu|)$$

shows that

$$\sigma_\nu(A_\nu^{-1}\psi_\nu(x)) \to 0 \quad \text{as } \nu \to \infty$$

uniformly on compact subsets in D. Since $\rho(\varphi_\nu(x)) < 0$, it follows that $\mathrm{Re}\, \psi_\nu^n(x) < 0$ for all sufficiently large ν. By Montel's theorem we may assume that

$$h_\nu(x) := \frac{\psi_\nu^n(x) + 1}{\psi_\nu^n(x) - 1} \to h(x)$$

uniformly on compact subsets in D.

Let $o \in \mathbb{H}_n$ be the point with coordinates $w_k(o) = 0 \ (k < n), \ w_n(o) = -1$. By the definition of ψ_ν we have

$$\psi_\nu(p) = o \quad \text{for all} \quad \nu.$$

In particular,

$$\psi_\nu^n(p) = -1 \quad \text{for all} \quad \nu,$$

whence $h(p) = 0$. Therefore, by the Maximum Principle, h maps D into the unit disk Δ^1. Consequently

$$\psi_\nu^n(x) \to \psi^n(x) := \frac{h(x) + 1}{h(x) - 1}$$

uniformly on compact subsets in D. In view of (3) each sequence

$$\{\psi_\nu^k(x)\}_{\nu=1,2,\ldots}, \qquad k = 1, 2, \ldots, n-1,$$

is uniformly bounded on compact subsets in D. Thus, again by Montel's theorem, we may assume that each of these sequences is convergent. Let $\psi = (\psi^1, \ldots, \psi^n) \in \mathrm{Hol}(D, \mathbb{C}^n)$ be the limit mapping. Since $\mu_\nu \to 1/2$, it follows from (3) that

$$\mathrm{Re}\, \psi^n(x) + \sum_{k=1}^{n-1} |\psi^k(x)|^2 < 0.$$

In other words, ψ maps D into $\mathbb{H}_n$.

Now, the inverse mapping $w \mapsto \chi_\nu(w) := \psi_\nu^{-1}(w) = \varphi_\nu^{-1}(U_\nu^{-1}A_\nu^{-1}w + q_\nu) \in D$ is well-defined for all sufficiently large ν, provided w remains in a relatively compact subdomain of $\mathbb{H}_n$. Since D is bounded, we may assume by Montel's theorem that the sequence $\{\chi_\nu\}$ has a limit mapping $\chi : \mathbb{H}_n \to \overline{D} \subset \mathbb{C}^n$. Observe that $\chi_\nu(o) = p$ for all ν so that $\chi(o) = p$. Since $\psi(\chi(w)) = w$ in a neighborhood of o, the Jacobian j_χ does not vanish in this neighborhood. From Lemma 2 of §2.6 it follows that j_χ does not vanish in $\mathbb{H}_n$. Therefore, for any point $w \in \mathbb{H}_n$ there exists a neighborhood $W \subset\subset \mathbb{H}_n$ such that $\chi|_W$ is homeomorphic. By Lemma 1 of §2.6 we have $\chi(w) \in \chi_\nu(W) \subset D$ for all sufficiently large ν. Thus χ maps $\mathbb{H}_n$ in D. It follows that $\psi \circ \chi = \mathrm{id}_{\mathbb{H}_n}$ and $\chi \circ \psi = \mathrm{id}_D$. By Lemma 1 the domain D is isomorphic to $\mathbb{B}_n$. □

Remark For each compact Lie group G one can construct a strictly pseudoconvex bounded domain D with real analytic boundary, such that $\mathrm{Aut}(D) = G$, see [SaZa], [BeDa].

2.9 Bounded domains with compact quotient $D/\mathrm{Aut}(D)$

Let D be a bounded domain in $\mathbb{C}^n$ and $K(z)$ the *Bergman kernel function* of D. Recall that

$$K(z) = \sum_{k=1}^{\infty} |f_k(z)|^2,$$

where $\{f_k\}_{k=1,2,\dots}$ is an orthonormal basis of the Hilbert space $L^2(D) \cap \mathcal{O}(D)$. Since $z_1, \dots, z_n \in L^2(D)$, it follows that $K(z)$ is a strictly plurisubharmonic function. The transformation law for $K(z)$ says that

$$K(\varphi(z)) \cdot |j_\varphi(z)|^2 = K(z), \tag{$*$}$$

where φ is an automorphism of D.

Let $G := \mathrm{Aut}(D)$. We have seen in §2.6 that the action of G on D is proper. By (iv) of the lemma in §2.5 the quotient D/G is Hausdorff. In this section we assume that the quotient D/G is compact. This assumption is fulfilled at least in two important cases: a) the domain D is homogeneous ; b) there exists a discrete subgroup $\Gamma \subset G$ with compact fundamental region.

Proposition 1 *If D/G is compact then D is a Stein domain.*

Proof Let $q \in \partial D$ be any boundary point and let $\{p_\nu\} \subset D$ be a sequence converging to q. It suffices to prove that $K(p_\nu) \to \infty$ when $\nu \to \infty$. Then D is holomorphically convex by Oka's theorem (see e.g. [GuRo], Ch. 9, D).

By assumption there exists a compact subset $C \subset D$ such that

$$p_\nu = \varphi_\nu(x_\nu), \qquad \text{where } \varphi_\nu \in G,\ x_\nu \in C.$$

By Montel's theorem we may assume that $\varphi_\nu \to \varphi \in \mathrm{Hol}(D, \mathbb{C}^n)$. Since C is compact, we may also assume that $x_\nu \to x \in D$. Then $\varphi(x) = q$ so that, according to the Compactness Principle, $\varphi(D) \subset \partial D$. It follows that $j_\varphi = 0$ everywhere in D. In particular, $j_\varphi(x) = 0$ showing that $j_{\varphi_\nu}(x_\nu) \to 0$. By ($*$) we have

$$K(p_\nu) \cdot |j_{\varphi_\nu}(x_\nu)|^2 = K(x_\nu).$$

Since

$$K(x_\nu) \geq \min_{x \in C} K(x) > 0$$

for all ν, it follows that $K(p_\nu) \to \infty$. □

Proposition 2 *If D is a bounded domain with C^2 boundary and D/G is compact, then D is isomorphic to $\mathbb{B}_n$.*

Proof The boundary of D is strictly pseudoconvex at least at one point $q \in \partial D$. In fact, consider a ball B such that $D \subset B$ and $\partial B \cap \partial D \neq \emptyset$. Let $q \in \partial B \cap \partial D$. In a neighborhood Ω of q we have $D \cap \Omega = \{x \in \Omega \mid \rho < 0\}$, where ρ is a C^2 function with $d\rho \neq 0$. One can choose ρ so that its real Hessian with respect to $x_k = \operatorname{Re} z_k$, $y_k = \operatorname{Im} z_k$ is positive definite. Then ρ is strictly plurisubharmonic. As in the proof of Proposition 1, one can find a sequence of automorphisms $\varphi_\nu : D \to D$ and a point $x \in D$ such that $\varphi_\nu \to \varphi \in \operatorname{Hol}(D, \mathbb{C}^n)$ and $\varphi(x) = q$. Now apply Theorem 1 of §2.8. □

Remark For bounded domains with piecewise C^2 boundary the compactness of D/G implies that D is isomorphic to the product $\mathbb{B}_{m_1} \times \ldots \times \mathbb{B}_{m_p}$, $m_1 + \ldots + m_p = n$, see [Pi1].

3 Compact Homogeneous Manifolds

In this chapter we study geometric properties of compact homogeneous complex manifolds. It is natural to begin with flag manifolds, which are defined as the coset spaces S/P, where S is a connected complex semisimple Lie group, $P \subset S$ a parabolic subgroup. Their description requires some work with roots systems, after which we prove that a flag manifold admits an equivariant projective embedding. Furthermore, flag manifolds can be characterized as projective homogeneous manifolds, which are rational and/or simply connected. We also discuss their automorphism groups, though the proof of one important theorem stated here will be given later in Chapter 4.

Next we consider parallelizable compact complex manifolds. Any such manifold is homogeneous and can be written in the Klein form $X = G/\Gamma$, where G is a complex Lie group, Γ a discrete uniform subgroup of G.

The role of the above two classes of homogeneous complex manifolds is explained by the normalizer theorem. Namely, if H is a closed complex Lie subgroup of a connected complex Lie group G and if $X = G/H$ is compact, then the normalizer of the connected component $H^\circ \subset H$ is a parabolic subgroup of G. Geometrically speaking, X admits a fibration, called the Tits fibration, whose base is a flag manifold and whose fiber is a parallelizable manifold. After giving a proof of the normalizer theorem, we consider some special cases. For example, we show that if X is simply connected then the fiber of the Tits fibration is a torus. More generally, if $\pi_1(X)$ is nilpotent (resp. solvable) then the fiber is homogeneous under a nilpotent (resp. solvable) complex Lie group. We also discuss the influence of the topology of X on the algebraic properties of a complex Lie group G acting transitively and effectively on X. It turns out that if $\pi_1(X)$ is nilpotent then G is locally a direct product of a semisimple group and a nilpotent group. For $\pi_1(X)$ solvable we construct an example, in which the adjoint action of the Levi subgroup of G on the radical is non-trivial. We also show (without any assuption on $\pi_1(X)$) that dim G is bounded from above by some number depending only on dim X. In the last section we prove the theorem of A.Borel and R.Remmert about compact homogeneous Kähler manifolds.

3.1 Flag manifolds

In this section G is a connected complex linear algebraic group. A G-action on a complex algebraic variety X is called *algebraic*, if the corresponding mapping $G \times X \to X$ is a morphism of algebraic varieties. For the theory of algebraic transformation groups the following theorem is of fundamental importance.

Fixed Point Theorem (A.Borel) *A connected solvable linear algebraic group, acting algebraically on a non-empty complete algebraic variety X, has a fixed point in X.*

The proof can be found in [Bo4] (Theorem 10.4), see also [Hum2], [OnVi].

For any Zariski closed subgroup $H \subset G$ the coset space G/H can be given the structure of a quasiprojective algebraic variety so that the natural action of G on G/H is algebraic. On the other hand, G/H can be considered as a complex G-homogeneous manifold. In this setting we shall use the same notation for algebraic varieties and for the corresponding complex spaces.

A maximal connected solvable algebraic subgroup of G is called a *Borel subgroup*. An algebraic subgroup of G, containing a Borel subgroup, is called a *parabolic subgroup*.

Proposition 1 *The following properties of the pair (G, H) are equivalent:*
(i) *G/H is a compact complex manifold;*
(ii) *G/H is a projective algebraic variety;*
(iii) *H is a parabolic subgroup of G.*

Proof (i) $\Longleftrightarrow$ (ii) is obvious because G/H is always quasiprojective, and (ii) $\Longrightarrow$ (iii) follows from the Fixed Point Theorem.

In order to prove (iii) $\Longrightarrow$ (ii) we may assume that G is a Zariski closed subgroup of $\mathrm{GL}(V)$, where V is a complex vector space of some dimension n. It suffices to show that G/B is a projective variety if B is a Borel subgroup of G. Denote by $\mathbb{F}(V)$ the projective variety whose points are identified with sequences of linear subspaces $\{0\} = V_0 \subset V_1 \subset V_2 \subset \ldots \subset V_j \subset \ldots \subset V_{n-1} \subset V_n = V$, where $\dim V_j = j$. To each $g \in \mathrm{GL}(V)$ there corresponds a transformation of $\mathbb{F}(V)$ sending $\{V_j\}$ to $\{gV_j\}$, and the arising action

$$\mathrm{GL}(V) \times \mathbb{F}(V) \to \mathbb{F}(V)$$

is algebraic and transitive. The isotropy subgroup is isomorphic to the subgroup of all upper triangular matrices. In particular, this subgroup is solvable. Now, G acts on $\mathbb{F}(V)$ as a subgroup of $\mathrm{GL}(V)$. Thus, for each $x \in \mathbb{F}(V)$ the isotropy subgroup G_x is also solvable. On the other hand, at least one G-orbit on $\mathbb{F}(V)$ is Zariski closed. It follows that for some solvable algebraic subgroup $A \subset G$ the quotient G/A is a projective variety. Replacing A by its identity component, which has finite index in A, we may assume that A is connected. But then there exists a Borel subgroup of G containing A. Therefore the quotient G/B is also a projective variety. □

Corollary 1 *Any two Borel subgroups of G are conjugate.*

Proof Since G/B is a projective variety, any other Borel subgroup $B_1 \subset G$ has a fixed point on G/B. Therefore $B_1 \subset gBg^{-1}$ for some $g \in G$. In view of the maximality property of B_1 one has the equality $B_1 = gBg^{-1}$. □

The radical of G is contained in any parabolic subgroup of G. Thus, if G acts effectively on some projective quotient space, then G is in fact semisimple. In other words, the class of all G-homogeneous projective varieties coincides with the class of all S-homogeneous projective varieties, where S is a semisimple algebraic group. Until the end of the section we assume without loss of generality that a transitive transformation group is semisimple. It should be mentioned that a connected semisimple complex Lie group S has the unique structure of a linear algebraic group, compatible with its complex Lie group structure. A homogeneous manifold S/P, where P is a parabolic subgroup of a connected semisimple complex group S, is called a *flag manifold* (of S).

Example 1 Let $S = \mathrm{SL}(V)$, where V is a complex vector space, dim $V = n$. Fix a sequence of integers $d_1, d_2, \ldots, d_k$ so that

$$0 < d_1 < d_2 < \ldots < d_k < n.$$

A flag of type $\{d_1, d_2, \ldots, d_k\}$ in V is by definition a sequence of linear subspaces $V_1 \subset V_2 \subset \ldots \subset V_j \subset \ldots \subset V_k$ of V such that dim $V_j = d_j$ for $j = 1, 2, \ldots, k$. Denote by $\mathbb{F}_{d_1,d_2,\ldots,d_k}(V)$ the projective variety whose points are identified with flags of type $\{d_1, d_2, \ldots, d_k\}$. The natural algebraic action

$$S \times \mathbb{F}_{d_1,d_2,\ldots,d_k}(V) \to \mathbb{F}_{d_1,d_2,\ldots,d_k}(V)$$

is transitive. Let $P_{d_1,d_2,\ldots,d_k}$ be the isotropy subgroup of some point. Then one can choose a basis of V so that $P_{d_1,d_2,\ldots,d_k}$ is identified with the subgroup of unimodular matrices having square blocks of dimensions $d_1, d_2 - d_1, \ldots, d_k - d_{k-1}, n - d_k$ along the main diagonal, zeros below these blocks, and arbitrary elements above them. The latter group contains the subgroup of all upper triangular unimodular matrices, which is a Borel subgroup in $\mathrm{SL}(n, \mathbb{C})$. Therefore $P_{d_1,d_2,\ldots,d_k}$ is a parabolic subgroup of S. Conversely, one can show that any parabolic subgroup of S is conjugate to one of the subgroups $P_{d_1,d_2,\ldots,d_k}$. Thus, $\mathbb{F}_{d_1,d_2,\ldots,d_k}(V) = S/P_{d_1,d_2,\ldots,d_k}$ is a flag manifold of S and any flag manifold of $\mathrm{SL}(V)$ may be obtained in this manner.

In particular, we have the specializations $\mathbb{F}_{1,2,\ldots,n-1}(V) = \mathbb{F}(V)$, $\mathbb{F}_m(V) = \mathbb{G}_m(V)$ (the *Grassmann manifold* of m-dimensional linear subspaces), and $\mathbb{F}_1(V) = \mathbb{G}_1(V) = \mathbb{P}(V)$.

Example 2 Consider a non-degenerate symmetric bilinear form on V, denote by S the corresponding *special orthogonal group* isomorphic to $\mathrm{SO}(n, \mathbb{C})$, and fix an integer m, $m \le n/2$, where $n = \dim V$. Let $Y \subset \mathbb{G}_m(V)$ be the algebraic subvariety whose points are identified with isotropic linear subspaces of V of dimension m. Then S acts on Y in the natural manner. By Witt's theorem, the action is transitive if $m < n/2$ and Y consists of two isomorphic S-orbits (each of them being a connected component of Y) if n is even and $m = n/2$. We set $\mathbb{IG}_m(V) := Y$ in the first case and denote by $\mathbb{IG}_m(V)$ one of the two S-orbits on Y in the second case. Then $\mathbb{IG}_m(V)$ is a flag manifold of S. In particular, for $m = 1$, $n > 2$ we obtain the quadric in $\mathbb{P}(V)$.

If V is the standard vector space $\mathbb{C}^n$ then we write $\mathbb{G}_m(n)$ and $\mathbb{IG}_m(n)$ instead of $\mathbb{G}_m(V)$ and $\mathbb{IG}_m(V)$ respectively. The quadric in $\mathbb{P}_{n-1}$ defined by the equation $\sum_{i=1}^n z_i^2 = 0$ is denoted by $\mathbb{Q}(n)$.

Let dim $V = 2m$ and let $V' \subset V$ be a linear subspace of codimension 1 such that the symmetric bilinear form, defined on V, is non-degenerate on V'. Then the mapping

$$\mathbb{IG}_m(V) \to \mathbb{IG}_{m-1}(V'), \quad s \mapsto s \cap V',$$

is an isomorphism. (The inverse mapping assigns to an $s' \in \mathbb{IG}_{m-1}(V')$ the unique maximal isotropic subspace s of V which is contained in the chosen orbit of $\mathrm{SO}(2m, \mathbb{C})$ and has the property that $s' \subset s \subset (s')^{\perp}$, where $\perp$ denotes the orthogonal complement in V.)

In other words, the subgroup $S = \mathrm{SO}(2m-1,\mathbb{C}) \subset \mathrm{SO}(2m,\mathbb{C})$ is transitive on $\mathbb{IG}_m(2m)$. Similarly, a simple observation shows that if n is even then the *symplectic group* $S = \mathrm{Sp}(n,\mathbb{C})$ is transitive on $\mathbb{P}_{n-1}$. Finally, the simple complex group S of type G_2 admits a 7-dimensional irreducible representation. The image of S is contained in $\mathrm{SO}(7,\mathbb{C})$ and is transitive on $\mathbb{Q}(7)$.

In these three cases a flag manifold of a simple group S can be written as a flag manifold of a larger simple group. Usually this does not happen and the above cases are the only exceptions (see Theorem 2 of §3.3).

The Lie algebra of a Borel (resp. parabolic) subgroup of G is called a *Borel* (resp. *parabolic*) *subalgebra* of $\mathfrak{g}$. We now proceed to the description of parabolic subalgebras in terms of roots and root vectors. All necessary facts concerning the structure of semisimple Lie algebras can be found in [Hum1] or [Se4].

Fix a maximal algebraic torus $T \subset S$ and denote by $\mathfrak{t}$ the corresponding *Cartan subalgebra* of $\mathfrak{s}$. Let Δ be the *root system* of $\mathfrak{s}$ with respect to $\mathfrak{t}$. Then we have the *root decomposition*

$$\mathfrak{s} = \mathfrak{t} \oplus \oplus_{\alpha\in\Delta} \mathfrak{s}_\alpha,$$

where each $\mathfrak{s}_\alpha$ has dimension 1. A subset $\Psi \subset \Delta$ is called *closed* if

$$\alpha, \beta \in \Psi, \alpha+\beta \in \Delta \implies \alpha+\beta \in \Psi.$$

A linear subspace $\mathfrak{h} \subset \mathfrak{s}$ containing $\mathfrak{t}$ is a Lie subalgebra if and only if

$$\mathfrak{h} = \mathfrak{t} \oplus \oplus_{\alpha\in\Psi} \mathfrak{s}_\alpha,$$

where $\Psi \subset \Delta$ is a closed subset.

Choose a base Π of the root system Δ and denote by Δ^+ (resp. Δ^-) the set of all positive (resp. negative) roots with respect to Π. The elements of Π are called *simple roots*. The subsets $\Delta^+, \Delta^- \subset \Delta$ are closed, so that

$$\mathfrak{b}^+ := \mathfrak{t} \oplus \oplus_{\alpha\in\Delta^+} \mathfrak{s}_\alpha \quad \text{and} \quad \mathfrak{b}^- := \mathfrak{t} \oplus \oplus_{\alpha\in\Delta^-} \mathfrak{s}_\alpha$$

are Lie subalgebras of $\mathfrak{s}$. It is easy to check that $\mathfrak{b}^+$ and $\mathfrak{b}^-$ are maximal solvable Lie subalgebras.

Proposition 2 *Let $\mathfrak{p}$ be any Lie subalgebra of $\mathfrak{s}$ containing $\mathfrak{b}^+$. Let $\mathfrak{h}$ be any Lie subalgebra of $\mathfrak{s}$ satisfying*

$$[\mathfrak{p},\mathfrak{p}] \subset \mathfrak{h} \subset \mathfrak{p}.$$

Denote by $\mathfrak{n}(\mathfrak{h})$ the normalizer of $\mathfrak{h}$ in $\mathfrak{s}$, i.e.,

$$\mathfrak{n}(\mathfrak{h}) := \{\mathrm{A} \in \mathfrak{s} \mid (\mathrm{ad\ A}) \cdot \mathfrak{h} \subset \mathfrak{h}\}.$$

Then $\mathfrak{n}(\mathfrak{h}) = \mathfrak{p}$.

Proof Since $\mathfrak{n}(\mathfrak{h}) \supset \mathfrak{p}$, we only have to show that each root subspace $\mathfrak{s}_{-\alpha} \subset \mathfrak{n}(\mathfrak{h})$ with $\alpha \in \Delta^+$ is in fact contained in $\mathfrak{p}$. But $\mathfrak{s}_\alpha \subset [\mathfrak{b}^+,\mathfrak{b}^+] \subset [\mathfrak{p},\mathfrak{p}] \subset \mathfrak{h}$. Therefore $[\mathfrak{s}_{-\alpha},\mathfrak{s}_\alpha] \subset \mathfrak{h}$ so that

$$\mathfrak{s}_{-\alpha} = [\mathfrak{s}_{-\alpha},[\mathfrak{s}_{-\alpha},\mathfrak{s}_\alpha]] \subset \mathfrak{h} \subset \mathfrak{p}.$$

□

Proposition 3 *Let $\mathfrak{p}$ be a Lie subalgebra of $\mathfrak{s}$ containing $\mathfrak{b}^+$. Then any Lie subgroup $P \subset S$ with Lie algebra $\mathfrak{p}$ is algebraic.*

Proof The adjoint representation $S \to \mathrm{GL}(\mathfrak{s})$ gives rise to an algebraic S-action on the Grassmann manifold $\mathbb{G}_k(\mathfrak{s})$, where $k := \dim \mathfrak{p}$. Let $o \in \mathbb{G}_k(\mathfrak{s})$ be the point corresponding to $\mathfrak{p}$. The subgroup

$$N(\mathfrak{p}) := \{s \in S \mid (\mathrm{Ad}\ s) \cdot \mathfrak{p} = \mathfrak{p}\}$$

is algebraic as the isotropy subgroup of o. According to Proposition 2 the Lie subalgebra of $N(\mathfrak{p})$ coincides with $\mathfrak{p}$. Since $P \subset N(\mathfrak{p})$, the subgroup P is also algebraic. □

Denote by B^+ and B^- the connected Lie subgroups of S with Lie algebras $\mathfrak{b}^+$ and $\mathfrak{b}^-$ respectively. By Proposition 3 these subgroups are algebraic. Since $\mathfrak{b}^+$ and $\mathfrak{b}^-$ are maximal solvable subalgebras of $\mathfrak{s}$, it follows that B^+ and B^- are Borel subgroups of S. The commutator subgroups $U^+ := (B^+)'$ and $U^- := (B^-)'$ are maximal unipotent subgroups of S.

Consider now a parabolic subgroup $P \subset S$. Replacing P if necessary by a conjugate subgroup, we may assume $P \supset B^+$. Let

$$\Delta_P := \{\alpha \in \Delta \mid \mathfrak{s}_\alpha \subset \mathfrak{p}\} \tag{1}$$

be the closed subset of Δ corresponding to $\mathfrak{p}$. We have

$$\Delta_P = \Delta_P^s \bigcup \Delta_P^a,$$

where

$$\Delta_P^s := \Delta_P \bigcap (-\Delta_P), \quad \Delta_P^a := \Delta_P - \Delta_P^s.$$

Since $\mathfrak{p} \supset \mathfrak{b}^+$, it follows that $\Delta_P^a \subset \Delta^+$. It is clear that Δ_P^s and Δ_P^a are closed subsets of Δ. Moreover,

$$\alpha \in \Delta_P,\ \beta \in \Delta_P^a,\ \alpha + \beta \in \Delta \Longrightarrow \alpha + \beta \in \Delta_P^a.$$

Proposition 4 *There exists a subset $\Phi_P \subset \Pi$ such that for $\gamma = \Sigma_{\alpha \in \Pi}\, k_\alpha \cdot \alpha \in \Delta$ one has:*

$$\gamma \in \Delta_P^s \iff k_\alpha = 0 \quad \textit{for all} \quad \alpha \notin \Phi_P.$$

Proof It is enough to prove that all simple roots which occur in the decompositions of elements of Δ_P^s are contained in Δ_P^s. For this it suffices to show that

$$\beta \in \Delta_P^s,\ \beta = \gamma + \delta,\ \gamma, \delta \in \Delta^+ \Longrightarrow \gamma, \delta \in \Delta_P^s.$$

One can rewrite the given equality in the form

$$-\delta = \gamma - \beta.$$

Since $\gamma \in \Delta^+$, one has $\gamma \in \Delta_P$. But $\beta \in \Delta_P^s$ so that $-\beta$ is also in Δ_P. Since Δ_P is closed, it follows that $-\delta \in \Delta_P$ and therefore $\delta \in \Delta_P^s$. The same argument applies to γ. □

Let $\mathfrak{c}_P$ denote the linear subspace of $\mathfrak{t}$, defined by the equations $\alpha(\mathrm{H}) = 0$, where $\mathrm{H} \in \mathfrak{t}$ and $\alpha \in \Delta_P^s$.

Corollary 2 *The same subspace is given by the equations $\alpha(\mathrm{H}) = 0$, where $\alpha \in \Phi_P$. In particular,*

$$\dim \mathfrak{c}_P = r - \operatorname{card} \Phi_P,$$

where $r = \dim \mathfrak{t}$ is the rank of S. □

Define three linear subspaces of $\mathfrak{s}$ by

$$\mathfrak{l}_P := \mathfrak{t} \oplus \oplus_{\alpha \in \Delta_P^s} \mathfrak{s}_\alpha, \quad \mathfrak{u}_P^+ := \oplus_{\alpha \in \Delta_P^a} \mathfrak{s}_\alpha, \quad \mathfrak{u}_P^- := \oplus_{\alpha \in (-\Delta_P^a)} \mathfrak{s}_\alpha. \tag{2}$$

It is clear that $\mathfrak{l}_P$, $\mathfrak{u}_P^+$, and $\mathfrak{u}_P^-$ are Lie subalgebras. Moreover,

$$[\mathfrak{l}_P, \mathfrak{u}_P^+] \subset \mathfrak{u}_P^+, \quad [\mathfrak{l}_P, \mathfrak{u}_P^-] \subset \mathfrak{u}_P^-.$$

Proposition 5 *The centralizer of $\mathfrak{c}_P$ in $\mathfrak{s}$ coincides with $\mathfrak{l}_P$. One has the decomposition*

$$\mathfrak{l}_P = \mathfrak{c}_P \oplus [\mathfrak{l}_P, \mathfrak{l}_P].$$

The commutator algebra $[\mathfrak{l}_P, \mathfrak{l}_P]$ is semisimple.

Proof A root subspace $\mathfrak{s}_\beta$ is contained in the centralizer of $\mathfrak{c}_P$ if and only if the linear form β vanishes on $\mathfrak{c}_P$. In view of Proposition 4 this means that $\beta \in \Delta_P^s$, and the first assertion follows. The second assertion is an immediate consequence of the decomposition

$$\mathfrak{t} = \mathfrak{c}_P \oplus \oplus_{\alpha \in \Phi_P} [\mathfrak{s}_\alpha, \mathfrak{s}_{-\alpha}].$$

In order to prove the last assertion it is enough to show that an abelian ideal $\mathfrak{a} \lhd \mathfrak{l}_P$ is contained in $\mathfrak{c}_P$. Clearly, any ideal of $\mathfrak{l}_P$ is spanned as a linear subspace by root vectors and by elements of $\mathfrak{t}$. If $\mathfrak{s}_\alpha \subset \mathfrak{a}$ then $[\mathfrak{s}_{-\alpha}, \mathfrak{s}_\alpha] \subset \mathfrak{a}$. But $[\mathfrak{s}_{-\alpha}, \mathfrak{s}_\alpha]$ does not commute with $\mathfrak{s}_\alpha$. Since $\mathfrak{a}$ is abelian, we get a contradiction. Finally, if $\mathrm{H} \in \mathfrak{t} \cap \mathfrak{a}$ and $\alpha(\mathrm{H}) \neq 0$ for some $\alpha \in \Phi_P$ then $\mathfrak{s}_\alpha = [\mathrm{H}, \mathfrak{s}_\alpha] \subset \mathfrak{a}$ and we again have a contradiction. Thus $\mathfrak{a} \subset \mathfrak{c}_P$. □

Denote by L_P, C_P, U_P^+, and U_P^- the associated connected Lie subgroups of S. Then C_P is an algebraic torus, L'_P a semisimple subgroup, and $L_P = C_P \cdot L'_P$ a reductive algebraic subgroup.

Now, $U_P^+ \subset U^+$ and $U_P^- \subset U^-$ are unipotent algebraic subgroups. Furthermore, the definition of $\mathfrak{l}_P$ and $\mathfrak{u}_P$ implies that

$$\mathfrak{p} = \mathfrak{l}_P \dotplus \mathfrak{u}_P^+. \tag{3}$$

Consequently U_P^+ is the unipotent radical of P. We write U_P and $\mathfrak{u}_P$ instead of U_P^+ and $\mathfrak{u}_P^+$ respectively.

Proposition 6 *A flag manifold S/P is a rational algebraic variety.*

Proof We have $\mathfrak{s} = \mathfrak{u}_P^- + \mathfrak{p}$ by the definition of $\mathfrak{u}_P^-$. This means that the orbit $O := U_P^-(e \cdot P) \subset S/P$ is open. But U_P^- is a unipotent algebraic group. Therefore O is isomorphic to an affine space. □

Proposition 7 *A flag manifold S/P is simply connected.*

Proof We keep the above notation. The complement E to the orbit O is an algebraic subvariety in S/P. Therefore

$$\dim_{\mathbb{R}} E \leq \dim_{\mathbb{R}} S/P - 2.$$

In this situation any continuous loop in S/P is homotopic to a continuous loop in O (see e.g. [He], Ch. 7, Prop. 12.4). Since O is simply connected, S/P is also simply connected. □

Corollary 3 *A parabolic subgroup of any connected linear algebraic group is connected.* □

Remark In particular, P in Proposition 3 is uniquely determined by $\mathfrak{p}$.

Corollary 4 *The normalizer* $\mathrm{Norm}_S(P)$ *of a parabolic subgroup $P \subset S$ coincides with P.*

Proof Taking $\mathfrak{h} = \mathfrak{p}$ in Proposition 2, we see that the Lie algebra of $\mathrm{Norm}_S(P)$ coincides with $\mathfrak{p}$. By Corollary 3 the groups P and $\mathrm{Norm}_S(P)$ are connected, hence $\mathrm{Norm}_S(P) = P$. □

Proposition 8 *Let $P \subset S$ be a parabolic subgroup. Then*

$$P = L_P \cdot U_P \tag{4}$$

and

$$P/P' \simeq (\mathbb{C}^*)^k, \tag{5}$$

where $k = r - \mathrm{card}\ \Phi_P$.

Proof Since P is connected, (4) follows from (3). Since $U_P \subset P'$, we have

$$P/P' \simeq L_P/L'_P \simeq C_P/\Gamma,$$

where Γ is finite. Thus $P/P' \simeq (\mathbb{C}^*)^k$ by Corollary 2. □

Theorem *Let S be a connected semisimple complex Lie group, $B \subset S$ be a Borel subgroup, and T a maximal torus in B. Choose the base Π of the root system Δ so that $B = B^+$. Then one has the natural bijections:*

$$\left\{\begin{matrix} \textit{conjugacy classes} \\ \textit{of parabolic} \\ \textit{subgroups of } S \end{matrix}\right\} \longleftrightarrow \left\{\begin{matrix} \textit{parabolic subgroups} \\ \textit{of } S \textit{ containing } B \end{matrix}\right\} \longleftrightarrow \{\textit{subsets } \Phi \subset \Pi\}.$$

The second bijection is given by

$$P \mapsto \Phi_P.$$

Proof We have to prove two things:

(a) for each parabolic subgroup there is only one conjugate subgroup, which contains B;

(b) each subset $\Phi \subset \Pi$ is of the form $\Phi = \Phi_P$, where P is a parabolic subgroup containing B.

Proof of (a) Assume that P_1 and P_2 contain B and that $P_1 = gP_2g^{-1}$. Then gBg^{-1} and B are two Borel subgroups in P_1. By Corollary 1 they are conjugate in P_1, i.e., $pgBg^{-1}p^{-1} = B$ for some $p \in P_1$. By Corollary 4 we have $pg \in B$, hence $g \in P_1$ showing that $P_1 = P_2$.

Proof of (b) Put

$$\Delta_P^s := \{\gamma = \sum_{\alpha \in \Pi} k_\alpha \cdot \alpha \in \Delta \mid k_\alpha = 0 \quad \text{for all} \quad \alpha \notin \Phi\}.$$

After that put $\Delta_P^a := \Delta^+ - \Delta_P^s$ and define $\mathfrak{l}_P$ and $\mathfrak{u}_P$ by (2). Finally, $\mathfrak{p}$ and P are recovered from (3) and (4). □

Later on we shall need the following property of parabolic subalgebras.

Proposition 9 *Let $\mathfrak{p} \neq \mathfrak{s}$ be a subalgebra of $\mathfrak{s}$ containing $\mathfrak{b}^+$. Then there exists an element $\mathrm{A} \in \mathfrak{p}$ such that*

$$\mathrm{tr}(\mathrm{ad}\ \mathrm{A})|_{\mathfrak{p}} \neq 0.$$

Proof Define Δ_P by (1) and take any vector $\mathrm{A} \in \mathfrak{t}$ such that

$$\alpha(\mathrm{A}) > 0 \quad \text{for all } \alpha \in \Pi.$$

Then

$$\mathrm{tr}(\mathrm{ad}\ \mathrm{A})|_{\mathfrak{p}} = \sum_{\alpha \in \Delta_P} \alpha(\mathrm{A}) = \sum_{\alpha \in \Delta_P^a} \alpha(\mathrm{A}) > 0.$$ □

3.2 Equivariant projective embeddings

Let X be a complex manifold, $L(X)$ the group of isomorphism classes of holomorphic line bundles over X with respect to tensor product, and $\mathcal{O}^* = \mathcal{O}_X^*$ the subsheaf (of abelian groups) of multiplicatively invertible elements of $\mathcal{O}_X$. The cohomology group $H^1(X, \mathcal{O}^*)$ is isomorphic to $L(X)$ and is called the *Picard group* of X, see [Hi], Ch.1, §3 or [GrHa], Ch.1, §1. Both $L(X)$ and $H^1(X, \mathcal{O}^*)$ are acted on by $\mathrm{Aut}(X)$. The isomorphism $L(X) \xrightarrow{\sim} H^1(X, \mathcal{O}^*)$, assigning to a line bundle the cocycle of its transition functions, commutes with the action of $\mathrm{Aut}(X)$. For a holomorphic line bundle $\mathbb{L}$ on X we denote by $\mathcal{O}_{\mathbb{L}}$ the sheaf of germs of local holomorphic sections of $\mathbb{L}$. The divisor of a meromorphic section $s : X \to \mathbb{L}$ is denoted by (s). The following embedding theorem is due to A.Blanchard, see [Bl], Théorème principal I.

Theorem 1 *Let X be a non-singular projective variety and G a group acting on X by holomorphic transformations. Assume that the induced action of G on $H^1(X, \mathcal{O}^*)$ is trivial. Then X admits a G-equivariant projective embedding.*

Proof Let V_0 be a vector space. We suppose that X is embedded into $\mathbb{P}(V_0^*)$, where V_0^* is the dual space. The embedding is denoted by φ_0. We identify X with $\varphi_0(X)$ and assume, without loss of generality, that X is not contained in a hyperplane of $\mathbb{P}(V_0^*)$. Consider the pull-back $\mathbb{L} := \varphi_0^*(\mathbb{L}_0)$, where $\mathbb{L}_0$ is the line bundle on $\mathbb{P}(V_0^*)$ associated with a hyperplane. The restriction map

$$V_0 = H^0(\mathbb{P}(V_0^*), \mathcal{O}_{\mathbb{L}_0}) \to V := H^0(X, \mathcal{O}_{\mathbb{L}})$$

is injective. In what follows V_0 is identified with its image in V.

Since G acts trivially on $H^1(X, \mathcal{O}^*)$, the line bundle $g^*\mathbb{L}$, induced from $\mathbb{L}$ by an automorphism $g \in G$, is isomorphic to $\mathbb{L}$. Let $s \in H^0(X, \mathcal{O}_{\mathbb{L}})$, $s \neq 0$. Then (s) and $g \cdot (s)$ are linearly equivalent for all $g \in G$. Thus, for g and s fixed, there exists a meromorphic function f on X such that $g \cdot (s) = (s) + (f)$. Therefore the meromorphic section fs of $\mathbb{L}$ is in fact holomorphic. In other words, $g \cdot (s)$ is the divisor of some $s' \in H^0(X, \mathcal{O}_{\mathbb{L}})$. Clearly, s and g determine s' uniquely up to a constant factor.

Denote by $[s]$ the image of $s \in V - \{0\}$ under the canonical projection $V - \{0\} \to \mathbb{P}(V)$. The above argument shows that the mapping $(g, [s]) \mapsto g \cdot [s] := [s']$ is a G-action on $\mathbb{P}(V)$. Therefore we also have the associated dual G-action on $\mathbb{P}(V^*)$. Assign to a point $x \in X$ a hyperplane $A_x \subset \mathbb{P}(V)$. Namely, put

$$A_x := \{[s] \mid s \in V - \{0\},\ s(x) = 0\}.$$

Then A_x can be regarded as a point of $\mathbb{P}(V^*)$. We claim that the mapping

$$X \ni x \overset{\varphi}{\mapsto} A_x \in \mathbb{P}(V^*)$$

is an embedding. In order to show this, consider the projection map $\pi : \mathbb{P}(V^*) \to \mathbb{P}(V_0^*)$. (Recall that V_0 is a linear subspace of V.) Then π is regular outside Y, where Y is identified with the set of all hyperplanes in V containing V_0. It is easy to see that $\varphi(X) \cap Y = \emptyset$ and $\pi \circ \varphi = \varphi_0$. Since φ_0 is an embedding, so is φ.

By the definition of the dual action we have

$$g \cdot A_x = \{[s] \mid g^{-1} \cdot [s]\ (x) = 0\}.$$

Therefore

$$\begin{aligned} g \cdot A_x = \{[s] \mid x \in g^{-1} \cdot (s)\} &= \{[s] \mid gx \in (s)\} = \\ = \{[s] \mid s(gx) = 0\} &= A_{gx}, \end{aligned}$$

showing that the embedding φ is G-equivariant. □

Remark In the coordinate form φ is given by

$$X \ni x \mapsto (s_0(x) : s_1(x) : \ldots : s_n(x)) \in \mathbb{P}_n,$$

where $s_0, s_1, \ldots, s_n$ is a basis of $H^0(X, \mathcal{O}_{\mathbb{L}})$.

Corollary *Let X be a non-singular projective variety with $H^1(X, \mathcal{O}_X) = 0$ and let G be a connected (real) Lie group acting continuously on X. Then X admits a G-equivariant projective embedding $X \hookrightarrow \mathbb{P}_n$.*

Proof Consider the exponential sequence of sheaves

$$0 \to \mathbb{Z} \to \mathcal{O} \to \mathcal{O}^* \to 1$$

and let $c : H^1(X, \mathcal{O}^*) \to H^2(X, \mathbb{Z})$ be the homomorphism from the corresponding exact cohomology sequence. Then c is injective because $H^1(X, \mathcal{O}) = 0$. Note that c commutes with the action of $\mathrm{Aut}(X)$ on the cohomology groups. Since G acts trivially on $H^2(X, \mathbb{Z})$ by the homotopy argument, the G-action on $H^1(X, \mathcal{O})$ is also trivial, and the theorem applies. □

Proposition 1 *For an algebraic variety $X \subset \mathbb{P}_n$ the group*

$$\mathrm{Aut}(\mathbb{P}_n, X) := \{g \in \mathrm{Aut}(\mathbb{P}_n) \simeq \mathrm{PGL}(n+1, \mathbb{C}) \mid g \cdot X = X\}$$

is linear algebraic.

Proof Let $I(X)$ be the ideal of X in the coordinate ring $\mathbb{C}[x_0, x_1, \ldots, x_n]$ of $\mathbb{P}_n$. Denote by $\mathbb{C}[x_1, x_2, \ldots, x_n]_d$ the subspace of homogeneous polynomials of degree d. Then

$$I(X) = \bigoplus_{d=0}^{\infty} I_d(X),$$

where $I_d(X) := I(X) \bigcap \mathbb{C}[x_0, x_1, \ldots, x_n]_d$. The group $\mathrm{Aut}(\mathbb{P}_n)$ acts algebraically on each Grassmann manifold $\mathbb{G}_k(\mathbb{C}[x_0, x_1, \ldots, x_n]_d)$. Let $k_d := \dim I_d(X)$. In case $k_d > 0$ denote by o_d the point of $\mathbb{G}_{k_d}(\mathbb{C}[x_0, x_1, \ldots, x_n]_d)$ corresponding to the linear subspace $I_d(X)$. Then

$$\mathrm{Aut}(\mathbb{P}_n, X) = \bigcap_{d:\ k_d > 0} \mathrm{Aut}(\mathbb{P}_n)_{o_d}, \qquad (*)$$

showing that the group on the left hand side is linear algebraic. □

Remark Although it is formally unnecessary for the proof, one can replace the intersection in (*) by a finite one. Namely, by Hilbert's theorem $I(X)$ is generated by a finite number of homogeneous polynomials $p_1, \ldots, p_m$. It follows that

$$\mathrm{Aut}(\mathbb{P}_n, X) = \bigcap_{j=1}^{m} \mathrm{Aut}(\mathbb{P}_n)_{o_{d_j}},$$

where $d_j := \deg p_j$, $j = 1, \ldots, m$.

The homomorphism

$$\tau : \mathrm{Aut}(\mathbb{P}_n, X) \to \mathrm{Aut}(X), \quad \tau(g) := g|_X,$$

is injective if and only if X is not contained in the union of linear subspaces $L_1, \ldots, L_k$ with the property that any linear subspace containing all L_i has dimension $\geq \sum_i (\dim L_i + 1) - 1$. In particular, if X is connected this means that X is not contained in a hyperplane. The image $\tau(\mathrm{Aut}(\mathbb{P}_n, X))$ is the subgroup of automorphisms of X, which are induced by automorphisms of the ambient projective space. If the embedding $X \hookrightarrow \mathbb{P}_n$ is equivariant with respect to some subgroup $G \subset \mathrm{Aut}(X)$ then $\tau(\mathrm{Aut}(\mathbb{P}_n, X))$ contains G.

Proposition 2 *Let X be a connected algebraic variety in $\mathbb{P}_n$ not contained in a hyperplane, $G := \mathrm{Aut}^\circ(X)$, and assume that the embedding $X \hookrightarrow \mathbb{P}_n$ is G-equivariant. Then the connected components of $\mathrm{Aut}(X)$ and $\mathrm{Aut}(\mathbb{P}_n, X)$ are isomorphic.*

Proof Since $\mathrm{Ker}\ \tau = \{e\}$ and $G \subset \mathrm{Im}\ \tau$, the assertion is clear. □

We now apply these results to homogeneous complex manifolds.

Theorem 2 *A flag manifold X admits a projective embedding, which can be chosen equivariant with respect to the connected automorphism group $\mathrm{Aut}^\circ(X)$. For a connected homogeneous complex manifold X the following conditions are equivalent:*

(i) *X is a rational projective variety;*

(ii) *X is a simply connected projective variety;*

(iii) *X is a flag manifold.*

Proof From Propositions 6 and 7 of §3.1 it follows that (iii) $\Longrightarrow$ (i) and (iii) $\Longrightarrow$ (ii). Assume now that X is a projective variety. Then $H^1(X, \mathbb{C}) = H^1(X, \mathcal{O}_X) \oplus H^0(X, \Omega_X)$ and $H^1(X, \mathcal{O}_X) \simeq H^0(X, \Omega_X)$, where Ω_X $(= \Omega_X^1)$ is the sheaf of germs of local holomorphic 1-forms on X. The number $q(X) = \dim H^0(X, \Omega_X)$, called the irregularity of X, is a birational invariant (see e.g. [GrHa], Ch. 4, §2). Therefore, if X is subject to (i) or (ii) then $H^1(X, \mathcal{O}_X) = 0$. By the above corollary there exists a projective embedding $X \hookrightarrow \mathbb{P}_n$ equivariant with respect to $\mathrm{Aut}^\circ(X)$. In order to deduce (iii) from (i) or (ii), identify X with its image in $\mathbb{P}_n$. By Chow's theorem [Ch] (see also [GR3], Ch. 9, §5) X is an algebraic variety in $\mathbb{P}_n$. The linear algebraic group $\mathrm{Aut}(\mathbb{P}_n, X)$ acts algebraically on X. Since each automorphism from $\mathrm{Aut}^\circ(X)$ is induced by some automorphism of $\mathbb{P}_n$, the action of $\mathrm{Aut}(\mathbb{P}_n, X)$ on X is transitive. Since X is connected, the connected component $\mathrm{Aut}^\circ(\mathbb{P}_n, X)$ is also transitive on X. By the Fixed Point Theorem the isotropy subgroup is parabolic and X is a flag manifold. □

3.3 Automorphism groups of flag manifolds

Lemma 1 *Let $X = S/P$ be a flag manifold of some connected semisimple complex Lie group S. Let G be a connected complex Lie group, acting on X holomorphically, transitively and effectively. Then G is semisimple and the action of G on X is algebraic.*

Proof Consider a projective embedding $X \hookrightarrow \mathbb{P}_n$ equivariant with respect to the connected automorphism group $\mathrm{Aut}^\circ(X)$. We may assume that X is not contained in a hyperplane of the ambient projective space. By Proposition 2 of §3.2 we can identify G with a complex Lie subgroup of the linear algebraic group $\mathrm{Aut}^\circ(\mathbb{P}_n, X)$. Let R be the radical of G and let $\hat{R}$ be the Zariski closure of R in $\mathrm{Aut}^\circ(\mathbb{P}_n, X)$. Then $\hat{R}$ is a connected solvable algebraic subgroup of $\mathrm{Aut}(\mathbb{P}_n)$ which leaves X invariant. By the Fixed Point Theorem $\hat{R}$ has a fixed point on X. Since R is a normal subgroup of a transitive group G, it follows that R acts on X trivially. The effectivity of the G-action on X implies $R = \{e\}$. Thus G is semisimple. Since any holomorphic homomorphism of a connected semisimple complex group to an algebraic group is rational, it follows that the G-action on X is algebraic. □

Lemma 2 *A flag manifold S/P of a simple group S cannot be represented as a product $S/P = X_1 \times X_2$, where X_1 and X_2 are complex manifolds of positive dimension.*

Proof Assume that $S/P = X_1 \times X_2$, where X_1 and X_2 are complex manifolds. We have to prove that either X_1 or X_2 is one point. Denote by π_1, π_2 the projection mappings onto the factors. By Proposition 1 of §2.4 any $s \in S$ interchanges the fibers of π_1 and also the fibers of π_2. Observe that the induced S-actions on X_1 and on X_2 are transitive. More precisely, let $o_1 \in X_1$ and $o_2 \in X_2$ be arbitrary points. Then $P_1 := S_{o_1}$, $P_2 := S_{o_2}$ are parabolic subgroups of S and X_1, X_2 can be written as $X_1 = S/P_1$, $X_2 = S/P_2$. Suppose that $x, y \in \pi_1^{-1}(o)$ and $sx = y$ for some $s \in S$. Then $s \in P_1$, and so P_1 acts transitively on the fiber $\pi_1^{-1}(o_1)$, isomorphic to X_2. By Lemma 1 the radical of P_1 acts on X_2 trivially. Therefore the S-action on X_2 has ineffectivity kernel of positive dimension. Since S is a simple group, this implies that S acts on X_2 trivially so that $X_2 = S/P_2$ is one point. □

If $S = S_1 \cdot \ldots \cdot S_m$ is the decomposition of a semisimple group S into simple factors, then any parabolic subgroup $P \subset S$ is of the form $P = P_1 \cdot \ldots \cdot P_m$ and $P_j := P \cap S_j$ is a parabolic subgroup of S_j for all j, $j = 1, \ldots, m$. Let $\pi_j : \tilde{S}_j \to S_j$ be the universal covering and let $\tilde{P}_j := \pi_j^{-1}(P_j)$. Since a flag manifold is simply connected, each subgroup $\tilde{P}_j \subset \tilde{S}_j$ is connected and $S_j/P_j = \tilde{S}_j/\tilde{P}_j$. It follows that

$$S/P = (S_1/P_1) \times \ldots \times (S_m/P_m).$$

We shall assume that the action of S on S/P is effective. This means that P_j is a proper parabolic subgroup of S_j for all j.

Theorem 1 *Let $S = S_1 \cdot \ldots \cdot S_m$ and $P = P_1 \cdot \ldots \cdot P_m$ as above. Assume that P_j is a proper parabolic subgroup of S_j for all j. Then*

$$\mathrm{Aut}^\circ(S/P) = \mathrm{Aut}^\circ(S_1/P_1) \times \ldots \times \mathrm{Aut}^\circ(S_m/P_m),$$

where each factor on the right hand side is a simple group.

Proof The decomposition is established in Proposition 2 of §2.4. By Lemma 1 each group on the right hand side is semisimple. We have to prove that if S is simple then $G := \mathrm{Aut}^\circ(S/P)$ is also simple. Assume the contrary and let $G = G_1 \cdot \ldots \cdot G_l$, $l \geq 2$, be the decomposition into simple factors. Denote by Q the isotropy subgroup of the point $e \cdot P$ in G. Then $G/Q = S/P$ and the subgroup Q is parabolic. Therefore

$$S/P = G/Q = G_1/Q_1 \times \ldots \times G_l/Q_l,$$

where Q_j is a proper parabolic subgroup of G_j for all j, $j = 1, \ldots, l$. By Lemma 2 we get a contradiction. □

According to Theorem 1, the group $\mathrm{Aut}^\circ(X)$ is easily determined for an arbitrary flag manifold X, provided $\mathrm{Aut}^\circ(X)$ is known for all flag manifolds of simple groups. Our next result solves the problem for $X = S/P$, where S is simple. Let r be the rank of S so that $\Pi = \{\alpha_1, \ldots, \alpha_r\}$. We adopt the conventions of [Bou] for indexing the simple roots. We have seen in §3.1 that a parabolic subgroup of S is uniquely up to conjugacy determined by a subset $\Phi \subset \Pi$. The parabolic subgroup associated with Φ is denoted here by $P_{i_1,\ldots,i_k}$, where $\Phi = \{\alpha_{i_1}, \ldots, \alpha_{i_k}\}$ and $1 \le i_1 < i_2 < \ldots < i_k \le r$. Specifically, P_i is a maximal parabolic subgroup and $P_{1,\ldots,r}$ is a Borel subgroup. For $S = \mathrm{SL}(V)$ this notation coincides with the notation of Example 1 of §3.1.

Observe that a parabolic subgroup $P \subset S$ contains the (finite) center of S. Thus, if S acts on S/P effectively then the center of S is trivial. Without loss of generality we assume that S has this property.

The special manifolds, arising in the following theorem, are introduced in §3.1 (see Example 2).

Theorem 2 *Let S be a simple complex Lie group with trivial center and let $X := S/P$ be a flag manifold. Then, as a rule, the connected automorphism group $\mathrm{Aut}^\circ(X)$ coincides with S. The only exceptions are as follows:*

1) *S is of type B_r, $P = P_r$, $X = \mathbb{IG}_{r+1}(2r+2)$, $\mathrm{Aut}^\circ(X)$ is of type D_{r+1}, $r \ge 3$;*

2) *S is of type C_r, $P = P_1$, $X = \mathbb{P}_{2r-1}$, $\mathrm{Aut}^\circ(X)$ is of type $A_{2r-1}, r \ge 2$;*

3) *S is of type G_2, $P = P_1$, $X = \mathbb{Q}(7)$, $\mathrm{Aut}^\circ(X)$ is of type B_3.*

The proof has to be postponed until §4.8 (Theorem 2).

Remark This result is mentioned in [Ti2] without proof (see Remark 6). A.L.Onishchik gave a proof based on the theory of topological transformation groups, see [On2] or [On4], Ch. 4, §15. Other proofs can be found in [Kan] and [St].

We now compute the full automorphism group of a flag manifold X. Let $G = \mathrm{Aut}^\circ(X)$. Then G is a semisimple complex Lie group, which is transitive on X. Let $o \in X$ be any point and $P := G_o \subset G$ the isotropy subgroup. Denote by $\mathrm{Aut}_{alg}(G)$ the group of all automorphisms of G as a complex Lie group. For each $g \in G$ we have the inner automorphism $\iota_g : G \to G$ defined by $\iota_g(x) = gxg^{-1}$. The subgroup of inner automorphisms $\mathrm{Int}(G)$ is normal in $\mathrm{Aut}_{alg}(G)$. Since G has trivial center, $\mathrm{Aut}_{alg}(G)$ and $\mathrm{Int}(G)$ are isomorphic to the automorphism group of the Lie algebra $\mathfrak{g}$ and to the group of inner automorphisms of $\mathfrak{g}$ respectively. In particular, $\mathrm{Aut}_{alg}(G)$ has a natural structure of a complex Lie group. Furthermore, $\mathrm{Int}(G)$ is isomorphic to G and $\mathrm{Int}(G) = \mathrm{Aut}^\circ_{alg}(G)$.

Fix a Borel subgroup of $B \subset P$ and a maximal torus $T \subset B$. Let E denote the subgroup of $\mathrm{Aut}_{alg}(G)$ consisting of those automorphisms which preserve B and T. Then E is a finite group isomorphic to the automorphism group of the corresponding Dynkin diagram, and

$$\mathrm{Aut}_{alg}(G) = E \ltimes \mathrm{Int}(G) \simeq E \ltimes G,$$

see [OnVi], Ch. 4, §4. Choose a base Π of the root system $\Delta = \Delta(\mathfrak{g}, \mathfrak{t})$ so that $B = B^+$, denote by Φ the subset of Π associated with P (see §3.1), and put

$$E_\Phi := \{\gamma \in E \mid \gamma(P) = P\}.$$

If E is realized as the automorphism group of the Dynkin diagram, then E_Φ is the subgroup preserving the subset of vertices corresponding to the elements of Φ.

Theorem 3 $\mathrm{Aut}(X) \simeq E_\Phi \ltimes G$.

Proof For any $\varphi \in \mathrm{Aut}(X)$ define $\tilde{\varphi} \in \mathrm{Aut}_{alg}(G)$ by

$$\tilde{\varphi}(g) \cdot x = \varphi(g \cdot \varphi^{-1}(x)), \quad g \in G = \mathrm{Aut}^\circ(X),\ x \in X.$$

Then $\varphi \mapsto \tilde{\varphi}$ is a homomorphism of $\mathrm{Aut}(X)$ into $\mathrm{Aut}_{alg}(G)$. Assuming that $\tilde{\varphi} = \mathrm{id}$, we obtain $\varphi(gx) = g \cdot \varphi(x)$ for all $g \in G, x \in X$. In particular, if $gx = x$ then $g \cdot \varphi(x) = \varphi(x)$. By Corollary 4 of §3.1 the normalizer in G of the isotropy subgroup G_x coincides with G_x. Thus G_x has only one fixed point on X. Hence $\varphi(x) = x$ and, since x is arbitrary, $\varphi = \mathrm{id}$.

We see that the homomorphism $\varphi \mapsto \tilde{\varphi}$ is injective. In order to understand its image, consider first an element $\varphi \in \mathrm{Aut}(X)$ such that $\varphi(o) = o$. Then the definition of $\tilde{\varphi}$ shows that $\tilde{\varphi}(P) = P$. Conversely, let $\gamma \in \mathrm{Aut}_{alg}(G)$ be an automorphism preserving P. Then we can define $\varphi \in \mathrm{Aut}(X)$ by $\varphi(g \cdot o) := \gamma(g) \cdot o$. One checks easily that this definition makes sense and that $\tilde{\varphi} = \gamma$.

Observe that if $\varphi \in \mathrm{Aut}^\circ(X) = G$ then $\tilde{\varphi}$ is an inner automorphism of G. It follows that

$$\{\tilde{\varphi} \mid \varphi \in \mathrm{Aut}(X)\} = \{\gamma \in \mathrm{Aut}_{alg}(G) \mid \gamma(P) = P\} \cdot \mathrm{Int}(G).$$

Thus, to finish the proof, it remains to show that the first group on the right hand side decomposes as $E_\Phi \ltimes P$. In other words, given $\gamma = \epsilon \cdot \iota_g \in \mathrm{Aut}_{alg}(G)$, where $\epsilon \in E$ and $\gamma(P) = P$, we have to prove that $g \in P$ and $\epsilon \in E_\Phi$. But $\epsilon^{-1}(P) = gPg^{-1}$ is a parabolic subgroup containing B. As we have seen in §3.1, such a subgroup is unique. Thus $gPg^{-1} = P$ and, consequently, $g \in P$. Finally, since $\epsilon(P) = \gamma(g^{-1}Pg) = P$, we conclude that $\epsilon \in E_\Phi$. □

Corollary 1 $\mathrm{Aut}(X)/\mathrm{Aut}^\circ(X) \simeq E_\Phi$. □

Let φ be a diffeomorphism of a compact oriented manifold M,

$$(\varphi^*)^k : H^k(X, \mathbb{R}) \to H^k(X, \mathbb{R})$$

the induced linear operator, and

$$L(\varphi) := \sum_k (-1)^k \ \mathrm{tr}\ (\varphi^*)^k$$

the *Lefschetz number* of φ. According to the Lefschetz formula, $L(\varphi) \neq 0$ implies that φ has a fixed point, see [GrHa], Ch. 3, §4.

Corollary 2 *Each automorphism of a flag manifold has at least one fixed point.*

Proof By Theorem 3 each automorphism of G/P lifts to G/B, and so we may assume that $X = G/B$. Since G acts trivially on $H^*(X, \mathbb{R})$ by the homotopy argument, it is enough to prove that $L(\varphi) \neq 0$ for those automorphisms which correspond to the elements of E. Recall that X is a cell complex with the cells defined by the Bruhat decomposition of G, see [Hum2], Ch. 10, § 28. Namely, each cell is of the form $B(no)$, where $n \in \mathrm{Norm}_G(T)$, and thus is a locally closed algebraic submanifold isomorphic to some $\mathbb{C}^k$. In particular, $H^k(X, \mathbb{R}) = 0$ for odd k and, consequently, $L(\varphi) = \sum_k \mathrm{tr}(\varphi^*)^k$. Now, if $\epsilon \in E$ and if φ is an automorphism of X defined by $\varphi(go) = \epsilon(g)o$, then $\varphi(Bno) = B\epsilon(n)o$ and $\epsilon(n) \in \mathrm{Norm}_G(T)$. This shows that φ interchanges the cells. Therefore φ acts on $H^*(X, \mathbb{R})$ by permutation of the base elements. The trace of such an operator in each $H^k(X, \mathbb{R})$ is non-negative. Since $\mathrm{tr}(\varphi^*)^0 = 1$, the proof is complete. □

Remark Applying this to an automorphism from $\mathrm{Aut}^\circ(X)$, we obtain the well-known fact that any $g \in G$ is contained in a Borel subgroup, see [OnVi], Ch. 3, §3 or [Hum2], Ch. 8, § 22.

3.4 Parallelizable manifolds

A complex manifold X is called *parallelizable* if its holomorphic tangent bundle TX is trivial. A discrete subgroup Γ of a Lie group G is called *uniform* if the quotient G/Γ is compact.

Theorem (H.-C.Wang [Wa3]) *A connected compact complex manifold X is parallelizable if and only if X is homogeneous and X can be expressed in the Klein form $X = G/\Gamma$, where G is a connected complex Lie group and Γ a uniform discrete subgroup of G.*

Proof Let $n := \dim X$. Assume first that $X = G/\Gamma$. Take n holomorphic linearly independent tangent vectors at $e \in G$. Pushing these vectors by right translations, we obtain n right invariant holomorphic vector fields on G, which are linearly independent at every point $g \in G$. By definition the vector fields are projectable onto $X = G/\Gamma$, showing that X is parallelizable.

In order to prove the converse assume that $A_1, \ldots, A_n \in \mathcal{T}_X(X)$ are linearly independent at every point $x \in X$. Then the Lie algebra $\mathfrak{g} := \mathcal{T}_X(X)$ is n-dimensional and $A_1, \ldots, A_n$ is a basis of $\mathfrak{g}$. Indeed, if $A \in \mathfrak{g}$ then A can be written as

$$A = \sum_{i=1}^{n} c_i(x) \cdot A_i \ ,$$

where the $c_i(x)$ are holomorphic functions on X. Since X is compact, $c_i(x)$ are constant and

$$A = \sum_{i=1}^{n} c_i \cdot A_i \ , \quad \text{where } \ c_i \in \mathbb{C}.$$

Let $G := \mathrm{Aut}^\circ(X)$. Then G is a complex Lie group, acting holomorphically on X, and $\mathfrak{g}$ is the Lie algebra of G (see §2.3). The G-orbit of each point $x \in X$ is open because $A_1, \ldots, A_n$ generate the tangent space T_xX. Since X is connected, it follows that X is G-homogeneous. The equality $\dim X = \dim G$ shows that the isotropy subgroup is discrete. □

Corollary *A compact homogeneous complex manifold X is parallelizable if and only if*

$$\dim \mathrm{Aut}(X) = \dim X.$$

Proof Let $X = G/\Gamma$, where Γ is a uniform discrete subgroup of G. Then the above proof shows that any holomorphic vector field on X belongs to the image of $\mathfrak{g}$ under the Lie homomorphism. Thus G is locally isomorphic to $\mathrm{Aut}^\circ(X)$, and so X and $\mathrm{Aut}(X)$ have the same dimension. The converse is clear. □

Example 1 Let $\Gamma \subset G := \mathbb{C}^n$ be a lattice generated by $2n$ vectors, which are linearly independent over $\mathbb{R}$. Then G/Γ is a complex torus.

A complex torus is an abelian complex Lie group. Conversely, any connected compact complex Lie group L is a complex torus. In order to show this, it suffices to consider the adjoint representation

$$L \to \mathrm{GL}(\mathfrak{l}), \quad l \mapsto \mathrm{Ad}\, l.$$

Since $\mathrm{GL}(\mathfrak{l})$ is a holomorphically separable complex manifold and L is connected and compact, the image of L in $\mathrm{GL}(\mathfrak{l})$ is one point. This means that L is abelian and, consequently, $L = \mathbb{C}^n/\Gamma$, where Γ is a lattice of rank $2n$.

Example 2 Let G be the group of all $n \times n$ matrices of the form

$$\begin{pmatrix} 1 & * & * & \dots & * \\ 0 & 1 & * & \dots & * \\ 0 & 0 & 1 & \dots & * \\ \vdots & \vdots & \vdots & \ddots & \vdots \\ 0 & 0 & 0 & \dots & 1 \end{pmatrix}$$

where $*$ denotes arbitrary complex numbers. Let $\Gamma \subset G$ be the subgroup consisting of similar matrices, where $*$ denotes arbitrary Gaussian integers. Then G is a nilpotent complex Lie group and Γ is a uniform discrete subgroup of G.

Example 3 According to a theorem of A.Borel [Bo3], any semisimple complex Lie group G has a uniform discrete subgroup Γ.

Not every connected complex Lie group G is of the form $G = \mathrm{Aut}^\circ(X)$, where X is a compact complex parallelizable manifold. In fact, (i) of the following proposition is a simple necessary condition for G to be such a group.

Proposition *Let G be a connected (real) Lie group, $\Gamma \subset G$ a discrete subgroup, and suppose that G/Γ has finite volume. Then:*

(i) *G is unimodular, i.e.,* $\det \mathrm{Ad} g = 1$ *for all $g \in G$;*

(ii) *for any automorphism $\phi : G \to G$ such that $\phi(\Gamma) = \Gamma$, we have*

$$|\det\,(d\phi)_e| = 1.$$

Proof Fix an orientation of G and consider G/Γ with the induced orientation. Let ω be a right invariant volume form on G. Denote by $\tilde{\omega}$ the induced volume form on G/Γ. For any $g \in G$ let l_g (resp. r_g) be the corresponding left (resp. right) translation of G. The form $l_g^*\omega$ is also right invariant. Thus $l_g^*\omega = \tau(g) \cdot \omega$, where $\tau(g)$ is a constant. Applying the map $(r_g^{-1})^*$ to both sides of this equality, we get

$\tau(g) = \det \operatorname{Ad} g$. Therefore the associated transformation $g : G/\Gamma \to G/\Gamma$ has the property that $g^*\tilde{\omega} = (\det \operatorname{Ad} g)\, \tilde{\omega}$. Hence

$$\operatorname{vol}(G/\Gamma) = \int_{G/\Gamma} \tilde{\omega} = \int_{G/\Gamma} g^*\tilde{\omega} = \det \operatorname{Ad} g \cdot \operatorname{vol}(G/\Gamma),$$

so that $\det \operatorname{Ad} g = 1$ for all $g \in G$. Thus (i) is proved.

In order to prove (ii), observe that $\phi \circ r_g = r_{\phi(g)} \circ \phi$. This yields $r_g^*(\phi^*\omega) = \phi^*\omega$, so that $\phi^*\omega$ is a right invariant form. Thus $\phi^*\omega = c \cdot \omega$, where the constant c is easily determined, namely, $c = \det\, (d\phi)_e$. Let $\tilde{\phi} : G/\Gamma \to G/\Gamma$ be the transformation induced by ϕ. Then it is clear that $\tilde{\phi}^*\tilde{\omega} = \det\, (d\phi)_e \cdot \tilde{\omega}$. Therefore

$$\operatorname{vol}(G/\Gamma) = \int_{G/\Gamma} \tilde{\omega} = \epsilon \cdot \int_{G/\Gamma} \tilde{\phi}^*\tilde{\omega} = \epsilon \cdot \det\, (d\phi)_e \cdot \operatorname{vol}(G/\Gamma),$$

where $\epsilon = 1$ or -1 depending on whether ϕ preserves or changes the orientation of G. In both cases we obtain (ii). □

3.5 Tits fibration

A connected, compact, homogeneous complex manifold can be written in a Klein form $X = G/H$, where G is a connected complex Lie group and H is a closed, not necessarily connected, complex Lie subgroup of G. Our purpose is to show that X admits a holomorphic fibration whose base is a flag manifold and whose fiber is a parallelizable complex manifold. Though G is important for the construction, it turns out that the fibration itself does not depend of the Klein form.

It is reasonable to extend the definitions of a Borel subgroup and a parabolic subgroup to the complex analytic category. Let G be a connected complex Lie group, R the radical of G and $G = S \cdot R$ a Levi decomposition, where S is any maximal connected semisimple subgroup of G.

A maximal connected solvable Lie subgroup $B \subset G$ is called a *Borel* subgroup of G. It is easily seen that B is a closed complex Lie subgroup of G. Moreover, if G is a linear algebraic group, then B is also algebraic and the definition coincides with the definition in §3.1. A Borel subgroup $B \subset G$ is of the form $B = B_* \cdot R$, where B_* is a Borel subgroup in S. Let $P \subset G$ be a closed complex Lie subgroup. As a result of our considerations in the preceding sections, *the following conditions are equivalent:*

(i) *P contains a Borel subgroup of G;*
(ii) *$P = P_* \cdot R$, where P_* is an (algebraic) parabolic subgroup of S;*
(iii) *$X = G/P$ is a flag manifold of S;*
(iv) *$X = G/P$ is a flag manifold of some semisimple group.*

Proof Assuming (i) we immediately get $P = P_* \cdot R$, where $P_* \subset S$ is a closed complex Lie subgroup containing a Borel subgroup of S. The algebraicity of P_* follows from Proposition 3 of §3.1. This gives (i) ⇒(ii). The implications (ii)⇒(iii) and (iii)⇒(iv) are clear. In order to prove (iv)⇒(i) consider the ineffectivity kernel I of the G-action on X. By Lemma 1 of §3.3 the group G/I is semisimple and the induced action of G/I on X is algebraic. Fix a Levi decomposition $G = S \cdot R$. Then $I^\circ = S_1 \cdot R$, where S_1 is a connected normal subgroup of S. Let S_2 be the complementary connected normal subgroup, so that $S = S_1 \cdot S_2$ and $S_1 \cap S_2$ is finite. Then P/I is an (algebraic) parabolic subgroup of $G/I \simeq S_2/I \cap S_2$. It follows that there exists a Borel subgroup $B_2 \subset S_2$ contained in P. Let B_1 be an arbitrary Borel subgroup of S_1. Then $B := B_1 \cdot B_2 \cdot R$ is a Borel subgroup of G and $B \subset P$. □

A closed complex subgroup $P \subset G$ is called *parabolic* if P satisfies the equivalent conditions (i) - (iv). For G linear algebraic this means that P is a parabolic subgroup of G in the sense of the theory of algebraic groups (see §3.1). The Lie algebra of a parabolic subgroup (resp. a Borel subgroup) of G is called a *parabolic subalgebra* (resp. a *Borel subalgebra*) of $\mathfrak{g}$.

Normalizer Theorem (A.Borel - R.Remmert [BoRe], J.Tits [Ti2]) *Let G be a connected complex Lie group, let $H \subset G$ be a closed complex Lie subgroup, and assume that $X = G/H$ is compact. Denote by H° the connected component of H containing e and by N be the normalizer of H° in G. Then:*
(i) *N is a parabolic subgroup of G;*
(ii) *if $P \subset G$ is a parabolic subgroup and $P \supset H$, then $P \supset N$.*

Remark The first statement is proved in [BoRe] and [Ti2], the proof of the second one is given in [Ti2].

Proof of (i) Let $k := \dim H$. Let o be the point of the Grassmann manifold $\mathbb{G}_k(\mathfrak{g})$, which corresponds to the Lie subalgebra $\mathfrak{h} \subset \mathfrak{g}$. The linear representation $\varphi := \mathrm{Ad}_G : G \to \mathrm{GL}(\mathfrak{g})$ along with the natural action

$$\mathrm{GL}(\mathfrak{g}) \times \mathbb{G}_k(\mathfrak{g}) \to \mathbb{G}_k(\mathfrak{g})$$

defines an action of G on $\mathbb{G}_k(\mathfrak{g})$. The isotropy subgroup G_o coincides with N. Since $N \supset H$, the orbit $Z := G(o)$ is a compact complex manifold. By Chow's theorem Z is an algebraic subvariety of $\mathbb{G}_k(\mathfrak{g})$. Let B be any Borel subgroup of G and let A be the Zariski closure of $\varphi(B)$ in $\mathrm{GL}(\mathfrak{g})$. Since $\varphi(B)$ is a connected solvable Lie subgroup in $\mathrm{GL}(\mathfrak{g})$, it follows that A is also solvable and connected. We have seen in §3.2 that the stabilizer of a subvariety in $\mathbb{P}_n$ is an algebraic subgroup in $\mathrm{Aut}(\mathbb{P}_n)$. This implies that the subgroup

$$\{g \in \mathrm{GL}(\mathfrak{g}) \mid g \cdot Z = Z\} \subset \mathrm{GL}(\mathfrak{g})$$

is algebraic. Therefore A leaves Z invariant and, by the Fixed Point Theorem, A has a fixed point on Z. Thus $B \subset G$ also has a fixed point on Z. But Z is G-homogeneous. Therefore $g \cdot B \cdot g^{-1} \subset G_o = N$ for some $g \in G$, showing that N is a parabolic subgroup.

Proof of (ii) Let P be a parabolic subgroup containing H. Since the homogeneous manifold P/H is compact, (i) implies that $N \cap P$ is a parabolic subgroup in P. Therefore $N \cap P$ is a parabolic subgroup in G and, consequently, also in N. It follows that $L := (N \cap P)/H^\circ$ is a parabolic subgroup in N/H°. On the other hand,

$$(N \cap P)/H \simeq ((N \cap P)/H^\circ)/(H/H^\circ) = L/\Gamma, \quad \text{where } \Gamma := H/H^\circ.$$

Therefore, Γ is a uniform discrete subgroup in L and, as we have shown in §3.4, L is unimodular. According to Proposition 9 of §3.1, a parabolic subgroup of a semisimple group is unimodular if and only if this subgroup coincides with the whole group. The same assertion is true for a parabolic subgroup $P = P_* \cdot R$ of any connected complex Lie group $G = S \cdot R$ because the adjoint representation of $P_* \subset S$ on the radical $\mathfrak{r} \subset \mathfrak{g}$ is always unimodular. As a result we obtain that $L = N/H^\circ$, whence $N \cap P = N$ or, equivalently, $N \subset P$. □

Let X be a compact homogeneous complex manifold. Suppose we have a locally trivial holomorphic fiber bundle (X, Y, π), where X is the total space, Y is the base, and $\pi : X \to Y$ is the bundle projection. The mapping

$$\pi : X \to Y$$

is called a *Tits fibration* (of X) if Y is a flag manifold and the following universality property holds. For any similar bundle (X, Y', π') with the base Y' being a flag manifold, the projection $\pi' : X \to Y'$ can be represented in the form $\pi' = \varphi \circ \pi$, where $\varphi : Y \to Y'$ is some holomorphic mapping.

The universality property of the Tits fibration shows that this fibration is unique (if it exists). More precisely, the equivalence relation defined by π on X is uniquely determined. The existence follows from the Normalizer Theorem. Namely, we have the following result.

Corollary 1 *Let $X = G/H$ be any Klein form of a compact homogeneous complex manifold X, where G is a connected complex Lie group. As above, denote by N the normalizer of H° in G and let $Y = G/N$. Then the fibration*

$$\pi : X \to Y, \quad \pi(gH) = gN \quad (g \in G) ,$$

is the Tits fibration of X.

Proof By (i) of the Normalizer Theorem Y is a flag manifold. In order to prove the universality of π consider another fibration $\pi' : X \to Y'$ with Y' being a flag manifold. Since Y' is simply connected, the typical fiber of π' is connected. By Proposition 1 of §2.4 G acts holomorphically on Y' so that $g \cdot \pi'(x) = \pi'(gx)$ for all $g \in G$ and $x \in X$. Since the G-action on X is transitive, the induced G-action on Y' enjoys the same property. But then π' is given by

$$\pi'(gH) = gP \quad \text{for all} \quad g \in G ,$$

where P is a closed complex Lie subgroup of G containing H. Since Y' is a flag manifold, P is a parabolic subgroup of G. By (ii) of the Normalizer Theorem P contains N. Therefore $\pi' = \varphi \circ \pi$, where the mapping $\varphi : Y \to Y'$ is defined by $\varphi(gN) := gP$. □

Corollary 2 *Let X be a compact homogeneous complex manifold. The fiber of the Tits fibration of X is a parallelizable complex manifold. Conversely, if a locally trivial holomorphic fiber bundle (X, Y, π) over a flag manifold Y has parallelizable fibers, then $\pi : X \to Y$ is the Tits fibration.*

Proof Using Corollary 1 we can write the fiber of the Tits fibration as $N/H = L/\Gamma$, where $L := N/H^\circ$ is a connected complex Lie group and $\Gamma := H/H^\circ$ is a discrete subgroup of L. Thus N/H is parallelizable.

Conversely, let $\pi : X \to Y$ be a mapping with the above properties. Since Y is simply connected, π has connected fibers. Therefore, according to Proposition 1 of §2.4, the mapping π is equivariant with respect to any connected group of holomorphic automorphisms of X. Fixing a Klein form $X = G/H$ we can write π as a fibration of homogeneous manifolds. Namely, $Y = G/P$, where $P \supset H$, and $\pi(gH) = gP$, $g \in G$. By the universality property of the Tits fibration $G/H \to G/N$ we have $P \supset N$.

It remains to show that $P \subset N$. Recall that P/H is a parallelizable compact manifold. Thus, as we have seen in §3.4, the dimensions of P/H and $\mathrm{Aut}(P/H)$ are equal. Let $I \triangleleft P$ be the ineffectivity kernel of the P-action on P/H. Then $H^\circ \subset I \subset H$ showing that H° is a normal subgroup of P. Therefore $P \subset \mathrm{Norm}_G(H^\circ) = N$.

□

3.6 Manifolds fibered by tori

Proposition 1 *Let G be a connected complex Lie group, $H \subset G$ a closed complex Lie subgroup, and $N = \mathrm{Norm}_G(H^\circ)$. Assume that $X = G/H$ is compact. Then the following conditions are equivalent:*

(i) *the fiber N/H of the Tits fibration $G/H \to G/N$ is a torus;*

(ii) $H \supset N'$;

(iii) *H is a normal subgroup of N.*

Proof (i) $\Rightarrow$ (ii). If N/H is a torus then $\mathrm{Aut}^\circ(N/H)$ is abelian (see Example 2 of §2.3). Therefore N' is contained in the ineffectivity kernel of the N-action on N/H. In particular, $N' \subset H$.

(ii) $\Rightarrow$ (iii) is obvious.

(iii) $\Rightarrow$ (i). See Example 1 of §3.4. □

Corollary (H.-C. Wang [Wa2]) *Let X be a compact, simply connected, homogeneous complex manifold. Then X fibers over a flag manifold with a complex torus as fiber.*

Proof Write X in the Klein form $X = G/H$ and consider the Tits fibration $\pi : X = G/H \to Y = G/N$. Since X is simply connected, the isotropy subgroup H is connected, i.e., $H = H^\circ$. Hence $H \triangleleft N$ and the fiber is a torus. □

Denote by R the radical of G and let S be a maximal semisimple subgroup of S. Then one has the Levi decomposition $G = S \cdot R$. Since N contains the radical, $N = N_* \cdot R$, where $N_* = N \cap S$.

Proposition 2 (see [Ti2]) *Assume that G acts locally effectively on $X = G/H$. If the fiber N/H of the Tits fibration $G/H \to G/N$ is a torus, then the adjoint action of G on R is trivial, i.e., $grg^{-1} = r$ for all $g \in G$, $r \in R$.*

Conversely, assume that $G = S \cdot R$ is a locally direct product of a semisimple group S and an abelian group R. Let $P = P_* \cdot R$, where P_* is a parabolic subgroup of S, and let H be any closed complex Lie subgroup of G, such that $P' \subset H \subset P$

and P/H is compact. Then $P = \mathrm{Norm}_G(H^\circ)$ so that P/H is the fiber of the Tits fibration. In particular, this fiber is a torus.

Proof We use the notation introduced in §3.1. Without loss of generality assume that $N_* \supset B^+$.

Let V be a finite-dimensional irreducible $\mathfrak{s}$-module and let $v_0 \in V$ be an eigenvector of $\mathfrak{b}^-$. Then V is spanned as a vector space by v_0 and vectors of the form $\mathrm{X}_1 \cdot \ldots \cdot \mathrm{X}_k \cdot v_0$, where $\mathrm{X}_i \in \mathfrak{b}^+$, $i = 1, \ldots, k$, $k > 0$ (see e.g. [Se4], Ch. VII). If $\dim V > 1$ then the weight of v_0 (the "lowest weight") is a non-zero linear form on $\mathfrak{t}$, and it follows that

(*) *each vector $v \in V$ can be represented in the form*

$$v = \sum_i \mathrm{X}_i \cdot v_i,$$

where $v_i \in V$, $\mathrm{X}_i \in \mathfrak{b}^+$.

Consider the linear representation

$$\mathrm{Ad}_G|_S : S \to \mathrm{GL}(\mathfrak{g}),$$

where Ad_G is the adjoint representation of G. Then $\mathfrak{r} \subset \mathfrak{g}$ is an invariant linear subspace. The induced representation of S on $V := \mathfrak{r}$ is completely reducible. Thus

$$V = V_1 \oplus \ldots \oplus V_m \oplus V^S,$$

where $V^S \subset V$ is the subspace of fixed vectors and each V_i, $1 \le i \le m$, is an irreducible S-module of dimension > 1. By (*) we have $V_i = [\mathfrak{b}^+, V_i]$ for all i. Therefore

$$[\mathfrak{g}, \mathfrak{r}] = [\mathfrak{s}, \mathfrak{r}] + [\mathfrak{r}, \mathfrak{r}] = (V_1 \oplus \ldots \oplus V_l) + [\mathfrak{r}, \mathfrak{r}] = [\mathfrak{b}^+, \mathfrak{r}] + [\mathfrak{r}, \mathfrak{r}] \subset [\mathfrak{n}, \mathfrak{n}] \subset \mathfrak{h}.$$

But $[\mathfrak{g}, \mathfrak{r}]$ is an ideal of $\mathfrak{g}$. Since the G-action on X is locally effective, it follows that $[\mathfrak{g}, \mathfrak{r}] = \{0\}$. This proves the first assertion.

In order to prove the converse we only have to show that $N \subset P$, because the opposite inclusion is obvious. Let $\mathfrak{h}_* := (\mathfrak{h} + \mathfrak{r}) \cap \mathfrak{s}$. Then

$$[\mathfrak{p}_*, \mathfrak{p}_*] \subset \mathfrak{h}_* \subset \mathfrak{p}_*$$

and N_* normalizes $\mathfrak{h}_*$. By Proposition 2 of §3.1 we have $\mathfrak{n}_* \subset \mathfrak{p}_*$. Since N_* and P_* are connected, it follows that $N_* \subset P_*$ and therefore $N \subset P$. □

Example 1 Consider the infinite cyclic subgroup $\Gamma_d \subset \mathrm{Aut}(\mathbb{C}^n - \{0\})$ generated by the automorphism $z \mapsto \gamma_d(z) = d \cdot z$, where $d \in \mathbb{C}^*$ is a fixed number with $|d| \neq 1$. The action of Γ_d on $\mathbb{C}^n - \{0\}$ is proper and free. Therefore the quotient

$$X_d = (\mathbb{C}^n - \{0\})/\Gamma_d$$

is a complex manifold. It is easily seen that X_d is diffeomorphic to $S^{2n-1} \times S^1$.

Since γ_d commutes with every linear transformation, the action of $\mathrm{GL}(n, \mathbb{C})$ on $\mathbb{C}^n - \{0\}$ induces an action on X_d, which is also transitive. The manifold X_d is called a *homogeneous Hopf manifold.* If $n = 1$ then X_d is an elliptic curve. Assume now that $n > 1$. Then it is convenient to consider X_d with the transitive group $G := \mathrm{SL}(n, \mathbb{C})$. Denote by P the parabolic subgroup of G such that $G/P = \mathbb{P}_{n-1}$. Then we have a surjective homomorphism $\varphi : P \to \mathbb{C}^*$ with kernel P'. Letting

$$H_d := \varphi^{-1}(\{d^k \mid k \in \mathbb{Z}\}),$$

we can write X_d in the Klein form $X_d = G/H_d$. Since $P' \subset H_d \subset P$, the Tits fibration is

$$X_d = G/H_d \xrightarrow{P/H_d} G/P = \mathbb{P}_{n-1},$$

where the projection is given by the canonical mapping $\mathbb{C}^n - \{0\} \to \mathbb{P}_{n-1}$. The fiber is isomorphic to the elliptic curve $\mathbb{C}^*/\{d\}$.

Example 2 The product $S^{2m-1} \times S^{2n-1}$, where $m, n > 1$, can be also given a complex structure. Moreover, one can construct a complex manifold, which is homogeneous and diffeomorphic to this product [CalEck].

In order to see this consider the natural transitive action of $G := \mathrm{SL}(m, \mathbb{C}) \times \mathrm{SL}(n, \mathbb{C})$ on $\mathbb{P}_{m-1} \times \mathbb{P}_{n-1}$. Denote by P the isotropy subgroup and observe that P/P' is isomorphic to $\mathbb{C}^* \times \mathbb{C}^*$. Fix a surjective homomorphism

$$\varphi : P \to \mathbb{C}^* \times \mathbb{C}^*$$

with kernel P'. For any $\lambda \in \mathbb{C} \setminus \mathbb{R}$ the one-parameter subgroup

$$A_\lambda := \{(e^z, e^{\lambda z}) \mid z \in \mathbb{C}\} \subset \mathbb{C}^* \times \mathbb{C}^*$$

is closed and the quotient group $(\mathbb{C}^* \times \mathbb{C}^*)/A_\lambda$ is an elliptic curve, which can also be written as $\mathbb{C}/(\mathbb{Z} \oplus \lambda\mathbb{Z})$. Let $H_\lambda := \varphi^{-1}(A_\lambda)$ and $X_\lambda := G/H_\lambda$. Then $P' \subset H_\lambda \subset P$ so that the Tits fibration of X_λ is of the form

$$G/H_\lambda \xrightarrow{P/H_\lambda} G/P,$$

where

$$P/H_\lambda \simeq (\mathbb{C}^* \times \mathbb{C}^*)/A_\lambda \simeq \mathbb{C}/(\mathbb{Z} \oplus \lambda\mathbb{Z}).$$

We claim that X_λ is diffeomorphic to $S^{2m-1} \times S^{2n-1}$. Indeed, consider the maximal compact subgroup $K = \mathrm{SU}(m) \times \mathrm{SU}(n) \subset G$. Since the additive group $\mathbb{C}$ has no compact subgroups $\neq \{0\}$, it follows from the definition of H_λ that

$$K \cap H_\lambda = K \cap P' = \mathrm{SU}(m-1) \times \mathrm{SU}(n-1).$$

Thus the K-orbit of the point $e \cdot H_\lambda$ is diffeomorphic to $S^{2m-1} \times S^{2n-1}$. By a dimension argument this orbit is the whole manifold X_λ.

Example 3 Let $G = \mathrm{SL}(3, \mathbb{C})$ and $K = \mathrm{SU}(3)$. Consider the subgroup $H_\lambda \subset G$ whose elements are matrices of the form

$$\begin{pmatrix} e^z & * & * \\ 0 & e^{\lambda z} & * \\ 0 & 0 & e^{-z-\lambda z} \end{pmatrix},$$

where $z \in \mathbb{C}$ and $\lambda \in \mathbb{C} \setminus \mathbb{R}$. Then H_λ is a closed complex subgroup in G and the same argument as in the preceding example shows that $G/H_\lambda = K$ as C^∞ manifolds. Thus we have obtained a left invariant complex structure on the compact Lie group K. Similar structures exist on every compact Lie group of even dimension. Moreover, if K is a connected compact Lie group and $J \subset K$ a connected closed subgroup, then K/J admits a K-invariant complex structure if and only if K/J has even dimension and the semisimple part of J coincides with the semisimple part of the centralizer of a torus in K. A compact homogeneous complex manifold of this type is always fibered by tori over a flag manifold, see [Wa2].

3.7 The role of the fundamental group

We have seen that simply connected, compact, homogeneous complex manifolds $X = G/H$ are torus fibrations over flag manifolds. Furthermore, if G acts on X locally effectively then the structure of G is given by Proposition 2 of §3.6. In this section we consider the case when the fundamental group $\pi_1(X)$ is not necessarily trivial, but solvable or nilpotent. We discuss the impact which these conditions have on the geometry of X and on the properties of G.

A connected complex Lie group G is called a *Cousin group* if $\mathcal{O}(G) = \mathbb{C}$, i.e., every holomorphic function on G is constant.

Proposition 1 *Let G be a Cousin group, H any connected complex Lie group, and Z the center of H. Then:*

(i) *any holomorphic mapping $\varphi : G \to H$ has the property that $\varphi(G) \subset h \cdot Z$ for some $h \in H$; in particular, if φ is a homomorphism then $\varphi(G) \subset Z$;*

(ii) *G is abelian; moreover, $G \simeq \mathbb{C}^n/\Gamma$, where $\Gamma \subset \mathbb{C}^n$ is a discrete subgroup generating $\mathbb{C}^n$ as a complex vector space;*

(iii) *the set* $\mathrm{Hom}(G, H)$, *consisting of all holomorphic homomorphisms, is discrete in* $\mathrm{Hol}(G, H)$.

Proof As usual, let $\mathrm{Ad} : H \to \mathrm{GL}(\mathfrak{h})$ denote the adjoint representation. Since the adjoint group $\mathrm{Ad}\, H \subset \mathrm{GL}(\mathfrak{h})$ is a holomorphically separable complex manifold, the image of $\mathrm{Ad} \circ \varphi$ is one point. In other words, $\varphi(G)$ is contained in a coset of H modulo Z. In particular, if φ is a homomorphism then $(\mathrm{Ad} \circ \varphi)(G) = \{\mathrm{id}_\mathfrak{h}\}$ and $\varphi(G) \subset Z$, showing (i).

Applying (i) to the identity mapping $\mathrm{id}_G : G \to G$ we see that G is abelian. Then, of course, $G \simeq \mathbb{C}^n/\Gamma$, where $\Gamma \subset \mathbb{C}^n$ is some discrete subgroup. If Γ generates a proper complex vector subspace $V \subset \mathbb{C}^n$, then any non-constant holomorphic function on $\mathbb{C}^n/V$ lifts to a non-constant function on $\mathbb{C}^n/\Gamma$. Since this contradicts $\mathcal{O}(G) = \mathbb{C}$, we obtain (ii).

In order to show (iii), it suffices to consider the case of abelian H. Then $H \simeq \mathbb{C}^m/\Delta$, where $\Delta \subset \mathbb{C}^m$ is some discrete subgroup. A holomorphic homomorphism $G \to H$ is identified with a linear mapping $\alpha : \mathbb{C}^n \to \mathbb{C}^m$ such that $\alpha(\Gamma) \subset \Delta$. Since Γ generates $\mathbb{C}^n$, the mapping

$$\mathrm{Hom}(G, H) \to \mathrm{Hom}_\mathbb{Z}(\Gamma, \Delta), \quad \alpha \mapsto \alpha|_\Gamma,$$

is injective. □

For $\alpha = (\alpha_1, \ldots, \alpha_n) \in \mathbb{Z}^n$ let χ_α denote the character of $(\mathbb{C}^*)^n$ given by $\chi_\alpha(z) := z_1^{\alpha_1} \cdot \ldots \cdot z_n^{\alpha_n}$. The following characterization of Cousin groups is often useful.

Proposition 2 *A Cousin group can be written as $G = (\mathbb{C}^*)^n/\Gamma$, where Γ is a discrete subgroup of $(\mathbb{C}^*)^n$. The following conditions are equivalent:*

(i) $G = (\mathbb{C}^*)^n/\Gamma$ *is a Cousin group;*
(ii) Γ *is dense in the Zariski topology of* $(\mathbb{C}^*)^n$;
(iii) $\chi_\alpha|_\Gamma = 1$ *if and only if* $\alpha = 0$.

Proof The expression of a Cousin group as $(\mathbb{C}^*)^n/\Gamma$ follows from (ii) of Proposition 1. Let H be the Zariski closure of $\Gamma \subset (\mathbb{C}^*)^n$. Then $(\mathbb{C}^*)^n/H \simeq (\mathbb{C}^*)^k$, where $k \leq n$. If G is a Cousin group then $k = 0$, hence (i) $\Longrightarrow$ (ii). The implication (ii) $\Longrightarrow$ (iii) is obvious. Assuming (iii) consider the Laurent series

$$f(z) = \sum c_\alpha \chi_\alpha(z)$$

of a holomorphic function on $(\mathbb{C}^*)^n$. For $\gamma \in \Gamma$ we have

$$f(z\gamma) = \sum c_\alpha \chi_\alpha(\gamma)\chi_\alpha(z).$$

If f is obtained by lifting a function from G then $f(z\gamma) = f(z)$ for all $z \in (\mathbb{C}^*)^n$ and $\gamma \in \Gamma$, showing that

$$c_\alpha \neq 0 \Longrightarrow \chi_\alpha|_\Gamma = 1 \Longrightarrow \alpha = 0.$$

Thus $f = c_0$ and G is a Cousin group. □

For an arbitrary group G and for any two subgroups $G_1, G_2 \subset G$ the set of all commutators $g_1 g_2 g_1^{-1} g_2^{-1}$, where $g_1 \in G_1, g_2 \in G_2$, generates a subgroup of G, which is denoted by (G_1, G_2). Let

$$G^{(0)} := G_{(0)} := G, \quad G^{(k+1)} := (G, G^{(k)}), \quad G_{(k+1)} := (G_{(k)}, G_{(k)}).$$

The sequences $\{G^{(k)}\}$ and $\{G_{(k)}\}$ are called respectively the *descending central* and the *derived* series of G. The group G is called *nilpotent* (resp. *solvable*) if $G^{(l)} = \{e\}$ (resp. $G_{(l)} = \{e\}$) for some $l \geq 0$. For G nilpotent (resp. solvable) we shall denote by c_G (resp. d_G) the length of the descending central (resp. derived) series of G, i.e., the minimal number l such that $G^{(l)} = \{e\}$ (resp. $G_{(l)} = \{e\}$). The number c_G is called the *nilpotency class* of a nilpotent group G.

If G is a Lie group then all terms of its descending central and derived series are (not necessarily closed) Lie subgroups. For the Lie algebras of $G^{(k)}$ (resp. $G_{(k)}$) we use the notation $\mathfrak{g}^{(k)}$ (resp. $\mathfrak{g}_{(k)}$).

For any subgroup H of a complex Lie group G let

$$\mathcal{O}(G)^H := \{f \in \mathcal{O}(G) \mid f(gh) = f(g) \quad \text{for all} \quad g \in G, h \in H\}.$$

Observe that if H is a closed complex Lie subgroup then $\mathcal{O}(G)^H$ is canonically isomorphic to $\mathcal{O}(G/H)$.

Theorem 1 *Let G be a connected complex Lie group, $H \subset G$ an arbitrary subgroup, and $\mathcal{O}(G)^H = \mathbb{C}$. If $H^{(l)} = \{e\}$ then $G^{(l+1)} = \{e\}$. In other words, if H is nilpotent then so is G and $c_G \leq c_H + 1$.*

Remark W.Barth and M.Otte [BaOt1] proved that G is nilpotent without an estimate on c_G. Our proof differs from the proof in [BaOt1].

Proof (by induction on l) If $l = 0$ then $H = \{e\}$ and G is a Cousin group. In particular, G is abelian and $G^{(1)} = G' = \{e\}$.

Assume now that $l > 0$. For any $g \in G$, $h \in H$, and $k \in H^{(l-1)}$ we have

$$(gh)k(gh)^{-1}k^{-1} = g(hkh^{-1})g^{-1}k^{-1} = gkg^{-1}k^{-1}$$

because $hkh^{-1}k^{-1} \in H^{(l)} = \{e\}$. In particular, $\mathrm{Ad}(gkg^{-1}k^{-1})$ does not change if one replaces g by gh. Therefore, for any fixed k and for any holomorphic function f on $\mathrm{GL}(\mathfrak{g})$ the function

$$G \ni g \mapsto f(\mathrm{Ad}(gkg^{-1}k^{-1}))$$

is constant. Hence $\mathrm{Ad}(gkg^{-1}k^{-1}) = \mathrm{id}_{\mathfrak{g}}$ for all $g \in G$.

Let Z be the center of G. It follows that

$$\zeta(g,k) := gkg^{-1}k^{-1} \in Z \quad \text{for all} \quad g \in G, k \in H^{(l-1)}.$$

From this one easily deduces that

$$\zeta(g_1g_2, k) = \zeta(g_1, k) \cdot \zeta(g_2, k).$$

On the other hand,

$$\zeta(gh, k) = \zeta(g, k) \quad \text{for all} \quad h \in H,$$

due to the fact that $H^{(l)} = \{e\}$. Thus

$$\zeta_k : G \to Z, \quad g \mapsto \zeta(g, k),$$

is a holomorphic homomorphism with $\zeta_k(g)$ depending only on the coset $g \cdot H$. Let

$$F := \bigcap_{k \in H^{(l-1)}} \mathrm{Ker}\, \zeta_k.$$

Observe that F is a closed, normal, complex Lie subgroup in G, containing G' and H. Furthermore, we have a continuous family of homomorphisms $G/F \to Z$, induced by $\{\zeta_k\}$. By Proposition 1 (iii) each of these homomorphisms is trivial, and so we obtain

$$\zeta(g,k) = gkg^{-1}k^{-1} = e \quad \text{for all} \quad g \in G, k \in H^{(l-1)}.$$

Hence $H^{(l-1)} \subset Z$. Letting $G_\star := G/Z$, $H_\star := H/H \cap Z$, we get $H_\star^{(l-1)} = \{e\}$ and $\mathcal{O}(G_\star)^{H_\star} = \mathbb{C}$, so that $G_\star^{(l)} = \{e\}$ by the induction hypothesis. Since G is a central extension of $G_\star$, it follows that $G^{(l+1)} = \{e\}$. □

Theorem 2 (see [A2]) *Let X be a compact homogeneous complex manifold with nilpotent fundamental group. Let G be a connected complex Lie group acting on X holomorphically, transitively, and locally effectively. Denote by R the radical and by S a maximal semisimple subgroup of G. Then:*

(i) *the adjoint action of S on R is trivial;*

(ii) *R is a nilpotent Lie group, and $c_R \leq c_{\pi_1(X)} + 1$;*

(iii) *if $R \neq \{e\}$ then $c_R \geq c_{\pi_1(X)}$;*

(iv) *if $R = \{e\}$ then the fiber of the Tits fibration of X is a torus; in particular, $\pi_1(X)$ is abelian.*

Proof Write X in the Klein form $X = G/H$. Then $N := \mathrm{Norm}_G(H^\circ)$ is a parabolic subgroup by the Normalizer Theorem. Thus $N = N_* \cdot R$, where N_* is a parabolic subgroup of S. Without loss of generality assume that $N_* \supset B^+$. Let $\mathrm{Y} \in \mathfrak{r}$ be an eigenvector of $\mathfrak{b}^-$ with non-zero weight $\lambda : \mathfrak{t} \to \mathbb{C}$. Take a vector $\mathrm{A} \in \mathfrak{t}$ such that $\lambda(\mathrm{A}) \neq 0$. Then for any p, $p = 0, 1, 2, \ldots$, we have

$$\mathrm{Y} = \frac{[\mathrm{A}, \mathrm{Y}]}{\lambda(\mathrm{A})} = \frac{[\mathrm{A}, [\mathrm{A}, \mathrm{Y}]]}{\lambda^2(\mathrm{A})} = \ldots = \frac{[\mathrm{A}, [\mathrm{A}, \ldots, [\mathrm{A}, \mathrm{Y}] \ldots]]}{\lambda^p(\mathrm{A})}, \tag{1}$$

where in the numerator of the last expression A appears p times.

Writing G/H and N/H in the Klein forms

$$G/H = (G/H^\circ)/(H/H^\circ) \quad \text{and} \quad N/H = (N/H^\circ)/(H/H^\circ),$$

we see that H/H_0 is a nilpotent, uniform, discrete subgroup of N/H_0. By Theorem 1 the Lie group N/H° is also nilpotent. Moreover,

$$c_{N/H^\circ} \leq c_{H/H^\circ} + 1 \leq c_{\pi_1(X)} + 1. \tag{2}$$

Returning to (1), take $p := c_{N/H^\circ}$ so that $\mathfrak{n}^{(p)} \subset \mathfrak{h}$. Since $\mathrm{A}, \mathrm{Y} \in \mathfrak{n}$, it follows that $\mathrm{Y} \in \mathfrak{h}$. The adjoint S-module $V := \mathfrak{r}$ can be decomposed as

$$V = V_1 \oplus \ldots \oplus V_m \oplus V^S,$$

where $V^S \subset V$ is the subspace of fixed vectors and each V_j, $1 \leq j \leq m$, is an irreducible S-module of dimension > 1. If Y is chosen in V_j, then V_j is spanned as a vector space by Y and the vectors of the form $[\mathrm{X}_1, [\mathrm{X}_2, \ldots, [\mathrm{X}_k, \mathrm{Y}] \ldots]]$, where $\mathrm{X}_i \in \mathfrak{b}^+, i = 1, \ldots, k, k > 0$. Since $\mathrm{Y} \in \mathfrak{h}$ and since $\mathfrak{b}^+$ normalizes $\mathfrak{h}$, it follows that $V_j \subset \mathfrak{h}$ for all j. Now, consider the vector subspace

$$\mathfrak{a} := \sum_{j=1}^{m} \sum_{r \in R} \mathrm{Ad} r \cdot V_j \subset \mathfrak{r}.$$

For any $r \in R$ the operator Ad r preserves $\mathfrak{h}$. Since $V_j \subset \mathfrak{h}$ for all j, we have $\mathfrak{a} \subset \mathfrak{h}$. It is clear that $\mathfrak{a}$ is R-invariant. On the other hand,

$$\mathrm{Ad} s \cdot \mathrm{Ad} r \cdot V_j = \mathrm{Ad} r' \cdot V_j,$$

where $r \in R$, $s \in S$ are arbitrary and $r' = srs^{-1} \in R$. Consequently, $\mathfrak{a}$ is also S-invariant. Thus $\mathfrak{a}$ is an ideal of $\mathfrak{g}$ which is contained in $\mathfrak{h}$. Since the G-action on X is locally effective, we have $\mathfrak{a} = 0$. Hence $V = V^S$, and (i) follows.

In order to prove (ii), consider the ideal $\mathfrak{r}^{(p)} \triangleleft \mathfrak{g}$. Since $\mathfrak{n}^{(p)} \subset \mathfrak{h}$ and $\mathfrak{r} \subset \mathfrak{n}$, we have $\mathfrak{r}^{(p)} \subset \mathfrak{h}$. The same argument as above shows then that $\mathfrak{r}^{(p)} = 0$. Therefore $c_R \leq p = c_{N/H^\circ}$, and (ii) follows from (2).

From now on and until the end of the proof we assume without loss of generality that G is simply connected. Then $\pi_1(X) \simeq H/H^\circ$. According to (i), S is a normal subgroup of G. Thus the Levi decomposition of G is direct, i.e., $G = S \times R$. Consequently $N = N_* \times R$ and $N'_* \triangleleft N$. Observe that the descending central series of the parabolic subalgebra $\mathfrak{n}_* \subset \mathfrak{s}$ stabilizes at $[\mathfrak{n}_*, \mathfrak{n}_*]$. In particular,

$$\mathfrak{n}_*^{(q+1)} = [\mathfrak{n}_*, \mathfrak{n}_*],$$

where $q := c_{\pi_1(X)}$. Since we already know that $\mathfrak{n}^{(p)} \subset \mathfrak{h}$ and $p \leq q+1$, it follows that $\mathfrak{h} \supset [\mathfrak{n}_*, \mathfrak{n}_*]$. Therefore it is possible to write

$$N/H^\circ = (N/N'_*)/(H^\circ/N'_*). \tag{3}$$

We have $N/N'_* = (N_*/N'_*) \times R$. If $R \neq \{e\}$ then $c_{N/N'_*} = c_R$. Along with (3) this yields

$$c_{\pi_1(X)} = c_{H/H^\circ} \leq c_{N/H^\circ} \leq c_R,$$

and (iii) is proved.

Finally, if $R = \{e\}$ then $N = N_*$ and (3) shows that N/H° is abelian. This implies (iv). □

Example 1 The nilpotency class can really increase by one in going from $\pi_1(X)$ to R. Let $S := \mathrm{SL}(3, \mathbb{C})$ and let R be the group of all 3×3 complex matrices of the form

$$\begin{pmatrix} 1 & x & z \\ 0 & 1 & y \\ 0 & 0 & 1 \end{pmatrix}.$$

In other words, R is the space $\mathbb{C}^3$ with coordinates x, y, z, in which the multiplication is given by the formula

$$(x, y, z) \cdot (x', y', z') = (x + x', y + y', z + z' + xy').$$

In the direct product $G := S \times R$ consider the subgroup H consisting of all pairs (s, r), where

$$s = \begin{pmatrix} \exp 2\pi i z_1 & * & * \\ 0 & \exp 2\pi z_2 & * \\ 0 & 0 & \exp(-2\pi(iz_1 + z_2)) \end{pmatrix}$$

and

$$r = (\alpha + i\beta, \gamma + i\delta, z_1 + z_2)$$

with $\alpha, \beta, \gamma, \delta \in \mathbb{Z}$, $z_1, z_2 \in \mathbb{C}$.

It is easily seen that H is closed and that $R \cap H$ is a uniform discrete subgroup of R, namely, the subgroup of matrices with Gaussian coordinates. Since the projection of H onto S is the full upper triangular group, it follows that $X := G/H$ is compact. Further, S and R are simply connected so that

$$\pi_1(X) \simeq H/H^\circ \simeq \mathbb{Z} \oplus \mathbb{Z} \oplus \mathbb{Z} \oplus \mathbb{Z}.$$

Finally, if $H^\circ \supset I$, where I is a connected normal subgroup of G, then the projection of I onto S must be trivial, since S is simple. Therefore $I \subset R$, which is impossible because

$$H^\circ \cap R = \{(0, 0, \mu + i\nu) \mid \mu, \nu \in \mathbb{Z}\}$$

is discrete. Consequently, G is locally effective on X. In this example we have $c_{\pi_1(X)} = 1$ and $c_R = 2$. The fiber of the Tits fibration is $R/R \cap H$, and the base is the flag manifold $\mathbb{F}_{1,2}(\mathbb{C}^3)$.

We now turn our attention to homogeneous manifolds with solvable fundamental group. Without an additional assumption, a theorem analogous to Theorem 1 does not hold. However, one has the following result for uniform discrete subgroups.

Theorem 3 (see [BaOt1]) *Let G be a connected complex Lie group, $\Gamma \subset G$ a uniform discrete subgroup. If Γ is solvable then so is G and $d_G \leq d_\Gamma + 1$.*

Proof Let R be the radical of G and $\rho : G \to S := G/R$ the canonical projection mapping. Assume dim $S > 0$. Since S has no compact factors, $\rho(\Gamma)$ is a uniform discrete subgroup of S, see [Aus]. Denote by B the Zariski closure of $\rho(\Gamma)$ in S. Then B is also solvable and S/B is compact. Therefore B is a Borel subgroup of S. On the other hand, $\rho(\Gamma)$ is a uniform discrete subgroup of B. Thus, as we have seen in §3.4, the group B is unimodular. However, this contradicts Proposition 9 of §3.1. Therefore $R = G$.

In order to prove the estimate on d_G, consider the adjoint representation of G and denote by A_G (resp. A_Γ) the Zariski closure of Ad G (resp. Ad Γ) in GL($\mathfrak{g}$). Then A_G/A_Γ is an affine algebraic variety (see Theorem 4 of §5.6). Any non-constant regular function on A_G/A_Γ gives rise to a non-constant holomorphic function on G/Γ. Since G/Γ is compact, it follows that $A_G = A_\Gamma$.

Now, $(\mathrm{Ad}\, G)_{(k)}$ and $(\mathrm{Ad}\, \Gamma)_{(k)}$ are Zariski dense in $(A_G)_{(k)}$ for each k. Therefore $d_{\mathrm{Ad}\, G} = d_{\mathrm{Ad}\, \Gamma}$ and, consequently, $d_G \leq d_\Gamma + 1$. □

Let V be a finite-dimensional module over a Lie algebra $\mathfrak{t}$. We denote by $\chi(V)$ the *trace form* on $\mathfrak{t}$. By definition

$$\chi(V)(\mathrm{A}) := \mathrm{tr}\{\mathrm{A} : V \to V\}, \quad \mathrm{A} \in \mathfrak{t}.$$

In what follows $\mathfrak{t}$ will be a Cartan subalgebra of a semisimple Lie algebra $\mathfrak{s}$.

Lemma 1 *Let* $\mathfrak{s} = \mathfrak{sl}\,(2, \mathbb{C}) = \mathbb{C} \cdot \mathrm{E} + \mathbb{C} \cdot \mathrm{H} + \mathbb{C} \cdot \mathrm{F}$ *and* $\mathfrak{t} = \mathbb{C} \cdot \mathrm{H}$, *where* $[\mathrm{H}, \mathrm{E}] = 2\mathrm{E}$, $[\mathrm{H}, \mathrm{F}] = -2\mathrm{F}$, *and* $[\mathrm{E}, \mathrm{F}] = \mathrm{H}$. *Let V be a finite-dimensional $\mathfrak{s}$-module and $V_0 \subset V$ a linear subspace which is invariant under* H *and* E. *Then:*

(i) $\chi(V_0)(\mathrm{H}) \geq 0$;

(ii) *if V_0 does not contain non-zero $\mathfrak{s}$-submodules, then* dim $V_0 \leq \chi(V_0)(\mathrm{H})$.

Proof First assume V is an irreducible $\mathfrak{s}$-module, i. e.,

$$V = \mathbb{C}v_m \oplus \mathbb{C}v_{m-2} \oplus \ldots \oplus \mathbb{C}v_{-m},$$

where $m \geq 0$, $\mathrm{H}v_k = kv_k$, $\mathrm{E}(\mathbb{C}v_k) = \mathbb{C}v_{k+2}$, and $\mathrm{F}(\mathbb{C}v_k) = \mathbb{C}v_{k-2}$ (we put $v_{m+2} = v_{-m-2} = 0$). Then

$$V_0 = \mathbb{C}v_m \oplus \mathbb{C}v_{m-2} \oplus \ldots \oplus \mathbb{C}v_{m-2d},$$

where $d \le m$. Therefore $\chi(V_0)(\mathrm{H}) = (d+1)(m-d) \ge 0$. Also, if $V_0 \ne V$, i.e., if $m \ge d+1$, then $\chi(V_0)(H) \ge d+1 = \dim V_0$. Thus, (i) and (ii) hold if V is irreducible.

Now, suppose (i) and (ii) are proved for all proper $\mathfrak{s}$-submodules of V. Let $V = V_1 \oplus V_2$ be some $\mathfrak{s}$-invariant decomposition, such that V_2 is irreducible. Denote by ψ the $\mathfrak{s}$-epimorphism of V onto V_2 with $\mathrm{Ker}\,\psi = V_1$ and consider the commutative diagram

$$\begin{array}{ccccccccc} 0 & \to & V_1 & \to & V & \overset{\psi}{\to} & V_2 & \to & 0 \\ & & \bigcup & & \bigcup & & \bigcup & & \\ 0 & \to & V_0 \bigcap V_1 & \to & V_0 & \overset{\psi}{\to} & \psi(V_0) & \to & 0\,. \end{array}$$

By our hypothesis

$$\chi(V_0)(\mathrm{H}) = \chi(V_0 \cap V_1)(\mathrm{H}) \; + \; \chi(\psi(V_0))(\mathrm{H}) \; \ge \; 0,$$

and (i) follows. In order to prove (ii) choose V_2 to be any irreducible $\mathfrak{s}$-submodule of V of maximal dimension. Without loss of generality we may assume that the action on V_2 is non-trivial. Since V_0 contains no non-zero $\mathfrak{s}$-submodules, the same is true for $V_0 \cap V_1$. Applying (ii) to this subspace of V_1 we obtain

$$\dim\,(V_0 \cap V_1) \; \le \; \chi(V_0 \cap V_1)(\mathrm{H}).$$

We still have to show that $\psi(V_0) \; \ne \; V_2$. Then one can apply (ii) to the subspace $\psi(V_0) \subset V_2$, and the proof will be complete.

Let $v_2 \ne 0$ be a vector in V_2 such that $\mathrm{F}v_2 = 0$. Then v_2 is unique up to scalar multiplication and $\mathrm{H}v_2 = \nu v_2$, $\nu \in \mathbb{Z}$, $\nu < 0$.

Assume $\psi(V_0) = V_2$, so that $\psi(v_0) = v_2$ for some $v_0 \in V_0$. We have $v_0 = v_1 + v_2$, where $v_1 \in V_1$. Write v_1 in the form $v_1 = \hat{v} + \tilde{v}$, where $\hat{v}$ is a weight vector of weight ν and $\tilde{v}$ is the sum of weight vectors of weight $> \nu$ (note that, due to our choice of V_2, a weight $< \nu$ cannot occur). Then $v_0 = (v_2 + \hat{v}) + \tilde{v}$. Applying powers of H to this decomposition, we obtain from the H-invariance of V_0 that $v_2 + \hat{v} \in V_0$. Since V_0 does not contain non-zero $\mathfrak{s}$-submodules, we get a contradiction. □

Lemma 2 *Let $V_0 \subset V$ be as above. The equality $\chi(V_0)(\mathrm{H}) = 0$ holds if and only if V_0 is invariant under $\mathfrak{s}$.*

Proof Assume $\chi(V_0)(\mathrm{H}) = 0$. We will prove by induction on $\dim V$ that V_0 is $\mathfrak{s}$-invariant. According to (ii), either $V_0 = 0$ or V_0 contains a non-zero $\mathfrak{s}$-submodule, say, V'. In the first case the assertion is trivial. In the second case consider the subspace $V_0/V' \subset V/V'$. Obviously, $\chi(V_0/V')(\mathrm{H}) = 0$, so that by the induction hypothesis V_0/V' is $\mathfrak{s}$-invariant. It follows that V_0 has the same property. □

Theorem 4 (see [A2]) *Let G be a connected complex Lie group, $H \subset G$ a closed complex Lie subgroup, R the radical of G, S a maximal semisimple subgroup of G, and U a maximal unipotent subgroup of S. Assume that $X := G/H$ is compact and that G acts locally effectively on X. Consider the following conditions:*

(i) *the adjoint action of S on R is trivial;*

(ii) *$\pi_1(X)$ is solvable;*

(iii) $H \supset gUg^{-1}$ *for some* $g \in G$.
Then we have (i)&(ii) $\Longleftrightarrow$ (iii).

Proof We start with several general remarks. Recall that $N := \mathrm{Norm}_G(H^\circ)$ is a parabolic subgroup of G, i.e., $N = N_* \cdot R$, where N_* is a parabolic subgroup of S. Pick a Borel subgroup B_* of S in N_* and let $B := B_* \cdot R$. Without loss of generality assume that the Cartan subalgebra $\mathfrak{t} \subset \mathfrak{s}$ is contained in $\mathfrak{b}_*$ and that the ordering of the root system Δ is chosen so that $B_* = B^+$, $U = U^+ = (B^+)'$. In what follows the linear form $\chi(V)$ on $\mathfrak{t}$ is always considered with respect to a $\mathfrak{t}$-module V, which is the quotient of two $\mathfrak{t}$-invariant linear subspaces of $\mathfrak{g}$ with the $\mathfrak{t}$-action induced by the adjoint one.

(i)&(ii) $\Longrightarrow$ (iii). The group H/H° is contained in N/H° as a uniform discrete subgroup. Due to (ii), H/H° is solvable. Therefore N/H° is also solvable by Theorem 3. In other words, $N_{(k)} \subset H^\circ$ for some $k \geq 0$. In particular, any semisimple subgroup of N_* is contained in H°. It follows that

$$\chi(\mathfrak{n}_*/\mathfrak{n}_* \cap \mathfrak{h}) = \sum_{\alpha \subset \Psi} \alpha,$$

where

$$\Psi = \{\alpha \in \Delta^+ \mid \mathfrak{s}_\alpha \text{ is not contained in } \mathfrak{h}\}.$$

We want to show that $\Psi = \emptyset$ so that $U^+ \subset H$. Since one can always choose $\mathrm{A} \in \mathfrak{t}$ satisfying $\alpha(\mathrm{A}) > 0$ for all $\alpha \in \Delta^+$, this is equivalent to

$$\chi(\mathfrak{n}_*/\mathfrak{n}_* \cap \mathfrak{h}) = 0. \tag{4}$$

In order to prove this equality, observe that, according to the proposition of §3.4, N/H° is an unimodular Lie group. In particular, $\chi(\mathfrak{n}/\mathfrak{h}) = 0$. On the other hand, due to (i), the adjoint $\mathfrak{t}$-action on $\mathfrak{r}$ is trivial. Recall that $\mathfrak{n} = \mathfrak{n}_* + \mathfrak{r}$ and let $\psi : \mathfrak{n} \to \mathfrak{r}$ denote the projection map, which obviously commutes with the adjoint $\mathfrak{t}$-action. We have

$$\chi(\mathfrak{n}_*/\mathfrak{n}_* \cap \mathfrak{h}) = \chi(\mathfrak{n}/\mathfrak{h}) - \chi(\mathfrak{r}/\psi(\mathfrak{h})) = \chi(\mathfrak{n}/\mathfrak{h}),$$

and (4) follows.

We now make two assumptions, which do not restrict the generality and which will be kept until the end of the proof. First, we assume that G is simply connected so that $\pi_1(X) \simeq H/H^\circ$. Second, replacing H if necessary by a conjugate subgroup, we assume that $H \supset U^+$ and, consequently, $N_* \supset B^+$.

(iii)$\Longrightarrow$(ii). Suppose $\mathfrak{s}_{-\alpha} \subset \mathfrak{n}_*$ for some $\alpha \in \Delta^+$. Then

$$\mathfrak{s}_{-\alpha} = [\mathfrak{s}_{-\alpha}, [\mathfrak{s}_{-\alpha}, \mathfrak{s}_\alpha]] \subset [\mathfrak{s}_{-\alpha}, [\mathfrak{n}_*, \mathfrak{h}]] \subset [\mathfrak{s}_{-\alpha}, \mathfrak{h}] \subset \mathfrak{h}.$$

It follows that $[\mathfrak{n}_*, \mathfrak{n}_*] \subset \mathfrak{h}$ and $[\mathfrak{n}, \mathfrak{n}] \subset \mathfrak{h} + \mathfrak{r}$. The Lie algebra $\mathfrak{n}/\mathfrak{h}$ has the solvable ideal $(\mathfrak{h} + \mathfrak{r})/\mathfrak{h} \simeq \mathfrak{r}/\mathfrak{r} \cap \mathfrak{h}$, such that the quotient algebra $\mathfrak{n}/(\mathfrak{h} + \mathfrak{r})$ is abelian. Therefore $\mathfrak{n}/\mathfrak{h}$ is a solvable Lie algebra and N/H° is a solvable Lie group. Since $\pi_1(X) \simeq H/H^\circ \subset N/H^\circ$, we obtain (ii).

(iii)$\Longrightarrow$(i). Denote by $\tau : \mathfrak{n} \to \mathfrak{n}_*$ the projection map with kernel $\mathfrak{r}$. We have just seen that $[\mathfrak{n}_*, \mathfrak{n}_*] \subset \mathfrak{h}$. In particular, $\tau(\mathfrak{h}) \supset [\mathfrak{n}_*, \mathfrak{n}_*]$. Since the Lie group N/H° is unimodular, it follows that

$$\chi(\mathfrak{r}/\mathfrak{r} \cap \mathfrak{h}) = \chi(\mathfrak{n}/\mathfrak{h}) - \chi(\mathfrak{n}_*/\tau(\mathfrak{h})) = 0,$$

hence

$$\chi(\mathfrak{r}\cap\mathfrak{h}) = \chi(\mathfrak{r}) - \chi(\mathfrak{r}/\mathfrak{r}\cap\mathfrak{h}) = 0.$$

Note that $\mathfrak{r}\cap\mathfrak{h}$ is a $\mathfrak{b}^+$-invariant linear subspace of $\mathfrak{r}$. Applying Lemma 2 to each three-dimensional simple root subalgebra of $\mathfrak{s}$, we see that $\mathfrak{r}\cap\mathfrak{h}$ is $\mathfrak{s}$-invariant. In view of the local effectivity of the G-action on X we obtain $\mathfrak{r}\cap\mathfrak{h} = 0$. Due to this fact, the natural mapping $\mathfrak{r} \to \mathfrak{n}/\mathfrak{h}$ is an embedding. Since $U^+ \subset H^\circ \lhd N$, the adjoint action of U^+ on $\mathfrak{n}/\mathfrak{h}$ is trivial. Therefore the action of U^+ on $\mathfrak{r}$ and on R is also trivial. This implies (i). □

Proposition 3 *Let $X = G/H$ be as in Theorem 4 and assume that the R-orbits on X are closed. Then we have* (ii)$\Longrightarrow$ (i).

Proof (V.V.Gorbacevich) As above, we assume that G is simply connected so that each element $g \in G$ can be uniquely written as $g = sr$, where $s \in S$, $r \in R$. Let $\pi : G \to S$ denote the natural projection mapping. Then $\pi(H)$ is a closed complex Lie subgroup of S, the quotient $S/\pi(H)$ is compact, and (ii) implies that its fundamental group is solvable. Theorem 4 tells us that $\pi(H^\circ)$ contains a maximal unipotent subgroup of S.

Let us show that $R\cap H$ is discrete. Since $R/R\cap H$ is compact, $(R\cap H)^\circ \lhd R$ by the Normalizer Theorem. Consider the uniform discrete subgroup $(R\cap H)/(R\cap H)^\circ$ in $R/(R\cap H)^\circ$. We have the natural action of N on R by inner automorphisms. The induced action on $R/(R\cap H)^\circ$ has the property that $H \subset N$ preserves $(R\cap H)/(R\cap H)^\circ$. For $n \in N$ denote by $\lambda(n)$ the corresponding linear transformation of $\mathfrak{r}/\mathfrak{r}\cap\mathfrak{h}$. By (ii) of the proposition of §3.4 we have $|\det \lambda(h)| = 1$ for $h \in H$. Since N is a connected complex Lie group, it follows that $\det \lambda(n) = 1$ for all $n \in N$. In particular, $\chi(\mathfrak{r}/\mathfrak{r}\cap\mathfrak{h}) = 0$. As in the above proof of (iii)$\Longrightarrow$(i), this implies $\mathfrak{r}\cap\mathfrak{h} = 0$. In other words, $R\cap H$ is discrete.

The uniform discrete subgroup $\Gamma := R\cap H \subset R$ is stable under the action of H. Therefore $h\gamma h^{-1} = \gamma$ for all $h \in H^\circ, \gamma \in \Gamma$. In other words, $\mathrm{Ad}\,\gamma \cdot \mathrm{A} = \mathrm{A}$ for any $\mathrm{A} \in \mathfrak{h}$. From this it follows that the holomorphic mapping $R \ni r \mapsto \mathrm{Ad}\, r\cdot \mathrm{A} \in \mathfrak{g}$ is constant on each coset $r \cdot \Gamma$. Since R/Γ is compact, the image is one point, i.e., $\mathrm{Ad}\, r \cdot \mathrm{A} = \mathrm{A}$ for all $r \in R$, $\mathrm{A} \in \mathfrak{h}$. This means that H° acts trivially on R.

Suppose $h = sr \in H^\circ$, where $s \in S, r \in R$. Then for any $x \in R$ we have

$$sxs^{-1} = hr^{-1}xrh^{-1} = (hr^{-1}h^{-1})(hxh^{-1})(hrh^{-1}) = (hr^{-1}h^{-1})x(hrh^{-1}).$$

Thus each $s \in \pi(H^\circ)$ induces an inner automorphism of R. Recall that $\pi(H^\circ)$ contains a maximal unipotent subgroup of S. It follows that the whole group S acts on R by inner automorphisms. Since the group of inner automorphisms of R is solvable, the action of S on R is trivial. □

We now compute the dimension of the fiber of the Tits fibration in topological terms. Let Γ be an abstract solvable group. According to G.D.Mostow [Mo3], the *rank* of Γ is defined by

$$\mathrm{rank}\,\Gamma := \sum_{k=0}^{\infty} \mathrm{rank}\,\Gamma_{(k)}/\Gamma_{(k+1)},$$

where $\mathrm{rank}\,\Gamma_{(k)}/\Gamma_{(k+1)}$ denotes the usual rank of an abelian group. We shall need the following properties of $\mathrm{rank}\,\Gamma$.

(*) *Given an exact sequence of solvable groups*

$$1 \to \Gamma_1 \to \Gamma \to \Gamma_2 \to 1,$$

one has

$$\text{rank } \Gamma \;=\; \text{rank } \Gamma_1 \;+\; \text{rank } \Gamma_2$$

(if one side is finite then so is the other and they are equal).

For the proof see [Mo3].

(**) *If G is a connected, simply connected, solvable Lie group, $\Gamma \subset G$ a uniform discrete subgroup, then*

$$\text{rank } \Gamma \;=\; \dim G.$$

Proof (by induction on dim G) For G abelian the assertion is obvious. Assume that G is non-abelian and that (**) is proved for all groups of smaller dimension. One can always find a connected closed normal subgroup $F \triangleleft G$ such that $\dim F < \dim G$ and $F/F \cap \Gamma$ is compact. Indeed, if G is not nilpotent, then one can take as F the nilradical of G (see [Mo1]). If G is nilpotent then one can put $F := G'$ (see [Ma]).

Observe that F is simply connected and denote by $\psi : G \to G/F$ the canonical epimorphism. Then $\psi(\Gamma)$ is a uniform discrete subgroup in G/F. In view of (*), the induction hypothesis yields

$$\text{rank } \Gamma = \text{rank } F \cap \Gamma + \text{rank } \psi(\Gamma) = \dim F + \dim G/F = \dim G. \qquad \square$$

Let X be a connected topological space with solvable fundamental group. Assume that X has only finitely many homotopy groups of non-zero rank. Then the *homotopy characteristic* of X is defined by

$$h(X) := \sum_{k=1}^{\infty} (-1)^{k+1} \text{ rank } \pi_k(X).$$

For homogeneous spaces of compact Lie groups $h(X)$ can be expressed in terms of these groups as follows (see [On3], Theorem 1).

(***) *Let $L \subset K$ be connected compact Lie groups. Then $h(K/L) = r_K - r_L$, where r_K denotes the rank of a connected compact Lie group K.*

Combining (*), (**), and (***) with the properties of the Tits fibration, we get the following result.

Proposition 4 *Let X be a connected, compact, homogeneous complex manifold with solvable fundamental group, $\pi : X \to Y$ the Tits fibration of X, and F the fiber of π. Then*

$$2 \dim_{\mathbb{C}} F = h(X).$$

Proof Since F is parallelizable and $\pi_1(F)$ is solvable, we have

$$\pi_k(F) = 0 \;\; (k \geq 2), \quad h(F) = \text{rank } \pi_1(F).$$

From (**) it follows that rank $\pi_1(F) = \dim_{\mathbb{R}} F$. In particular, $h(F) < \infty$.

By (*) the rank of a solvable group is additive in exact sequences. Applying this to the exact homotopy sequence of the Tits fibration, we obtain

$$h(X) = h(Y) + h(F).$$

The flag manifold Y can be written as the quotient of a compact semisimple Lie group by the centralizer of a torus. By (***) we have $h(Y) = 0$, and the assertion follows. □

Example 2 Following [A2] and [OtPo], we will show here that condition (ii) of Theorem 4 does not in general imply (i). The construction is based on some classical results of algebraic number theory. We refer the reader to [BorSh], Ch.2, for a detailed exposition of the facts which we use below.

Let K be a totally real extension of the field $\mathbb{Q}$ of degree $r+1$, let $\mathcal{O} \subset K$ be the ring of algebraic integers of K, and let $\sigma_1, \ldots, \sigma_{r+1}$ be all embeddings of K in $\mathbb{R}$. Then

$$\Gamma := \left\{ \left(\sigma_1(a) + i\sigma_1(b), \ldots, \sigma_{r+1}(a) + i\sigma_{r+1}(b)\right) \;\middle|\; a, b \in \mathcal{O} \right\} \subset \mathbb{C}^{r+1}$$

is a uniform discrete subgroup.

For $x \in K$ denote by $\sigma(x)$ the diagonal matrix diag $(\sigma_1(x), \ldots, \sigma_{r+1}(x))$. If x belongs to the group of units of K, then obviously $\sigma(x)(\Gamma) = \Gamma$. Let Λ be the group of units of K with norm 1 and let $D \subset \mathrm{SL}(r+1, \mathbb{C})$ be the subgroup of diagonal matrices. Then $\sigma(\Lambda) \subset D$. Moreover, it follows from the proof of Dirichlet's theorem that $\sigma(\Lambda)$ is a uniform discrete subgroup of D. Since $\sigma : \Lambda \to D$ is a faithful representation, we can identify the semidirect product $\Lambda \ltimes \Gamma$ with some uniform discrete subgroup of $D \ltimes \mathbb{C}^{r+1}$.

Now, let S be a simply connected, simple, complex Lie group of rank r, $B_* \subset S$ a Borel subgroup, $T \subset B_*$ a maximal torus, and U the unipotent radical of B_*, so that $B_* = T \cdot U$. We assume as usual that $\mathfrak{b}_*$ is spanned by $\mathfrak{t}$ and by all root vectors corresponding to positive roots. Denote by $\alpha_1, \ldots, \alpha_r$ the simple roots and put $\alpha := \alpha_1 + \ldots + \alpha_r$. It is easy to verify (for example, using the tables from [OnVi]) that α is a dominant weight if and only if S is of type A_r or B_r. The irreducible representation with highest weight α is the adjoint and the simplest one respectively. Note hat these representations are equivalent to their dual ones. We shall assume that S has one of these types and denote by V the irreducible S-module with highest weight α. Let $\hat{\alpha}_j$ and $\hat{\alpha}$ be the characters of T with differentials α_j and α respectively. In V there is a unique (up to scalar multiplication) vector w_0 such that $t \cdot w_0 = \hat{\alpha}(t)^{-1} w_0$ for $t \in T$. (The lowest weight of V is $-\alpha$, and w_0 is the corresponding weight vector.) Denote by V_0 the linear subspace of V spanned by all weight vectors of T having weights different from $-\alpha$. Then V_0 is B_*-invariant, and $V = V_0 \oplus \mathbb{C} \cdot w_0$ is a T-invariant decomposition.

We now define G as the semidirect product $S \ltimes V$ with respect to the given linear S-action on V. Further, let $B := B_* \ltimes V$ and $H^\circ := U' \ltimes V_0$. Then H° is a closed normal subgroup of B and the quotient group B/H° is isomorphic to the semidirect product $T \ltimes \mathbb{C}^{r+1}$, where T acts on $\mathbb{C}^{r+1}$ by

$$(z_1, \ldots, z_r, z_{r+1}) \xrightarrow{t} (\hat{\alpha}_1(t) z_1, \ldots, \hat{\alpha}_r(t) z_r, \hat{\alpha}^{-1}(t)) \qquad (t \in T).$$

Define a homomorphism $\varphi : T \ltimes \mathbb{C}^{r+1} \to D \ltimes \mathbb{C}^{r+1}$ by

$$\varphi([t,z]) := \left[\operatorname{diag}((\hat{\alpha}_1(t),\ldots,\hat{\alpha}_r(t),\hat{\alpha}^{-1}(t)),\ z\right]$$

where $t \in T,\ z \in \mathbb{C}^{r+1}$. It is easily seen that this definition makes sense and that φ is an epimorphism, whose kernel coincides with the center of S. We now define H as the preimage of $\Lambda \ltimes \Gamma$ in B by the composition of homomorphisms

$$B \to B/H^\circ \simeq T \ltimes \mathbb{C}^{r+1} \overset{\varphi}{\to} D \ltimes \mathbb{C}^{r+1}.$$

Then H is closed, $X := G/H$ is compact, G is locally effective on X, and $\pi_1(X)$ is a solvable group. On the other hand, the action of S on $R := V$ is given by a non-trivial linear representation.

In connection with Proposition 3 we note that the R-orbits on X are one-dimensional and that their closures are $(r+1)$-dimensional complex tori (see [A2], Prop. 6.5).

3.8 An estimate of the dimension of Aut(X)

If X is a compact complex manifold of fixed dimension then, generally speaking, the dimension of Aut(X) can be arbitrarily big. For example, the dimension of the automorphism group of the surface F_n is equal to $n+5$ (see § 2.4, Example 2). Our goal here is to show that this cannot happen for homogeneous manifolds.

In this section we use the following conventions and notations. A compact homogeneous complex manifold X is written in the Klein form $X = G/H$ with G a connected complex Lie group, $H \subset G$ a closed complex Lie subgroup. We fix a Levi decomposition $G = S \cdot R$, where R is the radical and S a Levi subgroup of G. The normalizer N of the connected component H° is a parabolic subgroup of G. This means that $N = P \cdot R$, where P is a parabolic subgroup of S. A Cartan subalgebra $\mathfrak{t} \subset \mathfrak{s}$ is chosen in $\mathfrak{p}$. Furthermore, the base Π of the root system Δ is chosen so that $B^+ \subset P$. The sets of positive and negative roots are denoted by Δ^+ and Δ^- respectively. As usual, B^- is the opposite Borel subgroup, U^+ and U^- are the commutator subgroups of B^+ and B^- respectively, and the similar notation is used for Lie algebras. The subset $\Delta_P \subset \Delta$ and the decomposition $\Delta_P = \Delta_P^s \cup \Delta_P^a$ are defined in §3.1. The linear form $\chi(V)$ on $\mathfrak{t}$ is the trace of an operator in a finite-dimensional $\mathfrak{t}$-module V. This form will be considered for quotients of $\mathfrak{t}$-invariant subspaces of $\mathfrak{g}$. Let Λ_P and ϱ be the linear forms on $\mathfrak{t}$ defined by

$$\Lambda_P := \sum_{\alpha\in\Delta_P^a} \alpha\,, \qquad \varrho := \frac{1}{2}\sum_{\alpha\in\Delta^+} \alpha.$$

Lemma 1 $\chi(\mathfrak{h}) = \Lambda_P$.

Proof According to the proposition of §3.4, N/H° is a unimodular Lie group. In particular, $\chi(\mathfrak{n}/\mathfrak{h}) = 0$. On the other hand, $\chi(\mathfrak{r}) = 0$ (since $\mathfrak{r}$ is an $\mathfrak{s}$-module) and $\chi(\mathfrak{p}) = \Lambda_P$ (since $\sum_{\alpha\in\Delta_P^s}\alpha = 0$). Therefore $\chi(\mathfrak{h}) = \chi(\mathfrak{n}/\mathfrak{h}) + \chi(\mathfrak{h}) = \chi(\mathfrak{n}) = \chi(\mathfrak{p}) + \chi(\mathfrak{r}) = \Lambda_P$. □

Lemma 2 *Assume that G is effective on $X = G/H$. If* $\mathrm{Y} \in \mathfrak{h}$ *and* Y *is annihilated by* ad $\mathfrak{u}^-$, *then* $\mathrm{Y} = 0$.

Proof Denote by V the $\mathfrak{s}$-submodule of $\mathfrak{g}$ generated by Y. We have

$$\mathrm{Y} = \mathrm{Y}_1 + \mathrm{Y}_2 + \ldots + \mathrm{Y}_k \, ,$$

where $\mathrm{Y}_1, \ldots, \mathrm{Y}_k$ are eigenvectors of $\mathfrak{b}^-$ in V. Let $\lambda_1, \ldots, \lambda_k$ be the corresponding weights. We may assume that $\lambda_i \neq \lambda_j$ for $i \neq j$. Then any eigenvector of $\mathfrak{b}^-$ in V coincides, up to a scalar factor, with some Y_i. Since $\mathfrak{t}$ normalizes $\mathfrak{h}$, it follows that $\mathrm{Y}_i \in \mathfrak{h}$ for all i, $i = 1, \ldots, k$. Since $\mathfrak{u}^+$ also normalizes $\mathfrak{h}$, we obtain $V \subset \mathfrak{h}$. But then the linear subspace

$$\sum_{r \in R} \mathrm{Ad} r \cdot V$$

is an ideal of $\mathfrak{g}$, which is contained in $\mathfrak{h}$. In view of the effectivity condition we have $V = \{0\}$ and $\mathrm{Y} = 0$. □

Lemma 3 *Assume that G is effective on $X = G/H$ and let*

$$\mathfrak{g}^S = \{\mathrm{A} \in \mathfrak{g} \mid (\mathrm{Ad} s) \cdot \mathrm{A} = \mathrm{A} \text{ for all } s \in S\}.$$

Then $\dim \mathfrak{g}^S \leq \dim N/H$.

Proof Let $Y := G/N = S/P$ and consider the Tits fibration $\pi : X \to Y$. According to Proposition 1 of §2.4 we have a canonical homomorphism of complex Lie groups $\mathrm{Aut}^\circ(X) \to \mathrm{Aut}^\circ(Y)$. In other words, every holomorphic vector field on X is projectable onto Y. Let

$$\pi_* : \mathcal{T}_X(X) \to \mathcal{T}_Y(Y)$$

be the corresponding Lie algebra homomorphism.

Since the G-action on X is effective, the Lie homomorphism $\mathfrak{g} \to \mathcal{T}_X(X)$ is injective (see §1.7). We identify $\mathfrak{g}$ with a subalgebra of $\mathcal{T}_X(X)$.

We claim that every vector field $\mathrm{A} \in \mathfrak{g}^S$ is vertical or, equivalently, $\pi_*(\mathrm{A}) = 0$. In order to prove this, observe that $\mathfrak{g}^S \subset \mathfrak{r}$, where $\mathfrak{r}$ is the radical of $\mathfrak{g}$. Therefore $\pi_*(\mathfrak{g}^S)$ is contained in the radical of $\pi_*(\mathfrak{g})$. On the other hand, Lemma 1 of §3.3 shows that the algebra $\pi_*(\mathfrak{g})$ is semisimple. Thus $\pi_*(\mathrm{A}) = 0$.

Since N/H is a parallelizable manifold, we have the equality $\dim \mathrm{Aut}(N/H) = \dim N/H$ (see §3.4). Therefore, it suffices to show that the vector field $\mathrm{A} \in \mathfrak{g}^S$ is uniquely determined by its restriction to the fiber $\pi^{-1}(o)$, where $o \in Y$ is some point. Denote by $\mathrm{A}(x)$ the tangent vector of A at $x \in X$. Then

$$(ds)_x \cdot \mathrm{A}(x) = (\mathrm{Ad} s \cdot \mathrm{A})(sx) = \mathrm{A}(sx)$$

for all $s \in S$, $x \in X$. Each point of X is of the form sx for some $s \in S$, $x \in \pi^{-1}(o)$. Assuming that $\mathrm{A}(x) = 0$ for all $x \in \pi^{-1}(o)$, we see that $\mathrm{A} = 0$. □

Lemma 4 *Let S be a semisimple complex Lie group, acting holomorphically and effectively on a complex manifold X. Then* $\dim S \leq n(n+2)$, *where* $n = \dim X$.

Proof Take a point $x \in X$ and choose a maximal compact subgroup $K \subset S$. By the Identity Theorem (see §2.2) the isotropy subgroup K_x is isomorphic to a subgroup of the unitary group $\mathrm{U}(n)$. Therefore

$$\dim_{\mathbb{C}} S = \dim_{\mathbb{R}} K = \dim_{\mathbb{R}} K(x) + \dim_{\mathbb{R}} K_x \leq 2n + \dim_{\mathbb{R}} \mathrm{U}(n) = n(n+2). \qquad \square$$

We now recall some basic facts from the representation theory of semisimple Lie algebras over $\mathbb{C}$, referring the reader to [Se4] or [Hum1] for a detailed exposition.

For each $\alpha \in \Delta$ there is a unique element $\mathrm{H}_\alpha \in [\mathfrak{s}_\alpha, \mathfrak{s}_{-\alpha}]$ such that $\alpha(\mathrm{H}_\alpha) = 2$. For any two roots α, β we have $\beta(\mathrm{H}_\alpha) \in \mathbb{Z}$. The element H_α is called the *coroot* associated with α. If $\Pi = \{\alpha_1, \dots, \alpha_r\}$, then $\{\mathrm{H}_{\alpha_1}, \dots, \mathrm{H}_{\alpha_r}\}$ is the base of the dual root system. Let $\mathcal{P}$ be the subgroup of $\mathfrak{t}^*$ consisting of all linear forms λ such that $\lambda(\mathrm{H}_\alpha) \in \mathbb{Z}$ for all $\alpha \in \Delta$ (or, equivalently, for all $\alpha \in \Pi$). It is clear that $\mathcal{P}$ is a free abelian group of rank r and that $\mathcal{P}$ is invariant under the Weyl group $\mathcal{W}$. The group $\mathcal{P}$ is often called the *weight lattice.* It is useful to observe that Δ is contained in $\mathcal{P}$ and that Δ generates a subgroup of finite index in $\mathcal{P}$ (the root lattice).

Further, let $\mathcal{P}^+$ be the subset of $\mathcal{P}$ defined by the inequalities $\lambda(\mathrm{H}_\alpha) \geq 0$ for all $\alpha \in \Delta^+$ (or, equivalently, for all $\alpha \in \Pi$). Then $\Delta \subset \mathcal{P}$ (in particular, $\Lambda_P \in \mathcal{P}$) and $\varrho \in \mathcal{P}^+$ (more precisely, $\varrho(\mathrm{H}_{\alpha_i}) = 1$ for all i, $i = 1, \dots, r$). For each $\lambda \in \mathcal{P}^+$ there is an irreducible finite-dimensional $\mathfrak{s}$-module having λ as *highest weight.* This $\mathfrak{s}$-module will be denoted by V^λ. Each irreducible finite-dimensional $\mathfrak{s}$-module is isomorphic to one and only one V^λ.

We shall consider $\mathfrak{g}$ as an $\mathfrak{s}$-module with the adjoint action. For $\lambda \in \mathcal{P}^+$ let m_λ be the multiplicity with which V^λ occurs in the dual module $\mathfrak{g}^*$. Denote by w_0 the unique element of $\mathcal{W}$ such that $w_0(\Delta^+) = \Delta^-$.

Theorem 1 *Assume that G is effective on $X = G/H$. Let $n = \dim X$, $d = \dim N/H$. Then :*

(i) $n\varrho + \Lambda_P + \sum_{\lambda \in \mathcal{P}^+} m_\lambda w_0(\lambda + \varrho) \in \mathcal{P}^+$;

(ii) $\sum_{\lambda \in \mathcal{P}^+} m_\lambda \leq n$;

(iii) $m_0 \leq d$.

Proof Statement (iii) is exactly Lemma 3. Let $\mathfrak{g} = V_1 \oplus V_2 \oplus \dots \oplus V_l$ be a decomposition of the adjoint $\mathfrak{s}$-module into irreducible submodules. In each V_i there is a unique (up to scalar multiplication) eigenvector of $\mathfrak{b}^-$. Call this vector v_i. By Lemma 2 the vectors $v_1, .., v_l$ are linearly independent modulo $\mathfrak{h}$. Therefore $l \leq n$, and (ii) follows. In order to prove (i), fix $\alpha \in \Delta^+$ and consider the associated simple three-dimensional subalgebra $\mathfrak{s}^{(\alpha)} \subset \mathfrak{s}$. Define $\hat{V}_i$ to be the $\mathfrak{s}^{(\alpha)}$-submodule of V_i generated by v_i and let $\hat{V} := \hat{V}_1 \oplus \hat{V}_2 \oplus \dots \oplus \hat{V}_l$. Since any lowest weight vector of $\mathfrak{s}^{(\alpha)}$ in $\hat{V}$ is a linear combination of v_i, it follows from Lemma 2 that $\hat{V} \cap \mathfrak{h}$ does not contain non-zero $\mathfrak{s}^{(\alpha)}$-submodules. By Lemma 1 (ii) of §3.7 we have $\dim\,(\hat{V} \cap \mathfrak{h}) \leq \chi(\hat{V} \cap \mathfrak{h})(\mathrm{H}_\alpha)$. On the other hand, if $\tau : \mathfrak{g} \to \mathfrak{g}/\hat{V}$ is the canonical projection map then

$$\chi(\hat{V} \cap \mathfrak{h})(\mathrm{H}_\alpha) \leq \chi(\hat{V} \cap \mathfrak{h})(\mathrm{H}_\alpha) + \chi(\tau(\mathfrak{h}))(\mathrm{H}_\alpha) = \chi(\mathfrak{h})(\mathrm{H}_\alpha)$$

by Lemma 1 (i) of §3.7. Therefore

$$\dim \hat{V} \;\leq\; n + \dim \hat{V} \cap \mathfrak{h} \;\leq\; n + \chi(\mathfrak{h})(\mathrm{H}_\alpha)$$

or, in view of Lemma 1,
$$\dim \hat{V} \le \Lambda_P(\mathrm{H}_\alpha).$$
Let $(-\mu_i) : \mathfrak{t} \to \mathbb{C}$ be the weight of v_i. Then $\dim \hat{V}_i = \mu_i(\mathrm{H}_\alpha) + 1$, and so we obtain

$$\sum_{i=1}^{l} (\mu_i(\mathrm{H}_\alpha) + 1) \le n + \Lambda_P(\mathrm{H}_\alpha).$$

Since $\varrho(\mathrm{H}_\alpha) \ge 1$ and $n \ge l$, it follows that

$$\sum_{i=1}^{l} (\mu_i + \varrho)(\mathrm{H}_\alpha) \le (n\varrho + \Lambda_P)(\mathrm{H}_\alpha)$$

for each $\alpha \in \Delta^+$. Hence

$$n\varrho + \Lambda_P + \sum_{\lambda \in \mathcal{P}^+} m_\lambda w_0(\lambda + \varrho) = n\varrho + \Lambda_P - \sum_{i=1}^{l} (\mu_i + \varrho) \in \mathcal{P}^+. \qquad \square$$

Theorem 2 (see [A6]) *Let*

$$\delta(n) = \max_{(S,P)} \prod_{\alpha \in \Delta^+} \frac{(n\varrho + \Lambda_P)(\mathrm{H}_\alpha)}{\varrho(\mathrm{H}_\alpha)},$$

where the maximum is taken over all parabolic subgroups P in all semisimple complex Lie groups S of dimension $\le n(n+2)$. Then

$$\dim \operatorname{Aut}(X) \le \delta(n)$$

for any compact homogeneous complex manifold X of dimension n.

Proof Let $G := \operatorname{Aut}^\circ(X)$. With the above notation the dual $\mathfrak{s}$-module $\mathfrak{g}^*$ is the sum of V^{μ_i}, $i = 1, 2, \ldots, l$. Therefore, the Weyl formula (see [Hum1] or [Bou], Chap. 8) yields

$$\dim G = \sum_{i=1}^{l} \dim V^{\mu_i} = \sum_{i=1}^{l} \prod_{\alpha \in \Delta^+} \frac{(\mu_i + \varrho)(\mathrm{H}_\alpha)}{\varrho(\mathrm{H}_\alpha)} \le$$

$$\le \prod_{\alpha \in \Delta^+} \sum_{i=1}^{l} \frac{(\mu_i + \varrho)(\mathrm{H}_\alpha)}{\varrho(\mathrm{H}_\alpha)} \le \prod_{\alpha \in \Delta^+} \frac{(n\varrho + \Lambda_P)(\mathrm{H}_\alpha)}{\varrho(\mathrm{H}_\alpha)},$$

where the last equality follows from Theorem 1. $\square$

Conjecture (R.Remmert) $\dim \operatorname{Aut}(X) \le n(n+2)$.

3.9 Compact homogeneous Kähler manifolds

The aim of this section is to prove the structure theorem for compact homogeneous Kähler manifolds, due to A.Borel and R.Remmert [BoRe]. For this we need the definition and some properties of the Albanese map. As usual, Ω^p_X denotes the sheaf of germs of local holomorphic p-forms on a complex manifold X. Throughout this section X is assumed to be a connected compact Kähler manifold. Therefore we have the canonical isomorphisms

$$H^i(X,\mathbb{C}) \simeq \oplus_{p+q=i} H^p(X,\Omega^q_X)$$

and

$$H^p(X,\Omega^q_X) \simeq H^q(X,\Omega^p_X).$$

Let $\omega_1,\ldots,\omega_g$ be a basis of $\Omega^1_X(X)$ and $\gamma_1,\ldots,\gamma_{2g}$ a basis of the free part of $H_1(X,\mathbb{Z})$. Then the vectors

$$p_j := \Big(\int_{\gamma_j}\omega_1, \ \ldots \ , \int_{\gamma_j}\omega_g\Big) \in \mathbb{C}^g, \quad j = 1,\ldots,2g,$$

are linearly independent over $\mathbb{R}$. Denote by Π the lattice in $\mathbb{C}^g$ generated by $p_1,\ldots,p_{2g}$. The complex torus $A(X) := \mathbb{C}^g/\Pi$ is called the *Albanese manifold* of X.

Let $x_0 \in X$ be some point which will be kept fixed. Given an arbitrary point $x \in X$ and a path γ from x_0 to x, put

$$a(x,\gamma) = \Big(\int_{x_0}^{x}\omega_1, \ \ldots \ , \int_{x_0}^{x}\omega_g\Big) \in \mathbb{C}^g,$$

where the integrals are taken along γ. Since all holomorphic forms on a compact Kähler manifold are closed, the coset of $a(x,\gamma)$ modulo Π does not depend on γ. Let $\alpha(x)$ denote this coset. Then $\alpha : X \to A(X)$ is a holomorphic map which is called the *Albanese map* of X. The Albanese map has the universality property:

for any holomorphic map $\psi : X \to T$, where T is a complex torus, there exists a unique affine map $A_\psi : A(X) \to T$ such that $\psi = A_\psi \circ \alpha$.

Proof Let $T = \mathbb{C}^m/\Delta$, where Δ is a lattice in $\mathbb{C}^m$, and let $\tilde{X} \to X$ be the universal covering. Then ψ is given by a system of holomorphic functions $\psi_j \in \mathcal{O}(\tilde{X})$, $j = 1,\ldots,m$, having the property that

$$(\psi_1,\ldots,\psi_m)(sy) = (\psi_1,\ldots,\psi_m)(y) \mod \Delta$$

for all $s \in \pi_1(X,x_0)$, $y \in \tilde{X}$. The differentials $d\psi_1,\ldots,d\psi_m$ are well-defined holomorphic 1-forms on X. Therefore

$$(d\psi_1,\ldots,d\psi_m) = (\omega_1,\ldots,\omega_g)\cdot C,$$

where $C = (c_{ij})$ is a complex $g \times m$-matrix.

We may assume that $\psi(x_0) = 0$ and that $\gamma_1, \ldots, \gamma_{2g}$ are represented by some elements $s_1, \ldots, s_{2g} \in \pi_1(X, x_0)$. Then

$$\left(\int_{\gamma_j} d\psi_1, \ldots, \int_{\gamma_j} d\psi_m\right) = (\psi_1(s_j y), \ldots, \psi_m(s_j y)) - (\psi_1(y), \ldots, \psi_m(y)) \in \Delta$$

for any $y \in \tilde{X}$. Therefore $p_j \cdot C \in \Delta$, $j = 1, \ldots, 2g$, so that the linear map

$$\mathbb{C}^g \ni z \mapsto z \cdot C \in \mathbb{C}^m$$

determines a homomorphism of complex tori $A(X) \to T$. Denoting this homomorphism by A_ψ, we see from its definition that $\psi(x) = A_\psi(\alpha(x))$ for all $x \in X$.

We now prove the uniqueness of A_ψ. Let $A^1_\psi, A^2_\psi : A(X) \to T$ be two affine maps such that $A^1_\psi \circ \alpha = A^2_\psi \circ \alpha$. The set $\{a \in A(X) | A^1_\psi(a) = A^2_\psi(a)\}$ is of the form $a_0 + S$, where $a_0 \in A(X)$ and $S \subset A(X)$ is a closed complex subgroup. If $A^1_\psi \neq A^2_\psi$ then dim $S < g$, and $\alpha(X)$ is contained in a proper subtorus of $A(X)$. This yields

$$c_1 \int_{x_0}^{x} \omega_1 + \ldots + c_g \int_{x_0}^{x} \omega_g = 0$$

for some $c_1, \ldots, c_g \in \mathbb{C}$ and for all x near x_0. Since $\omega_1, \ldots, \omega_g$ are linearly independent, we get a contradiction. Thus A_ψ is uniquely determined. □

The following result is a consequence of the universality of the Albanese map:

Aut(X) *acts holomorphically on* $A(X)$ *and* $\alpha : X \to A(X)$ *is an equivariant map.*

Proof For any $g \in \mathrm{Aut}(X)$ the map $\alpha \circ g : X \to A(X)$ can be represented as $A_g \circ \alpha$, where $A_g \in \mathrm{Aut}(A(X))$. The uniqueness of A_g implies that $A_g \circ A_h = A_{gh}$ and $A_e = \mathrm{id}_{A(X)}$. In order to show that the mapping $\mathrm{Aut}(X) \times A(X) \ni (g, a) \mapsto A_g(a) \in A(X)$ is holomorphic, it is enough to consider $A_g(x)$ for g in a small neighborhood of e. But then A_g is a translation, and so we obtain

$$A_g(a) = a + A_g(0) = a + A_g(\alpha(x_0)) = a + \alpha(gx_0),$$

showing that $A_g(a)$ depends holomorphically on g and a. □

Theorem (A.Borel - R.Remmert [BoRe]) *Let X be a connected, compact, homogeneous Kaehler manifold. Then X is isomorphic to the product $Y \times A(X)$, where Y is a flag manifold and $A(X)$ is the Albanese manifold. More precisely, Y is the base of the Tits fibration and the isomorphism is given by the product map $\pi \times \alpha$, where $\pi : X \to Y$ is the projection of the Tits fibration and $\alpha : X \to A(X)$ is the Albanese map.*

We start with several auxiliary propositions.

Lemma 1 *A connected, compact, parallelizable Kähler manifold X is a complex torus.*

Proof As we have seen in §3.4, the manifold X can be written in the Klein form $X = G/\Gamma$, where G is a connected complex Lie group, $\Gamma \subset G$ a uniform discrete subgroup. Let $\mu_1, \ldots, \mu_n$ be a basis of right-invariant holomorphic 1-forms on G. Since these forms can be regarded as holomorphic 1-forms on the Kähler manifold X, they are closed. Let $A_1, \ldots, A_n$ be a basis of right-invariant holomorphic vector fields on G. Then

$$\mu_i([A_j, A_k]) = -d\mu_i(A_j, A_k) = 0$$

for all i, j, k. Thus $[A_j, A_k] = 0$ for all j, k, and G is abelian. □

Lemma 2 (see [Oe2]) *Let X be a connected, compact, Kähler manifold with first Betti number $b_1(X) = 0$. Assume that X is almost homogeneous with respect to a complex Lie transformation group. Then:*
(i) $H^0(X, \Omega_X^q) = 0$ *for all* $q \geq 1$;
(ii) X *is projective algebraic.*

Proof Since dim $H^0(X, \Omega_X^1) = \frac{1}{2}b_1(X) = 0$, the first statement for q=1 is obvious. Let $n = \dim X$, $q > 1$, and $\omega \in \Omega_X^q(X)$. By assumption there exist n holomorphic vector fields $A_1, \ldots, A_n \in \mathcal{T}_X(X)$, which are linearly independent at some point. For any $i_1, \ldots, i_{q-1}$, $1 \leq i_1 < \ldots < i_{q-1} \leq n$, define $\omega_{i_1,\ldots,i_{q-1}} \in \Omega_X^1(X)$ by

$$(\omega_{i_1,\ldots,i_{q-1}})_x(\xi) := \omega_x(\xi, A_{i_1}(x), \ldots, A_{i_{q-1}}(x)), \quad \text{where} \quad x \in X, \xi \in T_x(X).$$

Since $\Omega_X^1(X) = 0$, we have $\omega_{i_1,\ldots,i_{q-1}} = 0$ for all $i_1, \ldots, i_{q-1}$. Hence

$$\omega_x(A_{i_1}(x), \ldots, A_{i_q}(x)) = 0$$

for all $i_1, \ldots, i_q$, $1 \leq i_1, \ldots, i_q \leq n$, and for all $x \in X$. On the other hand, $A_1(x), \ldots, A_n(x)$ is a basis of $T_x(X)$ for x in an open set. Thus ω is zero on this open set. Since ω is holomorphic, $\omega = 0$ everywhere on X, and (i) follows.

In order to prove (ii) we use the fundamental theorem of Kodaira [K]. Namely, a compact Kähler manifold is projective if and only if there exists a Kähler metric on X such that the real cohomology class of the associated closed $(1,1)$-form belongs to $H^2(X, \mathbb{Z})$. In our setting $H^0(X, \Omega_X^2) = 0$ by (i). Therefore, $H^2(X, \mathbb{C}) = H^1(X, \Omega_X^1)$ and each cohomology class in $H^2(X, \mathbb{R})$ is represented by a closed $(1,1)$-form. The integer cohomology classes in $H^2(X, \mathbb{R})$ form a lattice of maximal rank. The cohomology classes corresponding to Kähler metrics form a non-empty open cone with vertex 0. Since this cone obviously contains a point of the lattice, the manifold X is projective. □

Lemma 3 *Let X be a connected, compact, Kähler manifold. Assume that X is almost homogeneous with respect to a complex Lie transformation group. Then $\alpha(X) = A(X)$ and $\alpha : X \to A(X)$ is a fibration with connected fiber.*

Proof Let G be a connected complex Lie group acting on X with an open orbit. There is a homomorphism

$$\alpha_* : \mathrm{Aut}^\circ(X) \to \mathrm{Aut}^\circ(A(X)) \simeq A(X),$$

such that $\alpha(gx) = \alpha_*(g)\alpha(x)$ for all $g \in G$, $x \in X$. The subgroup $\alpha_*(G) \subset \mathrm{Aut}^\circ(A(X))$ acts on $A(X)$ freely. Therefore all G-orbits on $A(X)$ have equal dimension. Since $\alpha(X)$ is almost homogeneous with respect to G, it follows that G

is in fact transitive on $\alpha(X)$. In particular, $\alpha(X)$ is itself a torus. The proof of the universality property shows then that $\alpha(X) = A(X)$.

Since $A(X)$ is G-homogeneous, $(X, A(X), \alpha)$ is a locally trivial fiber bundle. In order to show that the fiber of α is connected, consider the Stein factorization $\alpha = \chi \circ \sigma$, where $\sigma : X \to Z$ is a surjective holomorphic map onto an irreducible complex space Z, such that $\mathcal{O}_Z = \sigma_* \mathcal{O}_X$, and $\chi : Z \to A(X)$ is a finite holomorphic map. Recall that all fibers of σ are connected (see e.g. [GR3], Ch. 10, §6). By Lemma 2 of §2.4, the group G acts holomorphically on Z and $\sigma : X \to Z$ is G-equivariant. But then $\chi : Z \to A(X)$ is also G-equivariant. Since G is transitive on $A(X)$, it follows that $\chi : Z \to A(X)$ is a finite unramified covering. Therefore, Z is a torus and χ is an isomorphism by the universality property of the Albanese map. □

We now proceed to the proof of the theorem.

Proof Let

$$X = G/H \xrightarrow{\pi} G/N = Y \quad \text{and} \quad X = G/H \xrightarrow{\alpha} G/M = A(X)$$

be the Tits and the Albanese fibrations. According to Lemma 1 the fiber of the first fibration is a torus. Since $\pi_1(Y) = \{e\}$, it follows that $\pi_1(X)$ is abelian.

By Lemma 3 the fiber of the second fibration is connected. As a consequence we have the exact sequence

$$\{e\} = \pi_2(A(X)) \to \pi_1(M/H) \to \pi_1(X) \to \pi_1(A(X)) \to \pi_0(M/H) = \{e\},$$

and the additivity of rank implies that

$$b_1(X) = \text{rank } \pi_1(X) = \text{rank } \pi_1(M/H) + \text{rank } \pi_1(A(X)) = \text{rank } \pi_1(M/H) + 2g.$$

On the other hand, $b_1(X) = 2g$. Therefore, $\pi_1(M/H)$ is finite and $b_1(M/H) = 0$. Applying Lemma 2, we see that M/H is a projective manifold. Furthermore, M/H is an orbit of a linear algebraic group in some $\mathbb{P}_n$ or, equivalently, a flag manifold (see §3.2).

We may assume without loss of generality that G is simply connected and that the G-action on X is locally effective. Then Proposition 2 of §3.6 tells us that $G = S \times R$, where S is a semisimple group and R is a vector group. Since S acts trivially on $A(X)$, it is clear that $M \supset S$ and S is a maximal semisimple subgroup of M. From Lemma 1 of §3.3 it follows that S acts on M/H transitively and algebraically. In other words, $M/H = S/Q$, where $Q := S \cap H$ is a parabolic subgroup in S.

Since S acts trivially on $A(X)$, the radical R is transitive on $A(X)$. Similarly, R acts trivially on Y and so the semisimple group S is transitive on Y. It follows that G is transitive on $Y \times A(X)$. In particular, the product map

$$\pi \times \alpha : X \to Y \times A(X)$$

is surjective. Let $x_0 := e \cdot H \in X$ and let $y_0 := \pi(x_0) = e \cdot N \in Y$. We still have to show that $\pi^{-1}(y_0) \cap \alpha^{-1}(0) = \{x_0\}$.

We can write Y in the Klein form $Y = S/P$, where $P := N \cap S$. Then $P \supset Q$ and P/Q is a flag manifold. Thus, $b_1(P/Q) = 0$ and $A(P/Q)$ is one point. On the other hand,

$$P/Q = \pi^{-1}(y_0) \cap \alpha^{-1}(0) \subset \pi^{-1}(y_0) = N/H.$$

Since N/H is a torus, the universality property of the Albanese map yields $P = Q$. This completes the proof. □

Remark 1 A proof of the Borel-Remmert theorem, based on the ideas of symplectic geometry, is given in [Hu1].

Remark 2 In the proof of the Borel-Remmert theorem we used the Klein form $X = G/H$, where G is a complex Lie group. In particular, if X is not a torus then G does not preserve a Kähler metric on X.

There is another classification problem concerning homogeneous Kähler manifolds. Namely, let X be a (not necessarily compact) Kähler manifold with a fixed Kähler metric and let G be a (real) Lie group acting transitively on X and preserving the metric. In 1967, S.G.Gindikin and E.B.Vinberg formulated the following conjecture:

X is the total space of a holomorphic fiber bundle, whose base is a homogeneous bounded domain $D \subset \mathbb{C}^n$ and whose fiber (with the induced Kähler metric) is the product of a flat homogeneous Kähler manifold T and a compact simply connected homogeneous Kähler manifold Y.

Note that Y is a flag manifold and $T = \mathbb{C}^m/\Gamma$, where $\Gamma \subset \mathbb{C}^m$ is a discrete additive subgroup. The fiber bundle is in fact holomorphically trivial so that X is biholomorphically equivalent to the product $D \times Y \times T$.

The conjecture was recently proved by J.Dorfmeister and K.Nakajima [DoNa]. Their result is the final step in the series of contributions made by many mathematicians (see [Bo2], [Kl], [Mat1], [Ha], [GPV2], [GV], [Shi]).

Remark 3 Let X be a connected, compact, almost homogeneous Kähler manifold. In the following example, taken from [BaOe], the Albanese fibration $X \to A(X)$ is topologically non-trivial. In general, this fibration may be holomorphically non-trivial for any finite unramified covering of X (see [Hu1] for a simple example). However, there exists a finite unramified covering $X^* \to X$ such that X^* is almost homogeneous and the Albanese fibration of X^* is *topologically* trivial (see [BaOe]).

Example Let $Z = \mathbb{C} \times \mathbb{P}_1 \times \mathbb{P}_1$ and let $\Gamma \simeq \mathbb{Z}^2$ be a transformation group of Z generated by

$$(z,p,q) \mapsto (z+1,p,q) \qquad \text{and} \qquad (z,p,q) \mapsto (z+\omega,q,p),$$

where $\omega \in \mathbb{C}-\mathbb{R}$. Put $\Lambda := \mathbb{Z}+\mathbb{Z}\omega$, $A := \mathbb{C}/\Lambda$, $X := Z/\Gamma$, and denote by $\alpha : X \to A$ the mapping induced by the projection of Z onto $\mathbb{C}$. Then X is a Kähler manifold and α is the Albanese map of X. The group $G := \mathbb{C} \times \mathrm{PSL}(2,\mathbb{C})$ acts on Z by

$$(z,p,q) \stackrel{(w,g)}{\longmapsto} (z+w, gp, gq),$$

where $w \in \mathbb{C}$, $g \in \mathrm{PSL}(2,\mathbb{C})$. Each transformation from $\mathbb{C} \times \mathrm{PSL}(2,\mathbb{C})$ commutes with each transformation from Γ. Since G has an open orbit on Z (namely, $\mathbb{C} \times ((\mathbb{P}_1 \times \mathbb{P}_1) - \Delta)$, where Δ is the diagonal), the induced action of G on X also has an open orbit and X is almost homogeneous with respect to G. The fiber Y of the Albanese map $\alpha : X \to A$ is isomorphic to $\mathbb{P}_1 \times \mathbb{P}_1$. However, X is not homemorphic to $A \times \mathbb{P}_1 \times \mathbb{P}_1$, because $H^2(X,\mathbb{Z}) = \mathbb{Z} \oplus \mathbb{Z}$, but $H^2(A \times \mathbb{P}_1 \times \mathbb{P}_1, \mathbb{Z}) = \mathbb{Z} \oplus \mathbb{Z} \oplus \mathbb{Z}$.

4 Homogeneous Vector Bundles

Suppose $\mathbb{V}$ is a holomorphic vector G-bundle on a complex manifold X, where G is a complex Lie group acting holomorphically on X. Let $\mathcal{O}_{\mathbb{V}}$ be the sheaf of germs of local holomorphic sections of $\mathbb{V}$. If X is compact then $H^q(X, \mathcal{O}_{\mathbb{V}})$ are finite-dimensional vector spaces. We show that the induced representations of G on these spaces are holomorphic. If X is homogeneous, $X = G/H$, then $\mathbb{V}$ is given by a holomorphic representation $\varphi : H \to \mathrm{GL}(V)$ and the induced representations are denoted by $\mathrm{I}^q\varphi : G \to \mathrm{GL}(H^q(X, \mathcal{O}_{\mathbb{V}}))$. The aim of this chapter is to give a proof and some applications of the theorem of R.Bott, which determines $\mathrm{I}^q\varphi$ in the case when H is a parabolic subgroup of a semisimple group G and φ is an irreducible representation. Namely, there is at most one number q, such that $H^q(X, \mathcal{O}_{\mathbb{V}}) \neq 0$, the representation $\mathrm{I}^q\varphi$ is again irreducible, and the theorem provides an algorithm for computing its highest weight. The proof below is due to M.Demazure. As an application, we consider the representations induced by characters of maximal parabolic subgrops. As another application, we prove that $H^q(X, \mathcal{T}) = 0$ $(q > 0)$ and determine the G-modules $H^0(X, \mathcal{T})$, where $\mathcal{T}$ is the tangent sheaf of a flag manifold $X = G/H$. It should be mentioned that, according to K.Kodaira and D.C.Spencer, the equality $H^1(X, \mathcal{T}) = 0$ implies that flag manifold are rigid, i.e., their complex structure cannot be altered by a small deformation.

4.1 Coherent analytic G-sheaves

Let $\mathcal{F}$ be a sheaf (of sets) on a topological space X and let $\pi : \mathcal{F} \to X$ denote the projection map. Suppose G is a group acting on X by homeomorphisms. Then $\mathcal{F}$ is called a *G-sheaf* if there is a G-action on $\mathcal{F}$, such that each element $g \in G$ acts as a homeomorphism $\sigma_g : \mathcal{F} \to \mathcal{F}$ and the diagram

$$\begin{array}{ccc} \mathcal{F} & \xrightarrow{\sigma_g} & \mathcal{F} \\ \pi \downarrow & & \downarrow \pi \\ X & \xrightarrow{g} & X \end{array}$$

commutes. For a G-sheaf $\mathcal{F}$ denote by $\sigma_{g,x}$ the restriction of σ_g to the stalk $\mathcal{F}_x$. Then

$$\sigma_{gh,x} = \sigma_{g,hx} \circ \sigma_{h,x}.$$

For an open set $U \subset X$ there is a bijection

$$\sigma_{g,U} : \mathcal{F}(U) \to \mathcal{F}(g \cdot U),$$

given by

$$[\sigma_{g,U}(s)]_x = \sigma_{g,g^{-1}x}(s_{g^{-1}x})$$

for all $s \in \mathcal{F}(U)$ and $x \in g \cdot U$.

Let $\mathcal{F}'$ and $\mathcal{F}''$ be two G-sheaves (of sets) on X. For any $g \in G$ let $\sigma'_g : \mathcal{F}' \to \mathcal{F}'$ and $\sigma''_g : \mathcal{F}'' \to \mathcal{F}''$ denote the corresponding homeomorphisms. A sheaf mapping $\varphi : \mathcal{F}' \to \mathcal{F}''$ is said to be a *mapping of G-sheaves* if $\sigma''_g \circ \varphi = \varphi \circ \sigma'_g$ for all $g \in G$.

A *G-sheaf of abelian groups* is a sheaf of abelian groups which is also a G-sheaf with the property that each mapping $\sigma_{g,x} : \mathcal{F}_x \to \mathcal{F}_{gx}$ is an isomorphism of abelian groups. In the same way one defines *G-sheaves of rings*, *G-sheaves of modules* (over G-sheaves of rings), and other *G-sheaves with algebraic structure.* If $\mathcal{F}'$ and $\mathcal{F}''$ are two G-sheaves with algebraic structure of the same type then a mapping of G-sheaves $\varphi : \mathcal{F}' \to \mathcal{F}''$ is called a *G-homomorphism* if for every $x \in X$ the mapping $\varphi_x : \mathcal{F}'_x \to \mathcal{F}''_x$ is a homomorphism of the corresponding algebraic objects.

Let G be a transformation group of a complex space $(X, \mathcal{O}_X)$. Then $\mathcal{O}_X$ is a G-sheaf of local $\mathbb{C}$-algebras with the operation of G on sections given by

$$\mathcal{O}(U) \ni f \mapsto f \circ g^{-1}|_U \in \mathcal{O}(g \cdot U).$$

In what follows we mainly consider G-sheaves of modules over $\mathcal{O}_X$, which we also call *analytic G-sheaves* . If $\mathcal{F}$ is such a sheaf then

$$\sigma_{g,U}(fs) = (f \circ g^{-1}|_U) \cdot \sigma_{g,U}(s) \qquad \text{for all } s \in \mathcal{F}(U), f \in \mathcal{O}(U).$$

Let $\varphi : \mathcal{F}' \to \mathcal{F}''$ be a G-homomorphism of G-sheaves of modules over $\mathcal{O}_X$ (or an *analytic G-homomorphism*). For any open $U \subset X$ we have the associated homomorphism of $\mathcal{O}(U)$-modules $\varphi_U : \mathcal{F}'(U) \to \mathcal{F}''(U)$. The definition of a G-homomorphism yields

$$\varphi_{g \cdot U} \circ \sigma'_{g,U} = \sigma''_{g,U} \circ \varphi_U .$$

From now on we assume that G is a complex Lie group acting holomorphically on a complex space X and that $\mathcal{F}$ is a coherent analytic sheaf on X. For any open set $U \subset X$ the vector space $\mathcal{F}(U)$ has a canonical Fréchet topology so that for each open subset $V \subset U$ the restriction map $\varrho_{U,V} : \mathcal{F}(U) \to \mathcal{F}(V)$ is continuous. We say that $\mathcal{F}$ is a *coherent analytic G-sheaf (under a holomorphic G-action)* if $\mathcal{F}$ is a G-sheaf of modules over $\mathcal{O}_X$ satisfying the following condition:

if $U, V \subset X$ and $W \subset G$ are open subsets such that $W \cdot U \subset V$ then the mapping

$$W \to \mathcal{F}(U), \qquad g \mapsto \sigma^{-1}_{g,U} \left[\varrho_{V,g \cdot U}(s)\right], \qquad (+)$$

is holomorphic for every $s \in \mathcal{F}(V)$.

First of all we check this condition for the structure sheaf.

If G is a complex Lie transformation group of X then $\mathcal{O}_X$ is a coherent analytic G-sheaf.

Proof Fix $f \in \mathcal{F}(V)$ and $g_0 \in W$. We have to show that the mapping $g \mapsto f \circ g|_U \in \mathcal{O}(U)$ is holomorphic in g_0. Let $N \subset W$ be a coordinate neighborhood of g_0 isomorphic to a polydisk in local coordinates $t_1, \ldots, t_n$, where $t_i(g_0) = 0$. For any open relatively compact subset $U' \subset\subset U$ the mapping $g \mapsto f \circ g|_{U'}$ is holomorphic in N by the definition of a holomorphic action (see §1.2). This means that for $g \in N$ one has

$$f \circ g|_{U'} = \sum a'_{i_1,\ldots,i_n} t_1^{i_1}(g) \cdot \ldots \cdot t_n^{i_n}(g), \qquad a'_{i_1,\ldots,i_n} \in \mathcal{O}(U'),$$

where the convergence is absolute with respect to any continuous seminorm on $\mathcal{O}(U')$. For another open relatively compact subset $U'' \subset\subset U$ there is a similar decomposition with coefficients $a''_{i_1,\dots,i_n} \in \mathcal{O}(U'')$, which converges everywhere in N. Since $a'_{i_1,\dots,i_n}|_{U'\cap U''} = a''_{i_1,\dots,i_n}|_{U'\cap U''}$, there exist holomorphic functions $a_{i_1,\dots,i_n} \in \mathcal{O}(U)$ whose restrictions to U' are equal to $a'_{i_1,\dots,i_n}$ for every $U' \subset\subset U$. The canonical topology on $\mathcal{O}(U)$ can be defined by a sequence of seminorms $\{p_k\}$, each of which is of the form $p_k = q_k \circ \varrho_k$, where $\varrho_k : \mathcal{O}(U) \to \mathcal{O}(U_k)$ is the restriction map, $U_k \subset\subset U$, and q_k is a continuous seminorm on $\mathcal{O}(U_k)$. It follows that

$$f \circ g|_U = \sum a_{i_1,\dots,i_n} t_1^{i_1}(g) \cdot \ldots \cdot t_n^{i_n}(g), \qquad g \in N,$$

where the convergence is absolute with respect to any continuous seminorm on $\mathcal{O}(U)$. □

Using this fact one gets many other examples of coherent analytic G-sheaves. Namely, if $\mathcal{J} \subset \mathcal{O}_X$ is a coherent ideal sheaf such that $f \circ g^{-1}|_U \in \mathcal{J}(g \cdot U)$ for any open $U \subset X$ and for all $f \in \mathcal{J}(U)$, $g \in G$, then $\mathcal{J}$ is itself a coherent analytic G-sheaf. In particular, if $A \subset X$ is a G-invariant analytic subset then $\mathcal{J}_A$ is a coherent analytic G-sheaf. It is also easy to see that the kernel, image and cokernel of an analytic G-homomorphism between coherent analytic G-sheaves are coherent analytic G-sheaves.

Remark For a real Lie transformation group of X and for $\mathcal{F} = \mathcal{O}_X$ the mapping (+) is not necessarily real analytic. (By the definition in §1.2 this should be so only if U is relatively compact in X and $W \cdot \overline{U} \subset V$.) For example, consider the action of $G = \mathbb{R}$ on the upper halfplane $X = \{z \in \mathbb{C} \mid \operatorname{Im} z > 0\}$ by translations $z \mapsto g_t(z) = z + t$, $t \in \mathbb{R}$. Taking $f = 1/z$ we obtain the mapping $t \mapsto f \circ g_t = \frac{1}{z+t} \in \mathcal{O}(X)$. It is clear that there is no such $\epsilon > 0$ that the series

$$\frac{1}{z+t} = \sum_{k=0}^{\infty} (-1)^k \frac{t^k}{z^{k+1}}$$

converges for all t, $|t| < \epsilon$, and for all $z \in X$.

In the next section we shall consider locally free analytic G-sheaves on homogeneous manifolds G/H. In this connection the following fact is worth mentioning.

If $\mathcal{F}$ is a coherent analytic G-sheaf over G/H then $\mathcal{F}$ is locally free.

Proof The set $B(\mathcal{F})$ of all points of G/H, where $\mathcal{F}$ is not locally free, is a nowhere dense analytic set (see [GR3], Ch.4, §4). Since $\mathcal{F}$ is a G-sheaf, we have $G \cdot B(\mathcal{F}) = B(\mathcal{F})$. Since G acts transitively, it follows that $B(\mathcal{F}) = \emptyset$. □

On the other hand, even in the homogeneous setting one cannot drop the coherence condition. More precisely, there are analytic G-sheaves on G/H, which are not locally finitely generated (and thus not coherent), but have the property that their stalks are finitely generated $\mathcal{O}_x$-modules for every point.

Example Consider the action of $G = \mathbb{C}$ on $X = \mathbb{C}$ by translations. Let $\mathcal{F}$ be the sheaf of germs of meromorphic functions on X having only simple poles. Then $\mathcal{F}$

is a G-sheaf and $\mathcal{F}_x$ is a free $\mathcal{O}_x$-module of rank 1 for every $x \in X$, but $\mathcal{F}$ is not locally finitely generated.

Let us return for a moment to our starting point. Namely, suppose that $\mathcal{F}$ is a G-sheaf of abelian groups over a topological space X and denote by $H^q(X, \mathcal{F})$ the q-th Čech cohomology group of X with coefficients in $\mathcal{F}$. We want to show that G acts on $H^q(X, \mathcal{F})$ by group automorphisms. In order to define this action consider an open covering $\mathfrak{U} = \{U_\iota\}$ of X and let

$$C^q(\mathfrak{U}, \mathcal{F}) := \prod_{\iota_0, \ldots, \iota_q} \mathcal{F}(U_{\iota_0, \ldots, \iota_q})$$

be the group of q-cochains of $\mathfrak{U}$ with coefficients in $\mathcal{F}$, where we use the usual notation $U_{\iota_0, \ldots, \iota_q} := U_{\iota_0} \cap \ldots \cap U_{\iota_q}$. Denote by $\delta^q_{\mathfrak{U}} : C^q(\mathfrak{U}, \mathcal{F}) \to C^{q+1}(\mathfrak{U}, \mathcal{F})$ the coboundary operators and put $Z^q(\mathfrak{U}, \mathcal{F}) := \operatorname{Ker} \delta^q_{\mathfrak{U}}$, $B^q(\mathfrak{U}, \mathcal{F}) := \operatorname{Im} \delta^{q-1}_{\mathfrak{U}}$ $(q \geq 1)$, $B^0(\mathfrak{U}, \mathcal{F}) := \{0\}$, $H^q(\mathfrak{U}, \mathcal{F}) := Z^q(\mathfrak{U}, \mathcal{F})/B^q(\mathfrak{U}, \mathcal{F})$. For each $\mathfrak{U}$ we have a canonical mapping $h^q_{\mathfrak{U}} : H^q(\mathfrak{U}, \mathcal{F}) \to H^q(X, \mathcal{F})$. If $\mathfrak{V} = \{V_\kappa\}$ is a refinement of $\mathfrak{U}$ and

$$h^q_{\mathfrak{U},\mathfrak{V}} : H^q(\mathfrak{U}, \mathcal{F}) \to H^q(\mathfrak{V}, \mathcal{F})$$

the associated homomorphism then $h^q_{\mathfrak{V}} \circ h^q_{\mathfrak{U},\mathfrak{V}} = h^q_{\mathfrak{U}}$ for all q, $q \geq 0$.

Let $g \in G$ be an arbitrary element. Then there is an associated covering $g \cdot \mathfrak{U} := \{g \cdot U_\iota\}$ with the same set of indices. The mappings

$$\sigma_{g, U_{\iota_0, \ldots, \iota_q}} : \mathcal{F}(U_{\iota_0, \ldots, \iota_q}) \to \mathcal{F}(g \cdot U_{\iota_0, \ldots, \iota_q})$$

determine the isomorphism

$$\sigma^q_{g,\mathfrak{U}} : C^q(\mathfrak{U}, \mathcal{F}) \to C^q(g \cdot \mathfrak{U}, \mathcal{F}).$$

Since the diagram

$$\begin{array}{ccc} C^q(\mathfrak{U}, \mathcal{F}) & \xrightarrow{\sigma^q_{g,\mathfrak{U}}} & C^q(g \cdot \mathfrak{U}, \mathcal{F}) \\ \delta^q_{\mathfrak{U}} \downarrow & & \downarrow \delta^q_{g \cdot \mathfrak{U}} \\ C^{q+1}(\mathfrak{U}, \mathcal{F}) & \xrightarrow{\sigma^{q+1}_{g,\mathfrak{U}}} & C^{q+1}(g \cdot \mathfrak{U}, \mathcal{F}) \end{array}$$

is obviously commutative, we obtain an isomorphism of the corresponding cohomology groups

$$\overline{\sigma}^q_{g,\mathfrak{U}} : H^q(\mathfrak{U}, \mathcal{F}) \to H^q(g \cdot \mathfrak{U}, \mathcal{F}).$$

Note that if $\mathfrak{V}$ is a refinement of $\mathfrak{U}$ then $g \cdot \mathfrak{V}$ is a refinement of $g \cdot \mathfrak{U}$ and we have

$$h^q_{g \cdot \mathfrak{U}, g \cdot \mathfrak{V}} \circ \overline{\sigma}^q_{g,\mathfrak{U}} = \overline{\sigma}^q_{g,\mathfrak{V}} \circ h^q_{\mathfrak{U},\mathfrak{V}}.$$

Given an element $\eta \in H^q(X, \mathcal{F})$, there exists a covering $\mathfrak{U}$ such that $h^q_{\mathfrak{U}}(\mu) = \eta$ for some $\mu \in H^q(\mathfrak{U}, \mathcal{F})$. Let us check that

$$h^q_{g \cdot \mathfrak{U}} \circ \overline{\sigma}^q_{g,\mathfrak{U}}(\mu) \in H^q(X, \mathcal{F})$$

depends only on η. Suppose that $\mathfrak{V}$ is another covering such that $h^q(\nu) = \eta$ for some $\nu \in H^q(\mathfrak{V}, \mathcal{F})$. Then, by the definition of the inductive limit, there exists a refinement $\mathfrak{W}$ of $\mathfrak{U}$ and $\mathfrak{V}$ with the property that

$$h^q_{\mathfrak{U},\mathfrak{W}}(\mu) = h^q_{\mathfrak{V},\mathfrak{W}}(\nu).$$

Therefore

$$h^q_{g\cdot\mathfrak{U}} \circ \overline{\sigma}^q_{g,\mathfrak{U}}(\mu) = h^q_{g\cdot\mathfrak{W}} \circ h^q_{g\cdot\mathfrak{U},g\cdot\mathfrak{W}} \circ \overline{\sigma}^q_{g,\mathfrak{U}}(\mu) = h^q_{g\cdot\mathfrak{W}} \circ \overline{\sigma}^q_{g,\mathfrak{W}} \circ h^q_{\mathfrak{U},\mathfrak{W}}(\mu) =$$

$$= h^q_{g\cdot\mathfrak{W}} \circ \overline{\sigma}^q_{g,\mathfrak{W}} \circ h^q_{\mathfrak{V},\mathfrak{W}}(\nu) = h^q_{g\cdot\mathfrak{W}} \circ h^q_{g\cdot\mathfrak{V},g\cdot\mathfrak{W}}(\nu) \circ \overline{\sigma}^q_{g,\mathfrak{V}} = h^q_{g\cdot\mathfrak{V}} \circ \overline{\sigma}^q_{g,\mathfrak{V}}(\nu),$$

and so we can put

$$g\cdot\eta := h^q_{g\cdot\mathfrak{U}} \circ \overline{\sigma}^q_{g,\mathfrak{U}}(\mu) \ \left(= h^q_{g\cdot\mathfrak{V}} \circ \overline{\sigma}^q_{g,\mathfrak{V}}(\nu)\right).$$

Since $\overline{\sigma}^q_{g_1,g_2\cdot\mathfrak{U}} \circ \overline{\sigma}^q_{g_2,\mathfrak{U}} = \overline{\sigma}^q_{g_1g_2,\mathfrak{U}}$, the mapping $g\times\eta \mapsto g\cdot\eta$ is an action of G on $H^q(X,\mathfrak{F})$. Since $\overline{\sigma}^q_{g,\mathfrak{U}}$ is a group isomorphism, it follows that each $g \in G$ acts as an automorphism of $H^q(X,\mathfrak{F})$.

Assume now again that X is a complex space and $\mathfrak{F}$ is an analytic G-sheaf on X. Then each cohomology group $H^q(X,\mathfrak{F})$ is a vector space over $\mathbb{C}$ and G acts on $H^q(X,\mathfrak{F})$ by $\mathbb{C}$-linear automorphisms. Recall that for X compact and $\mathfrak{F}$ coherent the Finiteness Theorem of H.Cartan and J.-P.Serre says that dim $H^q(X,\mathfrak{F}) < \infty$ $(q \geq 0)$. For the considerations of this chapter the following fact is of basic importance.

If X is compact and $\mathfrak{F}$ is a coherent analytic G-sheaf on X under a holomorphic G-action then the induced representation of G on $H^q(X,\mathfrak{F})$ is holomorphic.

Proof Suppose we have two coverings $\mathfrak{U} = \{U_\iota\}$ and $\mathfrak{V} = \{V_\iota\}$ with the same set of indices and assume that $U_\iota \subset V_\iota$ for every ι. Then the restriction maps

$$\varrho_{V_{\iota_0,\ldots,\iota_q},U_{\iota_0,\ldots,\iota_q}} : \mathfrak{F}(V_{\iota_0,\ldots,\iota_q}) \to \mathfrak{F}(U_{\iota_0,\ldots,\iota_q})$$

determine a map of q-cochains

$$\varrho^q_{\mathfrak{V},\mathfrak{U}} : C^q(\mathfrak{V},\mathfrak{F}) \longrightarrow C^q(\mathfrak{U},\mathfrak{F}).$$

If the set of indices is countable then $C^q(\mathfrak{U},\mathfrak{F})$ and $C^q(\mathfrak{V},\mathfrak{F})$ have natural Fréchet topology and $\varrho^q_{\mathfrak{V},\mathfrak{U}}$ is a continuous map of Fréchet spaces.

Since X is compact we can choose two finite Stein coverings $\mathfrak{U}$ and $\mathfrak{V}$ with the property that $U_\iota \subset\subset V_\iota$ for every ι. By the Leray theorem $h^q_{\mathfrak{U}}$ and $h^q_{\mathfrak{V}}$ are isomorphisms for all q, $q \geq 0$. Let $\eta \in H^q(X,\mathfrak{F})$ be an arbitrary cohomology class,

$$\nu := (h^q_{\mathfrak{V}})^{-1}(\eta) \in H^q(\mathfrak{V},\mathfrak{F}),$$

and $c \in Z^k(\mathfrak{V},\mathfrak{F})$ a q-cocycle representing ν. In what follows we assume that $g \in G$ is taken from a small neighborhood of e so that $\overline{U_\iota} \subset gV_\iota$ for every ι. From the definition of a coherent analytic G-sheaf it follows that the mapping

$$g \mapsto \sigma^q_{g,g^{-1}\cdot\mathfrak{U}}\,[\varrho^q_{\mathfrak{V},g^{-1}\cdot\mathfrak{U}}\,(c)] \in Z^q(\mathfrak{U},\mathfrak{F})$$

is holomorphic. Taking its composition with the canonical epimorphism $Z^q(\mathfrak{U},\mathfrak{F}) \to H^q(\mathfrak{U},\mathfrak{F})$, we obtain the mapping

$$g \mapsto \overline{\sigma}^q_{g,g^{-1}\cdot\mathfrak{U}} \circ h^q_{\mathfrak{V},g^{-1}\cdot\mathfrak{U}}\,(\nu) = h^q_{g\cdot\mathfrak{V},\mathfrak{U}} \circ \overline{\sigma}^q_{g,\mathfrak{V}}\,(\nu) \in H^q(\mathfrak{U},\mathfrak{F}),$$

which is holomorphic as well. Since

$$g \cdot \eta = h^q_{g \cdot \mathfrak{V}} \circ \overline{\sigma}^q_{g,\mathfrak{V}}(\nu) = h^q_{\mathfrak{U}} \, [h^q_{g \cdot \mathfrak{V},\mathfrak{U}} \circ \overline{\sigma}^q_{g,\mathfrak{V}}(\nu)],$$

it follows that the map $g \mapsto g \cdot \eta$ is also holomorphic. □

4.2 Holomorphic vector G-bundles

In this section we assume that X is a complex manifold with a holomorphic action of a complex Lie group G. Let $\mathbb{V}$ be a holomorphic vector bundle on X. We shall use the same notation $\mathbb{V}$ for the total space of the bundle and denote by $p : \mathbb{V} \to X$ the projection map. A holomorphic vector bundle is said to be a *G-bundle* if there is a holomorphic G-action on the total space $\mathbb{V}$ which commutes with p and has the property that for every pair $(g, x) \in G \times X$ the mapping of $p^{-1}(x)$ onto $p^{-1}(gx)$ induced by g is an isomorphism of vector spaces.

Let $\Gamma(U, \mathbb{V})$ denote the vector space of holomorphic sections of $\mathbb{V}$ over an open subset $U \subset X$. If $\mathbb{V}$ is a G-bundle then we have a bijection

$$\sigma_{g,U} : \Gamma(U, \mathbb{V}) \to \Gamma(g \cdot U \, , \mathbb{V})$$

defined by

$$\sigma_{g,U}(s)(x) = gs(g^{-1}x),$$

where $s \in \Gamma(U, \mathbb{V})$ and $x \in g \cdot U$.

As usual, one can associate with $\mathbb{V}$ a locally free analytic sheaf, which will be denoted by $\mathcal{O}_{\mathbb{V}}$. By definition, $\mathcal{O}_{\mathbb{V}}$ is the *sheaf of germs of local holomorphic sections of* $\mathbb{V}$. If s is a holomorphic section of $\mathbb{V}$ defined in a neighborhood U of $x \in X$ then the germ $[\sigma_{g,U}(s)]_{gx}$ depends only on s_x and is denoted by $\sigma_{g,x}(s_x)$. It is easily seen that the mappings $\sigma_{g,x} : (\mathcal{O}_{\mathbb{V}})_x \to (\mathcal{O}_{\mathbb{V}})_{gx}$ define a G-action on $\mathcal{O}_{\mathbb{V}}$ and that $\mathcal{O}_{\mathbb{V}}$ is a G-sheaf of modules over $\mathcal{O}_X$. Furthermore,

$\mathcal{O}_{\mathbb{V}}$ is a locally free analytic G-sheaf under a holomorphic G-action.

Proof The only thing we have to check is the condition (+) of the preceding section. Let $U, V \subset X$ and $W \subset G$ be open subsets such that $W \cdot U \subset V$ and let s be a fixed holomorphic section of $\mathbb{V}$ over V. Then we have to show that the section $s(g, \cdot) \in \Gamma(U, \mathbb{V})$ defined by $x \mapsto g^{-1}s(gx)$ depends holomorphically on $g \in W$. Let $g_0 \in W$ and let $N \subset W$ be a coordinate neighborhood of g_0 isomorphic to a polydisk in local coordinates $t_1, \ldots, t_n$, where $t_i(g_0) = 0$. Pick an arbitrary point $x_0 \in U$ and denote by U' a sufficiently small neighborhood of x_0 in U. As we have seen in §4.1, $\mathcal{O}_X$ is a G-sheaf under a holomorphic G-action. Applying the same argument now, we reduce our assertion to the following one: *the mapping* $g \mapsto s(g, \cdot)|_{U'} \in \Gamma(U', \mathbb{V})$ *is holomorphic at* g_0. In order to show this we may assume that $\mathbb{V}$ is trivial over U'. Choose a local frame $s_1, \ldots, s_m \in \Gamma(U', \mathbb{V})$, so that $s_1(x), \ldots, s_m(x)$ form a basis of the fiber $p^{-1}(x)$ for every $x \in U'$. Then

$$s(g, x) = f_1(g, x) \cdot s_1(x) + \ldots + f_m(g, x) \cdot s_m(x),$$

where $f_i \in \mathcal{O}(N \times U')$. Each f_i is the sum of a power series in $t_1, \ldots, t_n$ with coefficients in $\mathcal{O}(U')$, which is absolutely convergent with respect to any continuous seminorm on $\mathcal{O}(U')$. Therefore the mapping $N \to \Gamma(U', \mathbb{V})$, $g \mapsto s(g,.)|_{U'}$, is holomorphic. □

In what follows we shall be mainly concerned with *homogeneous holomorphic vector bundles*, i.e., holomorphic vector G-bundles over homogeneous manifolds of the form G/H, where $H \subset G$ is a closed complex Lie subgroup. If $\mathbb{V}$ is such a bundle over $X = G/H$ then there is a natural holomorphic linear representation of H on the fiber $V := p^{-1}(e \cdot H)$. Conversely, for a given holomorphic representation $\varphi : H \to \mathrm{GL}(V)$ it is easy to construct a homogeneous holomorphic vector bundle. Namely, consider the fiber product $\mathbb{V} = G \times_H V$, i.e., the quotient space of the direct product $G \times V$ by the action of H

$$(g, v) \overset{h}{\mapsto} (gh^{-1}, \varphi(h)v) \quad (g \in G, h \in H, v \in V),$$

and define the projection map by $p([g,v]) := g \cdot H$, where $[g,v]$ is the equivalence class of (g,v) in $G \times_H V$. The group G acts on $\mathbb{V}$ by $g \cdot [x,v] = [gx,v]$, where $g, x \in G, v \in V$. A direct verification shows that $\mathbb{V}$ is a homogeneous holomorphic vector bundle over G/H such that the representation of H on the fiber over $e \cdot H$ coincides with the given one. Thus we obtain the one-to-one correspondence

$$\left\{ \begin{array}{c} \text{holomorphic homogeneous} \\ \text{vector bundles over } G/H \end{array} \right\} \longleftrightarrow \left\{ \begin{array}{c} \text{holomorphic representations} \\ \text{of the subgroup } H \end{array} \right\}.$$

This correspondence has various natural properties. Namely, let $\mathbb{V}^{\varphi}$ be the homogeneous vector bundle corresponding to the representation $\varphi : H \to \mathrm{GL}(V)$. Then $\mathbb{V}^{\varphi_1 \oplus \varphi_2} = \mathbb{V}^{\varphi_1} \bigoplus \mathbb{V}^{\varphi_2}$, $\mathbb{V}^{\varphi_1 \otimes \varphi_2} = \mathbb{V}^{\varphi_1} \bigotimes \mathbb{V}^{\varphi_2}$, $\mathbb{V}^{\wedge^n \varphi} = \bigwedge^n \mathbb{V}^{\varphi}$, and $\mathbb{V}^{\varphi^*} = (\mathbb{V}^{\varphi})^*$. Here we use the standard notation for direct sum, tensor product, exterior power, and dual representation, and also the analogous notation for vector bundles.

Example 1 Let $\mathbb{T}$ be the holomorphic tangent bundle over G/H and let $\mathbb{K} = \wedge^n \mathbb{T}^*$ be the canonical bundle (here $n = \dim G/H$). The holomorphic tangent space at the point $e \cdot H \in G/H$ is identified with the quotient space $\mathfrak{g}/\mathfrak{h}$. The representation $\theta : H \to \mathrm{GL}(\mathfrak{g}/\mathfrak{h})$ defined by

$$\theta(h)(\mathrm{Y} + \mathfrak{h}) = \mathrm{Ad}_G(h)(\mathrm{Y}) + \mathfrak{h} \qquad (h \in H,\ \mathrm{Y} \in \mathfrak{g}),$$

is identified with the isotropy representation. Thus $\mathbb{T}$ is a homogeneous vector bundle corresponding to $\theta : H \to \mathrm{GL}(\mathfrak{g}/\mathfrak{h})$. It follows that $\wedge^n \mathbb{T}$ and $\mathbb{K} = \wedge^n \mathbb{T}^*$ are homogeneous line bundles associated with the characters $\det \theta$ and $\delta := (\det \theta)^{-1}$ respectively. We note that

$$\delta(h) = \det \mathrm{Ad}_H(h) \cdot (\det \mathrm{Ad}_G(h))^{-1} \quad (h \in H).$$

Assume now that $X = G/H$ is compact and let $\mathbb{V}$ be a holomorphic homogeneous vector bundle over X corresponding to a holomorphic representation $\varphi : H \to \mathrm{GL}(V)$. As we have seen in the preceding section, each *induced representation*

$$\mathrm{I}^q \varphi : G \to \mathrm{GL}(H^q(X, \mathcal{O}_{\mathbb{V}})), \quad q \geq 0,$$

is holomorphic.

For $q = 0$ the induced representation has a simple realization in a subspace of $\mathrm{Hol}(G, V)$. Namely, to each holomorphic section $s \in \Gamma(X, \mathbb{V})$ there corresponds a holomorphic mapping $f_s : G \to V$ defined by $f(g) = f_s(g) = g^{-1} \cdot s(gH)$. It is easy to see that

$$f(gh) = \varphi^{-1}(h)f(g) \qquad (*)$$

for all $g \in G, h \in H$. Conversely, given $f \in \mathrm{Hol}(G, V)$ satisfying $(*)$ one can define a section $s = s_f \in \Gamma(X, \mathbb{V})$ by $s(gH) = [gH, f(g)]$. The mappings $f \mapsto s_f$ and $s \mapsto f_s$ are inverse to each other. Quite often we shall identify $\Gamma(X, \mathbb{V})$ with the subspace of $\mathrm{Hol}(G, V)$ defined by $(*)$. The induced representation of G on this space is given by

$$(\mathrm{I}^0\varphi(g) \cdot f)(x) = f(g^{-1}x) \quad (x, g \in G).$$

Main Problem *For a given holomorphic representation* $\varphi : H \to \mathrm{GL}(V)$ *determine the induced representations* $\mathrm{I}^q\varphi : G \to \mathrm{GL}(H^q(X, \mathcal{O}_{\mathbb{V}}))$, $q \geq 0$.

The main result of this chapter (see §4.3) gives a solution of this problem under the following assumptions: (a) G is semisimple; (b) H is parabolic; (c) φ is irreducible. At the same time one obtains a geometric realization for the irreducible finite-dimensional representations of semisimple complex Lie groups.

Example 2 We calculate here the induced representations of $G = \mathrm{SL}(m+1, \mathbb{C})$ on the cohomology spaces of invertible sheaves over $\mathbb{P}_m$. Let P be the parabolic subgroup of G consisting of the matrices

$$p = \left(\begin{array}{ccc|c} & & & 0 \\ & A(p) & & \vdots \\ & & & 0 \\ \hline * & \cdots & * & a(p) \end{array}\right)$$

and let $\varphi_n : P \to \mathbb{C}^*$ be the character defined by $\varphi_n(p) := a(p)^{-n}$. Denote by $\mathbb{L}$ the homogeneous line bundle corresponding to φ_1. The n-th power of $\mathbb{L}$ is then associated with φ_n. Any line bundle over $\mathbb{P}_m = G/P$ is homogeneous with respect to G and is isomorphic to one of $\mathbb{L}^n$. Note that

$$\delta(p) = \frac{a(p)^m}{\det A(p)} = a(p)^{m+1},$$

and so we obtain $\mathbb{K} = \mathbb{L}^{-m-1}$ (see Example 1 for the definition of δ and $\mathbb{K}$). The usual notation for $\mathcal{O}_{\mathbb{L}^n}$ is $\mathcal{O}(n)$. The global sections of this sheaf are identified with the holomorphic functions on $\mathbb{C}^{m+1} - \{0\}$ satisfying

$$f(tz) = t^n f(z) \qquad (z \in \mathbb{C}^{m+1}, t \in \mathbb{C}^*).$$

Using Hartogs' Theorem, we conclude that f is a homogeneous polynomial of degree n for $n \geq 0$ and $f = 0$ for $n < 0$. Thus

$$H^0(\mathbb{P}_m, \mathcal{O}(n)) = \begin{cases} W_n & \text{if } n \geq 0 \\ 0 & \text{if } n < 0, \end{cases}$$

where W_n is the space of homogeneous polynomials of degree n on $\mathbb{C}^{m+1}$ with the natural action of $\mathrm{SL}(m+1,\mathbb{C})$ and the equality is understood as the equivalence of representations.

According to a theorem of J.-P. Serre [Se2], $H^k(\mathbb{P}_m, \mathcal{O}(n)) = 0$ if $0 < k < m$. Thus the only non-trivial induced representations are $\mathrm{I}^0\varphi_n$ and $\mathrm{I}^m\varphi_n$. In order to determine the latter one, we can use the Serre duality theorem. Namely, if $\mathbb{V}$ is a holomorphic vector G-bundle over $\mathbb{P}_m$ then $H^p(\mathbb{P}_m, \mathcal{O}_{\mathbb{V}})$ and $H^{m-p}(\mathbb{P}_m, \mathcal{O}_{\mathbb{V}^*\otimes\mathbb{K}})$ are dual G-modules. Since $(\mathbb{L}^n)^* \otimes \mathbb{K} = \mathbb{L}^{-(m+n+1)}$, we obtain that

$$H^m(\mathbb{P}_m, \mathcal{O}(n)) = H^0(\mathbb{P}_m, \mathcal{O}(-m-n-1)) = \begin{cases} W_{-m-n-1} & \text{if } m+n+1 \leq 0 \\ 0 & \text{if } m+n+1 > 0. \end{cases}$$

In particular, if $-(m+1) < n < 0$ then $H^k(\mathbb{P}_m, \mathcal{O}(n)) = 0$ for all k.

4.3 Theorem of R.Bott. Proof of the Borel-Weil theorem

We shall use freely the notation introduced in § 3.1. Let G be a connected semisimple complex Lie group, $T \subset G$ a maximal torus, $\mathfrak{t} \subset \mathfrak{g}$ the corresponding Cartan subalgebra, and Δ the root system of $\mathfrak{g}$ with respect to $\mathfrak{t}$. We note that in the space of an irreducible finite-dimensional representation of an algebraic group the unipotent radical of this group acts trivially. Therefore describing the irreducible representations of the whole group reduces to describing the irreducible representations of its reductive part. In particular, this applies to a parabolic subgroup $P \subset G$. It is convenient to consider parabolic subgroups which contain B^-. However, the notion of a *highest weight*, which we employ in the sequel, is defined with respect to B^+. Thus, a highest weight vector of a representation of G (resp. P) is an eigenvector of the subgroup B^+ (resp. $P \cap B^+$).

The weight lattice $\mathcal{P} \subset \mathfrak{t}^*$ and the subset $\mathcal{P}^+ \subset \mathcal{P}$ have been introduced in §3.8. We shall also need the subset $\mathcal{P}^{++} \subset \mathcal{P}^+$ defined by the strict inequalities $\lambda(\mathrm{H}_\alpha) > 0$, where $\alpha \in \Delta^+$. A linear form $\mu \in \mathcal{P}$ is called *singular* if $\mu(\mathrm{H}_\alpha) = 0$ for some $\alpha \in \Delta$, and *regular* otherwise.

The Weyl group is denoted by $\mathcal{W}$. Recall that $\mathcal{W}$ is generated by reflections s_α, where α is a simple root. The *length* of an element $w \in \mathcal{W}$ is the minimal number $l = l(w)$ such that w is the product of l reflections corresponding to simple roots. The length of an element $w \in \mathcal{W}$ is equal to the number of $\alpha \in \Delta^+$ such that $w(\alpha) \in \Delta^-$ (see [Bou], Ch. 8, §1 and also the remark in §4.5). If $\mu \in \mathcal{P}$ is regular then there exists a unique element $w = w_\mu \in \mathcal{W}$ such that $w(\mu) \in \mathcal{P}^{++}$. The length of w_μ is equal to the number of roots $\alpha \in \Delta^+$ for which $\mu(\mathrm{H}_\alpha) < 0$. As usual, the half-sum of the positive roots is denoted by ϱ. Recall that $\varrho(\mathrm{H}_\alpha) = 1$ for all simple roots α. Thus $\mu \in \mathcal{P}^{++}$ implies $\mu - \varrho \in \mathcal{P}^+$.

Theorem (R.Bott [B]) *Let $X = G/P$ be a flag manifold, $\mathbb{V} = \mathbb{V}^\varphi$ the homogeneous holomorphic vector bundle over X, associated with an irreducible holomorphic representation $\varphi : P \to \mathrm{GL}(V)$, and $\lambda \in \mathcal{P}$ the highest weight of φ. If the form $\lambda + \varrho$ is singular, then $H^k(X, \mathcal{O}_{\mathbb{V}}) = 0$ for all k. If the form $\lambda + \varrho$ is regular, let l denote the length of $w = w_{\lambda+\varrho} \in \mathcal{W}$. Then $H^k(X, \mathcal{O}_{\mathbb{V}}) = 0$ for $k \neq l$ and $H^l(X, \mathcal{O}_{\mathbb{V}})$ is an irreducible G-module with highest weight $\Lambda = w(\lambda + \varrho) - \varrho$.*

In addition to the original proof in [B], there are other proofs due to B.Kostant [Kt], F. Aribaud [Ar], and M.Demazure [De1], [De2]. The latter is reproduced in

§4.5. We begin with a special case of the above theorem, which plays an important role in its proof. Let $\mathrm{X}(L)$ be the group of holomorphic characters of a complex Lie group L. For $\varphi \in \mathrm{X}(L)$ we denote by $\dot{\varphi}$ the differential of φ at $e \in L$.

Theorem (A.Borel - A.Weil (1954), see [Sel]) *Let $\varphi \in \mathrm{X}(B^-)$ and let $\mathbb{L}^\varphi$ be the homogeneous line bundle over G/B^- associated with φ. If there exists $\lambda \in \mathcal{P}^+$ such that $\lambda = \dot{\varphi}|_{\mathfrak{t}}$ then $H^0(G/B^-, \mathcal{O}_{\mathbb{L}^\varphi})$ is an irreducible G-module with highest weight λ. Otherwise $H^0(G/B^-, \mathcal{O}_{\mathbb{L}^\varphi}) = 0$.*

Proof In §4.2 we have identified the space of holomorphic sections of a homogeneous holomorphic vector bundle $\mathbb{V}$ with a certain subspace of $\mathrm{Hol}(G, V)$. Thus, $\Gamma(G/B^-, \mathbb{L}^\varphi)$ is interpreted as a subspace of $\mathcal{O}(G)$ consisting of all functions f, such that $f(gb) = \varphi(b)^{-1} f(g)$ for all $g \in G, b \in B^-$. Assume $\Gamma(G/B^-, \mathbb{L}^\varphi) \neq 0$ and let $f_0 \in \Gamma(G/B^-, \mathbb{L}^\varphi)$ be a highest weight vector, i.e.,

$$f_0(b^{-1}g) = \chi(b) f_0(g), \qquad g \in G,\ b \in B^+,$$

where $\chi \in \mathrm{X}(B^+)$. Recall that U^+ (resp. U^-) is the unipotent radical of B^+ (resp. B^-). It is clear that $f_0(e) \neq 0$, for otherwise f_0 equals 0 on the open set $U^+TU^- \subset G$. Thus we may assume that $f_0(e) = 1$. Denote by u^+, t, and u^- arbitrary elements of U^+, T, and U^- respectively. Then

$$f_0(u^+tu^-) = f_0(tu^-) = f_0(e \cdot tu^-) = \varphi(t)^{-1} f_0(e) = \varphi(t)^{-1}$$

and, on the other hand,

$$f_0(u^+tu^-) = f_0(u^+t) = f_0(u^+t \cdot e) = \chi(t)^{-1} f_0(e) = \chi(t)^{-1}.$$

Thus $\chi|_T = \varphi|_T$ and $\dot{\chi}|_{\mathfrak{t}} = \dot{\varphi}|_{\mathfrak{t}}$. Since $\dot{\chi}|_{\mathfrak{t}} \in \mathcal{P}^+$, we obtain the second assertion.

In order to prove the first one, denote by V an irreducible G-module with highest weight $\lambda = \dot{\varphi}|_{\mathfrak{t}}$. Let V^* be the dual G-module, $\langle \cdot, \cdot \rangle : V^* \times V \to \mathbb{C}$ the canonical pairing, v_0 a highest weight vector in V, and ν_0 an eigenvector of B^- in V^*. Assume without loss of generality that $\langle \nu_0, v_0 \rangle = 1$ and note that $b\nu_0 = \varphi(b)^{-1}\nu_0$, where $b \in B^-$. For any $v \in V$ define a holomorphic function on G by

$$f_v(g) := \langle \nu_0, g^{-1}v \rangle = \langle g\nu_0, v \rangle.$$

One checks easily that $f_v \in \Gamma(G/B^-, \mathbb{L}^\varphi)$ and that the mapping $v \mapsto f_v$ is a homomorphism of G-modules. Since the orbit $G\nu_0$ generates V^* as a vector space, this homomorphism is injective.

We claim that $f_0 = f_{v_0}$. It suffices to prove the equality $f_0(g) = f_{v_0}(g)$ for $g = u^+tu^-$, where $u^+ \in U^+, t \in T, u^- \in U^-$. But

$$f_{v_0}(u^+tu^-) = \langle u^+tu^-\nu_0, v_0 \rangle = \langle tu^-\nu_0, (u^+)^{-1}v_0 \rangle =$$

$$= \varphi(t)^{-1} \langle \nu_0, v_0 \rangle = \varphi(t)^{-1} = f_0(u^+tu^-).$$

It follows that, up to multiplication by a non-zero number, f_{v_0} is the only highest weight vector in $\Gamma(G/B^-, \mathbb{L}^\varphi)$. Therefore $\Gamma(G/B^-, \mathbb{L}^\varphi)$ is an irreducible G-module isomorphic to V. □

Let B be any Borel subgroup of G containing T and let $U = B'$ be the corresponding maximal unipotent subgroup. Since $B = T \cdot U$, a character $\varphi \in \mathrm{X}(T)$ has a unique extension to B, denoted by the same letter, such that $\varphi_U = 1$. Let

$$\mathcal{P}_G := \{\lambda \in \mathcal{P} \mid \lambda = \dot{\varphi} \quad \text{for some} \quad \varphi \in \mathrm{X}(T)\}, \quad \mathcal{P}_G^+ := \mathcal{P}_G \cap \mathcal{P}^+.$$

The equivalence classes of irreducible holomorphic finite-dimensional G-modules are in one-to-one correspondence with elements of $\mathcal{P}_G^+$. Namely, if λ is the highest weight of such a G-module then $\lambda \in \mathcal{P}_G^+$.

Observe that G/U is a holomorphic principal T-bundle over G/B with the bundle projection given by $g \cdot U \mapsto g \cdot B$. We identify holomorphic functions on G/U with holomorphic functions on G, which are right invariant under U. As an immediate application of the Borel-Weil theorem, we want to prove the following interesting properties of the induced representation of G on $\mathcal{O}(G/U)$.

Proposition (a) *Each irreducible holomorphic finite-dimensional G-module occurs in $\mathcal{O}(G/U)$ exactly once.*

(b) *For $\lambda \in \mathcal{P}_G^+$ let V^λ denote the irreducible submodule of $\mathcal{O}(G/U)$ with highest weight λ. Then, for any $f \in \mathcal{O}(X)$, there is a uniquely defined decomposition*

$$f = \sum_{\lambda \in \mathcal{P}_G^+} f_\lambda,$$

where $f_\lambda \in V^\lambda$ and the series converges absolutely and uniformly on compact sets in G/U.

Proof It is convenient to assume that $B = B^-$ and $U = U^-$. For any $\lambda \in \mathcal{P}_G$ define a subspace $V^\lambda \subset \mathcal{O}(G/U)$ by

$$V^\lambda := \{f \in \mathcal{O}(G) \mid f(gb) = \varphi(b)^{-1} f(g)\}, \quad \lambda = \dot{\varphi},$$

where $\varphi \in \mathrm{X}(T)$ is extended to B as above. It is then clear that V^λ is a G-submodule of $\mathcal{O}(G/U)$ and that $V^\lambda \simeq H^0(G/B, \mathcal{O}_{\mathbb{L}^\varphi})$ as G-modules. For $\lambda \in \mathcal{P}^+$ the Borel-Weil theorem tells us that V^λ is an irreducible G-module with highest weight λ. Since λ is an arbitrary element of $\mathcal{P}_G^+$, it follows that each irreducible G-module occurs in $\mathcal{O}(G/U)$ at least once. On the other hand, let $f_1, f_2 \in \mathcal{O}(G/U)$ be two eigenvectors of B^+ with the same weight. Then the ratio f_1/f_2 is a B^+-invariant meromorphic function on G/U. Since the B^+-orbit of $e \cdot U \in G/U$ is open, it follows that f_1/f_2 is a constant. Therefore each irreducible G-module occurs in $\mathcal{O}(G/U)$ at most once, and we obtain (a).

For $f \in \mathcal{O}(G/U), \varphi \in \mathrm{X}(T)$, and $\lambda = \dot{\varphi}$ let

$$f_\lambda(x) = \int_{T_c} f(xt)\varphi(t)d\mu(t) \qquad (x \in G),$$

where T_c is the (unique) maximal compact subgroup in T and μ is the normalized Haar measure on T_c. One checks easily that $f_\lambda \in V^\lambda$. Applying again the Borel-Weil theorem, we see that $V^\lambda \neq 0$ only if $\lambda \in \mathcal{P}_G^+$. Thus $f_\lambda \neq 0$ only if $\lambda \in \mathcal{P}_G^+$.

We still have to prove the convergence and the uniqueness of the decomposition in (b). For this consider a more general setting. Namely, let X be an arbitrary holomorphic principal T-bundle over a complex manifold Y. The action of the structure group T on X is written as right multiplication. For $f \in \mathcal{O}(X)$ define f_λ by the above formula. Since the convergence is a local fact along the base, we may assume that the bundle is trivial, i.e., the total space is of the form $X = Y \times (\mathbb{C}^*)^r$. Then

$$f(y,z) = \sum f_\lambda(y,z), \quad \text{where} \quad \lambda = \dot{\varphi}, \ \varphi \in \mathrm{X}(T),$$

is a Laurent series in $z \in (\mathbb{C}^*)^r$ with coefficients depending on $y \in Y$, so that the convergence and the uniqueness of the decomposition are classical facts. □

Remark 1 For a generalization of Bott's theorem to non-algebraic homogeneous compact manifolds see [Gr], [A1].

Remark 2 It was already noted that Bott's theorem gives geometric realizations of the irreducible finite-dimensional representations of semisimple complex Lie groups, or equivalently, realizations of irreducible unitary representations of compact semisimple Lie groups. Now, let G be a connected real form of a semisimple complex Lie group $G_{\mathbb{C}}$ and let $B \subset G_{\mathbb{C}}$ be a Borel subgroup. Then G has finitely many orbits on $G_{\mathbb{C}}/B$, so that at least one orbit is open (see [W2]). We call an open G-orbit on $G_{\mathbb{C}}/B$ a *flag domain* (of G). Assume that a maximal compact subgroup of G has the same rank as the whole group G. According to a theorem of Harish-Chandra, this condition is necessary and sufficient for the group G to have unitary representations, whose matrix elements belong to $L^2(G)$. An irreducible unitary representation of this type is called a representation of *discrete series*. R.Langlands (1966) conjectured that the representations of discrete series are realized in the properly defined L^2-cohomology spaces of flag domains with coefficients in invertible $G_{\mathbb{C}}$-sheaves on $G_{\mathbb{C}}/B$. For a precise definition of the cohomology spaces in question see [GrSch]. The conjecture of R.Langlands was proved in final form by W.Schmid [Sch1], [Sch2]. A related result for Hermitian symmetric spaces was obtained in [NaOk].

4.4 Application of the Leray spectral sequence

Let (X, Y, π) be a locally trivial holomorphic fiber bundle, where the total space X and the base Y are complex manifolds. Denote by X_y the fiber over $y \in Y$ and assume that X_y is connected and compact. Let G be a connected complex Lie group acting holomorphically on X. By Lemma 2 of §2.4 there is an induced holomorphic G-action on Y so that π is G-equivariant. In what follows we assume that the induced G-action on Y is transitive. As usual, G_y is the isotropy subgroup of G at $y \in Y$.

Let $\mathbb{F}$ be a holomorphic vector G-bundle on X, $\mathbb{F}_y$ the restriction of $\mathbb{F}$ to X_y, and $\mathcal{F} := \mathcal{O}_{\mathbb{F}}$ (resp. $\mathcal{F}_y := \mathcal{O}_{\mathbb{F}_y}$) the corresponding locally free sheaves. Then $\mathbb{F}_y$ is a holomorphic G_y-bundle and $\mathcal{F}_y$ is a G_y-sheaf under a holomorphic G_y-action. Thus we obtain holomorphic representations $G_y \to \mathrm{GL}(H^k(X_y, \mathcal{F}_y))$. Let $\mathbb{H}^k$ be the associated homogeneous vector bundles on Y and let $\mathcal{H}^k := \mathcal{O}_{\mathbb{H}^k}$ be the corresponding sheaves.

An element $g \in G$ interchanges the fibers X_y and induces the sheaf isomorphisms $\mathcal{F}_y \to \mathcal{F}_{gy}$. Therefore $\dim H^k(X_y, \mathcal{F}_y)$ is independent of $y \in Y$. It is then known that all direct images $\pi_{(k)}(\mathcal{F})$ are locally free and that the natural maps

$$\pi_{(k)}(\mathcal{F})/\mathfrak{m}_y\pi_{(k)}(\mathcal{F}) \to H^k(X_y, \mathcal{F}_y)$$

are isomorphisms (see [GR3], Ch. 10, §5.5). Since these isomorphisms commute with the action of G_y, it follows that $\pi_{(k)}(\mathcal{F}) \simeq \mathcal{H}^k$ for all k, $k \geq 0$.

The foolowing result is a special case of the Leray theorem. The proof can be found in [Go] (see Ch. 2, Théorème 4.17.1).

Theorem *In the above notation there exists a spectral sequence* $\{E_r\}$ *with*

$$E_2^{p,q} = H^p(Y, \mathcal{H}^q),$$

whose final term is associated to $H^*(X, \mathcal{F})$.

Remark All terms of the spectral sequence are G-modules, all differentials are homomorphisms of G-modules, the filtration in $H^k(X, \mathcal{F})$ is G-invariant, and the corresponding graded G-module is isomorphic to $\oplus_{p+q=k} E_\infty^{p,q}$.

The spectral sequence $\{E_r\}$ is called the *Leray spectral sequence.* Returning to the set-up of the preceding section, we want to prove two propositions, which are direct consequences of the above theorem. The first of them reduces the proof of Bott's theorem to the special case of line bundles over G/B.

Proposition 1 *Assume that Bott's theorem is proven for all homogeneous line bundles over* G/B*, where* G *is an arbitrary connected semisimple complex Lie group,* $B \subset G$ *a Borel subgroup.*

Fix such a group G *and let* $P \subset G$ *be a parabolic subgroup containing* $B = B^-$. *Consider a homogeneous vector bundle* $\mathbb{V} = \mathbb{V}^\tau$ *over* $Y = G/P$*, where* $\tau : P \to$ GL(V) *is an irreducible holomorphic representation with highest weight* $\lambda \in \mathcal{P}$. *Denote by* φ *the character of* B *defined by* $\dot{\varphi}|_{\mathfrak{t}} = \lambda$ *and let* $\mathbb{L} = \mathbb{L}^\varphi$ *be the associated homogeneous line bundle over* $X = G/B$. *Then*

$$H^k(X, \mathcal{O}_{\mathbb{L}}) \simeq H^k(Y, \mathcal{O}_{\mathbb{V}}) \qquad \textit{for all } k, \ k \geq 0,$$

as G*-modules, proving Bott's theorem for* $\mathbb{V}$.

Proof Consider the holomorphic fiber bundle (X, Y, π), where $\pi : G/B \to G/P$ is the canonical projection. Let $F := P/B$ and let $\mathbb{L}_F$ be the restriction of $\mathbb{L}$ to F. We want to apply the Leray spectral sequence to the line bundle $\mathbb{L}$ on X. For this we have to calculate the representation of P on $H^k(F, \mathcal{O}_{\mathbb{L}_F})$.

There exists an algebraic torus $C \subset T$, whose centralizer in G is a reductive Levi subgroup of P. Moreover, one can choose C to be the connected center of its centralizer. The latter is then a locally direct product $S \cdot C$, where S is a connected semisimple group. It follows that $P = S \cdot C \cdot U_P$, where U_P is the unipotent radical of P.

Observe that $C \cdot U_P \subset B$ showing that F can be written in the Klein form $F = S/S \cap B$. Applying Bott's theorem to F, we obtain that $H^q(F, \mathcal{O}_{\mathbb{L}_F}) = 0$ for $q > 0$ and that $H^0(F, \mathcal{O}_{\mathbb{L}_F}) \simeq V$ as S-modules .

We claim that $H^0(F, \mathcal{O}_{\mathbb{L}_F}) \simeq V$ as P-modules. Since V is an irreducible S-module, we only have to show that C acts in V and in $H^0(F, \mathcal{O}_{\mathbb{L}_F})$ via the same character. As in §4.2, we can identify the sections of $\mathbb{L}_F$ with holomorphic functions on P satisfying an appropriate functional equation. Namely,

$$H^0(F, \mathcal{O}_{\mathbb{L}_F}) = \Gamma(F, \mathbb{L}_F) = \{f \in \mathcal{O}(P) \mid f(pb) = \varphi(b)^{-1} f(p) \quad (p \in P, b \in B)\}.$$

Let $a \in C$, $f \in \Gamma(F, \mathbb{L}_F)$, and write $p = scu$, where $s \in S$, $c \in C$, $u \in U_P$. Then

$$f(a^{-1}p) = f(a^{-1}scu) = f(a^{-1}sc) = f(sca^{-1}) = \varphi(a) f(sc) = \varphi(a) f(p).$$

This shows that C acts in $H^0(F, \mathcal{O}_{\mathbb{L}_F})$ via φ. On the other hand, it is clear that the representation of C on V is also given by

$$v \overset{a}{\mapsto} \varphi(a) \cdot v \qquad (v \in V, a \in C),$$

since this formula holds if v is a highest weight vector. Therefore the homogeneous vector bundle $\mathbb{H}^q$ is isomorphic to $\mathbb{V}$ for $q = 0$ and is the zero bundle for $q > 0$. Hence

$$E_2^{p,q} = H^p(Y, \mathcal{H}^q) = \begin{cases} H^p(Y, \mathcal{O}_{\mathbb{V}}) & \text{if } q = 0, \\ 0 & \text{if } q > 0. \end{cases}$$

It follows that all differentials in the spectral sequence are trivial and

$$H^k(X, \mathcal{O}_{\mathbb{L}}) \simeq E_\infty^{k,0} \simeq E_2^{k,0} = H^k(Y, \mathcal{O}_{\mathbb{V}}).$$

□

Let α be a simple root. Denote by $P^{(\alpha)}$ the parabolic subgroup of G with Lie algebra

$$\mathfrak{p}^{(\alpha)} = \mathfrak{t} \oplus \mathfrak{g}_\alpha \oplus \oplus_{\beta \in \Delta^-} \mathfrak{g}_\beta.$$

Proposition 2 *Let $\tau : B \to \mathrm{GL}(V)$ be a holomorphic linear representation of $B = B^-$ and let $\varphi : B \to \mathbb{C}^*$ be a holomorphic character. If τ extends to a linear representation of $P^{(\alpha)}$ on V and if $\dot{\varphi}(\mathrm{H}_\alpha) = -1$ then*

$$H^q(G/B, \mathcal{O}_{\mathbb{V} \otimes \mathbb{L}^\varphi}) = 0$$

for all q, $q \geq 0$.

Proof Consider the holomorphic fiber bundle (X, Y, π), where $X = G/B$, $Y = G/P^{(\alpha)}$, and $\pi : G/B \to G/P^{(\alpha)}$ is the canonical projection, and apply the Leray spectral sequence to the vector bundle $\mathbb{V} \otimes \mathbb{L}^\varphi$ on X. It suffices to show that $\mathcal{H}^q = 0$ for all q or, equivalently, that the restriction of $\mathbb{V} \otimes \mathbb{L}^\varphi$ to each fiber of π has trivial cohomology. The fiber is isomorphic to $\mathbb{P}_1$. The restriction of $\mathbb{V} \otimes \mathbb{L}^\varphi$ to the fiber is the tensor product of a trivial vector bundle and a line bundle of degree -1. Since $H^0(\mathbb{P}_1, \mathcal{O}(-1)) = H^1(\mathbb{P}_1, \mathcal{O}(-1)) = 0$, the proof is complete. □

4.5 Proof of the theorem of R.Bott

By Proposition 1 of §4.4 it is enough to prove the theorem for line bundles over G/B, where $B = B^-$. We simplify our previous notation as follows. Given a holomorphic B-module V, we denote by $\mathbb{V}$ the associated homogeneous vector bundle on G/B and write $\mathcal{V}$ for the corresponding locally free sheaf $\mathcal{O}_{\mathbb{V}}$. Without loss of generality we assume that G is simply connected. Under this assumption we can identify $\mathcal{P}$ with the character group $X(T)$ which, in its turn, is canonically isomorphic to $X(B)$. We denote by L_λ the one-dimensional B-module given by $\lambda \in \mathcal{P}$. For each simple root α we have defined a parabolic subgroup $P^{(\alpha)} \subset G$ containing B. Denote by s_α the reflection corresponding to α. Recall that s_α is an element of the Weyl group acting on $\mathfrak{t}^*$ by $s_\alpha(\lambda) = \lambda - \lambda(\mathrm{H}_\alpha)\alpha$.

Lemma 1 *Let $\lambda \in \mathcal{P}$ and suppose that $\lambda(\mathrm{H}_\alpha) \geq 0$ for a simple root α. Then there exists an irreducible linear representation $P^{(\alpha)} \to \mathrm{GL}(V_{\lambda,\alpha})$ such that one has an exact sequence of B-modules*

$$0 \to M \to V_{\lambda,\alpha} \to L_\lambda \to 0, \tag{1}$$

where $M = 0$ if $\lambda(\mathrm{H}_\alpha) = 0$ and $M = L_{s_\alpha(\lambda)}$ if $\lambda(\mathrm{H}_\alpha) = 1$. Moreover, if $\lambda(\mathrm{H}_\alpha) \geq 2$ then there is an exact sequence of B-modules

$$0 \to L_{s_\alpha(\lambda)} \to M \to V_{\lambda-\alpha,\alpha} \to 0. \tag{2}$$

Proof Take $\mathrm{E}_\alpha \in \mathfrak{g}_\alpha$ and $\mathrm{F}_\alpha \in \mathfrak{g}_{-\alpha}$ so that $[\mathrm{E}_\alpha, \mathrm{F}_\alpha] = \mathrm{H}_\alpha$. Let $S^{(\alpha)}$ be the connected three-dimensional Lie subgroup of G with Lie algebra $\mathfrak{s}^{(\alpha)} = \mathbb{C}\cdot \mathrm{E}_\alpha + \mathbb{C}\cdot \mathrm{H}_\alpha + \mathbb{C}\cdot \mathrm{F}_\alpha$. Further, let $C^{(\alpha)}$ be the algebraic torus in G with Lie algebra $\mathfrak{c}^{(\alpha)} = \{\mathrm{H} \in \mathfrak{t} \mid \alpha(\mathrm{H}) = 0\}$. Then $M^{(\alpha)} := C^{(\alpha)} \cdot S^{(\alpha)}$ is a reductive Levi subgroup of $P^{(\alpha)}$. We start by defining $V_{\lambda,\alpha}$ as an irreducible $\mathfrak{s}^{(\alpha)}$-module of dimension $m+1$, where $m = \lambda(\mathrm{H}_\alpha)$. Namely, let

$$V_{\lambda,\alpha} := \mathbb{C}v_m \oplus \mathbb{C}v_{m-2} \oplus \ldots \oplus \mathbb{C}v_{-m}, \qquad m = \lambda(\mathrm{H}_\alpha),$$

where $\mathrm{H}_\alpha v_k = k v_k$, $\mathrm{E}_\alpha(\mathbb{C}v_k) = \mathbb{C}v_{k+2}$, and $\mathrm{F}_\alpha(\mathbb{C}v_k) = \mathbb{C}v_{k-2}$ (we put $v_k = 0$ if $|k| > m$). Since G is simply connected, we have $\exp \pi i \mathrm{H}_\alpha \neq e$, for otherwise H_α can be written as a linear combination of coroots with even coefficients. Thus $S^{(\alpha)}$ is isomorphic to $\mathrm{SL}(2,\mathbb{C})$, and so the representation of $\mathfrak{s}^{(\alpha)}$ on $V_{\lambda,\alpha}$ can be integrated to a representation of $S^{(\alpha)}$. Recall that λ is the differential of a character $\varphi : T \to \mathbb{C}$. Define a representation of $C^{(\alpha)}$ on $V_{\lambda,\alpha}$ by

$$v \overset{a}{\mapsto} \varphi(a) \cdot v \qquad (a \in C^{(\alpha)}, v \in V_{\lambda,\alpha}).$$

Since $\varphi(\exp \pi i \mathrm{H}_\alpha) = e^{\pi i \lambda(\mathrm{H}_\alpha)} = (-1)^m$, it follows that the representations of $S^{(\alpha)}$ and $C^{(\alpha)}$ coincide on the intersection $S^{(\alpha)} \cap C^{(\alpha)}$. Therefore we get a representation $M^{(\alpha)} \to \mathrm{GL}(V_{\lambda,\alpha})$, which extends to a representation of $P^{(\alpha)}$.

Consider $V_{\lambda,\alpha}$ as a B-module and define a B-submodule M by

$$M = \bigoplus_{k=-m}^{m-2} \mathbb{C}v_k.$$

Since $tv_m = \varphi(t) \cdot v_m$ and $\dot{\varphi} = \lambda$, we obtain the exact sequence (1).

Note that if $m = \lambda(\mathrm{H}_\alpha) = 0$ then $M = 0$ by definition. It is easy to see that the weight of v_k is $\lambda - \frac{(m-k)}{2}\alpha$. In particular, in case $m = 1$ we have $M = \mathbb{C}v_{-1} \simeq L_{\lambda-\alpha} = L_{s_\alpha(\lambda)}$.

Finally, assume that $\lambda(\mathrm{H}_\alpha) \geq 2$. Then $\mathbb{C}v_{-m}$ is a B-submodule of M, which is isomorphic to $L_{\lambda-m\alpha} = L_{s_\alpha(\lambda)}$. Since the quotient $M/\mathbb{C}v_{-m}$ is isomorphic to $V_{\lambda-\alpha,\alpha}$, we get the exact sequence (2). □

In what follows it is convenient to complete the definition of the cohomology groups by $H^k(G/B, \mathcal{F}) = 0$ if $k < 0$.

Lemma 2 *Let $\lambda \in \mathcal{P}$ and suppose that $(\lambda + \varrho)(\mathrm{H}_\alpha) \geq 0$ for a simple root α. Then*

$$H^k(G/B, \mathcal{L}_\lambda) \simeq H^{k+1}(G/B, \mathcal{L}_{s_\alpha(\lambda+\varrho)-\varrho}), \qquad k \in \mathbb{Z},$$

as G-modules.

Proof Suppose that $(\lambda + \varrho)(\mathrm{H}_\alpha) \geq 2$. In the remaining cases $(\lambda + \varrho)(\mathrm{H}_\alpha) = 0$ and $(\lambda + \varrho)(\mathrm{H}_\alpha) = 1$ the proof admits an obvious modification.

By Lemma 1 we have the exact sequences of B-modules

$$0 \to M \to V_{\lambda+\varrho,\alpha} \to L_{\lambda+\varrho} \to 0$$

and

$$0 \to L_{s_\alpha(\lambda+\varrho)} \to M \to V_{\lambda+\varrho-\alpha,\alpha} \to 0.$$

Taking the tensor product with $L_{-\varrho}$, we obtain the exact sequences

$$0 \to N \to V_{\lambda+\varrho,\alpha} \otimes L_{-\varrho} \to L_\lambda \to 0$$

and

$$0 \to L_{s_\alpha(\lambda+\varrho)-\varrho} \to N \to V_{\lambda+\varrho-\alpha,\alpha} \otimes L_{-\varrho} \to 0,$$

where $N = M \otimes L_{-\varrho}$. To each of them there corresponds an exact sequence of G-sheaves. Since $\varrho(\mathrm{H}_\alpha) = 1$, we have

$$H^k(G/B, \mathcal{V}_{\lambda+\varrho,\alpha} \otimes \mathcal{L}_{-\varrho}) = H^k(G/B, \mathcal{V}_{\lambda+\varrho-\alpha,\alpha} \otimes \mathcal{L}_{-\varrho}) = 0, \qquad k \in \mathbb{Z},$$

by Proposition 2 of §4.4. Therefore the exact cohomology sequences yield the isomorphisms of G-modules

$$H^k(G/B, \mathcal{L}_\lambda) \simeq H^{k+1}(G/B, \mathcal{N}) \simeq H^{k+1}(G/B, \mathcal{L}_{s_\alpha(\lambda+\varrho)-\varrho}), \qquad k \in \mathbb{Z}. \quad □$$

Lemma 3 *Let $\lambda \in \mathcal{P}$, let w be an element of the Weyl group $\mathcal{W}$, and let l be the length of w. If $\lambda + \varrho \in \mathcal{P}^+$, then $H^k(G/B, \mathcal{L}_\lambda)$ and $H^{k+l}(G/B, \mathcal{L}_{w(\lambda+\varrho)-\varrho})$ are isomorphic as G-modules for all $k \in \mathbb{Z}$.*

Proof (by induction on l) If $l = 1$ then $w = s_\alpha$, where α is a simple root, and we apply Lemma 2. If $l > 1$ then there is a decomposition $w = s_1 \cdot s_2 \cdot \ldots \cdot s_l$, where s_i are the reflections corresponding to simple roots α_i (some of these roots may coincide). Put $\alpha := \alpha_1$ and $w' := s_2 \cdot \ldots \cdot s_l$. Then the length of w' is $l - 1$. We claim that $(w')^{-1}(\alpha)$ is a positive root. To show this write $(w')^{-1}(\alpha) = s_l \cdot \ldots \cdot s_2(\alpha)$ and recall that $s_k(\beta) \in \Delta^+ - \{\alpha_k\}$ for all $\beta \in \Delta^+ - \{\alpha_k\}$. Thus, if $(w')^{-1}(\alpha) \in \Delta^-$ then $s_{k-1} \cdot \ldots \cdot s_2(\alpha) = \alpha_k$ for some k, $2 \leq k \leq l$. This implies $(s_{k-1} \cdot \ldots \cdot s_2)s_1(s_2 \cdot \ldots \cdot s_{k-1}) = s_k$ showing that w can be written as

$w = s_2 \cdot \ldots s_{k-1} s_{k+1} \cdot \ldots \cdot s_l$. Since the length of this decomposition is $l-2$, we get a contradiction.

Put $\lambda' := w'(\lambda+\varrho)-\varrho$. By the induction hypothesis we have the isomorphisms

$$H^k(G/B, \mathcal{L}_\lambda) \simeq H^{k+l-1}(G/B, \mathcal{L}_{\lambda'}).$$

On the other hand, since $(w')^{-1}(\alpha) \in \Delta^+$ and $\lambda + \varrho \in \mathcal{P}^+$, we obtain

$$(\lambda' + \varrho)(\mathrm{H}_\alpha) = w'(\lambda + \varrho)(\mathrm{H}_\alpha) = (\lambda + \varrho)(\mathrm{H}_{(w')^{-1}(\alpha)}) \geq 0.$$

The desired isomorphisms follow now from Lemma 2 applied to λ'. □

Lemma 4 *If $\lambda + \varrho \in \mathcal{P}^+$ then $H^k(G/B, \mathcal{L}_\lambda) = 0$ for $k > 0$.*

Proof Let $w_0 \in \mathcal{W}$ be the element of maximal length. Then $w_0(\Delta^+) = \Delta^-$ and, consequently, $l(w_0) = \dim G/B$. Taking $w = w_0$ in Lemma 3, we obtain the result. □

Proof of the theorem There exists an element $w \in \mathcal{W}$ such that $w(\lambda + \varrho) \in \mathcal{P}^+$. Put $\Lambda := w(\lambda + \varrho) - \varrho$.

Suppose first that $\lambda + \varrho$ is a singular linear form. By Lemma 3 it suffices to show that $H^k(G/B, \mathcal{L}_\Lambda) = 0$ for all $k \in \mathbb{Z}$. Note that $\Lambda + \varrho$ is also singular. Thus there exists a simple root α such that $(\Lambda + \varrho)(\mathrm{H}_\alpha) = 0$. It follows that $\Lambda(\mathrm{H}_\alpha) = -1$, so that $H^k(G/B, \mathcal{L}_\Lambda) = 0$ for all $k \in \mathbb{Z}$ by Proposition 2 of §4.4.

Assume now that $\lambda + \varrho$ is regular. Then the element w is unique ($w = w_{\lambda+\varrho}$). Let l denote the length of w (and of w^{-1}). By Lemma 3 $H^k(G/B, \mathcal{L}_\Lambda) \simeq H^{k+l}(G/B, \mathcal{L}_\lambda)$ for all $k \in \mathbb{Z}$. Therefore the theorem is reduced to the following two assertions:

(a) $H^0(G/B, \mathcal{L}_\Lambda)$ is an irreducible G-module with highest weight Λ;

(b) $H^k(G/B, \mathcal{L}_\Lambda) = 0$ for $k > 0$.

Observe that (a) is the Borel - Weil theorem and that (b) is a consequence of Lemma 4. This completes the proof. □

Remark The proof of Lemma 3 shows also that $l = l(w)$ is equal to the number of $\alpha \in \Delta^+$ such that $w(\alpha) \in \Delta^-$. This is clear for $l = 1$. For $l > 1$ write $w = s_\alpha w'$ as above and assume by induction that $\beta_1, \ldots, \beta_{l-1} \in \Delta^+$, $\beta_i \neq \beta_j$, are all positive roots sent by w' into Δ^-. Since $(w')^{-1}(\alpha) \in \Delta^+$ and, in particular, $w'(\beta_i) \neq \alpha$, it follows that $w(\beta_i) = s_\alpha w'(\beta_i) \in \Delta^-$, $i = 1, \ldots, l-1$. Let $\beta_l := (w')^{-1}(\alpha)$. Then $w(\beta_l) = s_\alpha(\alpha) = -\alpha \in \Delta^-$ and $\beta_l \neq \beta_i$, $i < l$, because $w'(\beta_l) = \alpha \in \Delta^+$, but $w'(\beta_i) \in \Delta^-$. Finally, if $\gamma \in \Delta^+$ has the property that $w(\gamma) \in \Delta^-$, then either $\gamma = \beta_i$, $i < l$, or $w'(\gamma) \in \Delta^+$, and in the latter case one has $w'(\gamma) = \alpha$, so that $\gamma = \beta_l$.

4.6 Invertible sheaves on G/P for P maximal parabolic

As an application of Bott's theorem, we will generalize the result of Example 2 of §2 to arbitrary quotients G/P, where $P \subset G$ is a maximal parabolic subgroup. Since such a subgroup P contains all simple factors of G except one, we assume without loss of generality that G is simple. We also assume that G is simply connected, since it is always possible to replace G by the universal covering group. Although we do not use this fact in the sequel, we remark that every line bundle on G/P is G-homogeneous. In general, this is not the case if G is not simply connected.

Let Π be the base of Δ^+, $\alpha \in \Pi$ a fixed simple root, and ω_α the corresponding *fundamental weight.* Recall that ω_α is a linear form on $\mathfrak{t}$ defined by

$$\omega_\alpha(\mathrm{H}_\beta) = \begin{cases} 1 & \text{if } \beta = \alpha, \\ 0 & \text{if } \beta \in \Pi - \{\alpha\}. \end{cases}$$

Put

$$\Delta_\alpha := \{\gamma \in \Delta \mid \gamma = \sum_{\beta\in\Pi} k_\beta \cdot \beta, \quad k_\alpha \neq 0\},$$

$$\Delta_\alpha^\pm := \Delta_\alpha \cap \Delta^\pm, \quad \Delta'_\alpha := \Delta - \Delta_\alpha, \quad (\Delta'_\alpha)^\pm := \Delta'_\alpha \cap \Delta^\pm,$$

and denote by $P_{(\alpha)}$ the maximal parabolic subgroup of G having the Lie algebra

$$\mathfrak{p}_{(\alpha)} = \mathfrak{t} \oplus \oplus_{\gamma\in(\Delta'_\alpha)^+} \mathfrak{g}_\gamma \oplus \oplus_{\gamma\in\Delta^-} \mathfrak{g}_\gamma.$$

It is convenient to write ϱ in the form

$$\varrho = \varrho_\alpha + \varrho'_\alpha,$$

where

$$\varrho_\alpha := \frac{1}{2} \sum_{\beta\in\Delta_\alpha^+} \beta, \quad \varrho'_\alpha := \frac{1}{2} \sum_{\beta\in(\Delta'_\alpha)^+} \beta,$$

and to define an integer c_α by $c_\alpha := 2\varrho_\alpha(\mathrm{H}_\alpha)$. As $\varrho(\mathrm{H}_\beta) = 1$ for all $\beta \in \Pi$ and, by the same reason, $\varrho'_\alpha(\mathrm{H}_\beta) = 1$ for $\beta \in \Pi - \{\alpha\}$, we have

$$2\varrho_\alpha = c_\alpha\omega_\alpha.$$

Since $\beta(\mathrm{H}_\alpha) \leq 0$ for $\beta \in \Pi - \{\alpha\}$, we obtain that

$$c_\alpha = 2(\varrho(\mathrm{H}_\alpha) - \varrho'_\alpha(\mathrm{H}_\alpha)) = 2 - 2\varrho'_\alpha(\mathrm{H}_\alpha) \geq 2.$$

Since G is simply connected, there exists a character $\varphi_\alpha \in \mathrm{X}(T)$ with $\dot{\varphi}_\alpha = \omega_\alpha$. Obviously, φ_α has a unique extension to $P_{(\alpha)}$, which we denote by the same letter. Then $\mathrm{X}(P_{(\alpha)})$ is a cyclic group having φ_α as a generator. Let $\mathbb{L}$ be the homogeneous line bundle on $G/P_{(\alpha)}$ associated with φ_α. As in the case of $\mathbb{P}_m$, we write $\mathcal{O}(n)$ for $\mathcal{O}_{\mathbb{L}^n}$.

The tangent space to $G/P_{(\alpha)}$ at the point $o = e \cdot P_{(\alpha)}$ is identified with $\mathfrak{g}/\mathfrak{p}_{(\alpha)}$. As a T-module this quotient space is isomorphic to the sum of the root spaces $\mathfrak{g}_\beta$, where $\beta \in \Delta_\alpha^+$. Letting $m = m_\alpha := \dim G/P_{(\alpha)} = \operatorname{card} \Delta_\alpha^+$, we see that $\bigwedge^m \mathbb{T}$ (resp. $\mathbb{K} = (\bigwedge^m \mathbb{T})^*$) is isomorphic to $\mathbb{L}^{c_\alpha}$ (resp. $\mathbb{L}^{-c_\alpha}$).

Denote by $\mathcal{W}_\alpha$ the subgroup of the Weyl group $\mathcal{W}$, generated by the reflections s_β, where $\beta \in \Pi - \{\alpha\}$. Let w_0 (resp. $w_{0,\alpha}$) be the element of $\mathcal{W}$ (resp. $\mathcal{W}_\alpha$) having the maximal length.

Lemma *With the above notation one has:*

(i) $w_{0,\alpha}(\varrho'_\alpha) = -\varrho'_\alpha$;

(ii) $w_{0,\alpha}(\varrho_\alpha) = \varrho_\alpha$; $w_{0,\alpha}(\omega_\alpha) = \omega_\alpha$;
(iii) $l(ww_\alpha) = m_\alpha$.

Proof The element w_0 has the property that $w_0(\Delta^+) = \Delta^-$ Similarly,

$$w_{0,\alpha}((\Delta'_\alpha)^+) = (\Delta'_\alpha)^-.$$

This implies (i). On the other hand, if

$$\gamma = k_\alpha \cdot \alpha + \sum_{\beta \in \Pi - \{\alpha\}} k_\beta \cdot \beta \in \Delta_\alpha^+, \qquad k_\alpha > 0,$$

then

$$w_{0,\alpha}(\gamma) = k_\alpha \cdot \alpha + \sum_{\beta \in \Pi - \{\alpha\}} l_\beta \cdot \beta.$$

Hence $w_{0,\alpha}(\Delta_\alpha^+) = \Delta_\alpha^+$, and (ii) follows. Finally,

$$w_0 w_{0,\alpha}(\Delta_\alpha^+) = w_0(\Delta_\alpha^+) \subset \Delta^-$$

and

$$w_0 w_{0,\alpha}((\Delta'_\alpha)^+) = w_0((\Delta'_\alpha)^-) \subset \Delta^+,$$

showing (iii). □

Denote by V^λ an irreducible G-module with highest weight $\lambda \in \mathcal{P}^+$. Observe that, together with ω_α, the form $-w_0(\omega_\alpha)$ is also a fundamental weight.

Theorem (a) *If $n \geq 0$ then*

$$H^0(G/P_{(\alpha)}, \mathcal{O}(n)) \simeq V^{n\omega_\alpha}$$

as G-modules and

$$H^k(G/P_{(\alpha)}, \mathcal{O}(n)) = 0 \qquad \textit{for } k > 0.$$

(b) *If $n + c_\alpha \leq 0$ then*

$$H^{m_\alpha}(G/P_{(\alpha)}, \mathcal{O}(n)) \simeq V^{(n+c_\alpha)w_0(\omega_\alpha)}$$

as G-modules and

$$H^k(G/P_{(\alpha)}, \mathcal{O}(n)) = 0 \qquad \textit{for } k < m_\alpha.$$

(c) *If $-c_\alpha < n < 0$ then $H^k(G/P_{(\alpha)}, \mathcal{O}(n)) = 0$ for all k.*

Proof (a) is a direct consequence of Bott's theorem. In order to prove (b), write

$$w_0 w_{0,\alpha}(n\omega_\alpha + \varrho) = nw_0(\omega_\alpha) + w_0(\varrho_\alpha - \varrho'_\alpha) =$$

$$= nw_0(\omega_\alpha) + 2w_0(\varrho_\alpha) - w_0(\varrho) = (n + c_\alpha)w_0(\omega_\alpha) + \varrho,$$

where we used (i) and (ii) of the above lemma. This shows that the element $w_0 w_{0,\alpha} \in \mathcal{W}$ brings the form $n\omega_\alpha + \varrho$ into $\mathcal{P}^{++}$. Therefore (b) follows from (iii) of the lemma and from Bott's theorem. We still have to prove that in case (c) the form $\mu := n\omega_\alpha + \varrho$ is singular. Observe that

$$\mu(\mathrm{H}_\alpha) = n + 1 \leq 0$$

and

$$\mu(\mathrm{H}_{w_{0,\alpha}(\alpha)}) = w_{0,\alpha}^{-1}(\mu)(\mathrm{H}_\alpha) = (n\omega_\alpha + \varrho_\alpha - \varrho'_\alpha)(\mathrm{H}_\alpha) = n + c_\alpha - 1 \geq 0,$$

by (i) and (ii) of the lemma.

Assume $\mu(\mathrm{H}_\gamma) \neq 0$ for all $\gamma \in \Delta$. Then, in particular,

$$\mu(\mathrm{H}_\alpha) < 0, \qquad \mu(\mathrm{H}_{w_{0,\alpha}(\alpha)}) > 0.$$

A coroot H_γ is said to be indecomposable if $\mu(\mathrm{H}_\gamma) > 0$ and there are no such $\delta, \epsilon \in \Delta$ that $\mathrm{H}_\gamma = \mathrm{H}_\delta + \mathrm{H}_\epsilon$, $\mu(\mathrm{H}_\delta) > 0$, and $\mu(\mathrm{H}_\epsilon) > 0$. The set of all indecomposable coroots is a base of the dual root system (see e.g. [Se4], Ch. 5, §8). Since $\mu(\mathrm{H}_\beta) = 1$ for $\beta \in \Pi - \{\alpha\}$, all such H_β are indecomposable. Thus, there exists exactly one indecomposable coroot H_γ, where $\gamma \in \Pi - \{\alpha\}$. Consider two possible cases: $\gamma \in \Delta^+$ and $\gamma \in \Delta^-$.

1) $\gamma \in \Delta^+$. Then

$$\mathrm{H}_\gamma = k \cdot \mathrm{H}_\alpha + \sum_{\beta \in \Pi - \{\alpha\}} k_\beta \cdot \mathrm{H}_\beta, \qquad k > 0,\ k_\beta \geq 0. \tag{1}$$

On the other hand, since $\mu(\mathrm{H}_\alpha) < 0$, we have

$$-\mathrm{H}_\alpha = \mathrm{H}_{-\alpha} = l \cdot \mathrm{H}_\gamma + \sum_{\beta \in \Pi - \{\alpha\}} l_\beta \cdot \mathrm{H}_\beta\,, \qquad l \geq 0,\ l_\beta \geq 0. \tag{1'}$$

Substituting (1) into (1′), we see that $lk = -1$. This is a contradiction.

2) $\gamma \in \Delta^-$. Then

$$\mathrm{H}_\gamma = k \cdot \mathrm{H}_\alpha + \sum_{\beta \in \Pi - \{\alpha\}} k_\beta \cdot \mathrm{H}_\beta, \qquad k < 0,\ k_\beta \leq 0. \tag{2}$$

On the other hand, since $\mu(\mathrm{H}_{w_{0,\alpha}(\alpha)}) > 0$, we have

$$\mathrm{H}_{w_{0,\alpha}(\alpha)} = l \cdot \mathrm{H}_\gamma + \sum_{\beta \in \Pi - \{\alpha\}} l_\beta \cdot \mathrm{H}_\beta, \qquad l \geq 0,\ l_\beta \geq 0. \tag{2'}$$

Substituting (2) into (2′) and taking into account that $\omega_\alpha(\mathrm{H}_{w_{0,\alpha}(\alpha)}) = 1$, we obtain $kl = 1$. This contradiction completes the proof. □

In closing, we calculate the numbers c_α for all simple groups. Let r be the rank of G and let $\Pi = \{\alpha_1, \ldots, \alpha_r\}$. We adopt the conventions of [Bou] for indexing the simple roots and write c_i instead of c_{α_i}.

Proposition *The numbers c_i for simple groups are given as follows:*

type A_r: $c_i = r + 1 \quad (1 \leq i \leq r)$;

type B_r: $c_i = 2r - i \quad (1 \le i \le r-1), \quad c_r = 2r;$
type C_r: $c_i = 2r - i + 1 \quad (1 \le i \le r);$
type D_r: $c_i = 2r - i - 1 \quad (1 \le i \le r-2), \quad c_{r-1} = c_r = 2r - 2;$
type E_6: $c_1 = 12,\ c_2 = 11,\ c_3 = 9,\ c_4 = 7,\ c_5 = 9,\ c_6 = 12;$
type E_7: $c_1 = 17,\ c_2 = 28,\ c_3 = 11,\ c_4 = 8,\ c_5 = 10,\ c_6 = 13,\ c_7 = 18;$
type E_8: $c_1 = 23,\ c_2 = 17,\ c_3 = 13,\ c_4 = 9,\ c_5 = 11,\ c_6 = 14,\ c_7 = 19,\ c_8 = 29;$
type F_4: $c_1 = 8,\ c_2 = 5,\ c_3 = 7,\ c_4 = 11;$
type G_2: $c = 5,\ c_2 = 3.$

Proof Put $\omega_i := \omega_{\alpha_i}$ and write

$$2\varrho = p_1\alpha_1 + \ldots + p_r\alpha_r,$$
$$\omega_i = q_{i1}\alpha_1 + \ldots + q_{ir}\alpha_r \quad (1 \le i \le r).$$

From the definition of c_i it follows that

$$c_i = \frac{p_i}{q_{ii}} \qquad (1 \le i \le r).$$

The numbers p_i, q_{ij} are known (see [Bou]). The result is obtained by a direct calculation. □

4.7 Computations in root systems

This section contains preparatory material for the next one. Given a reduced root system Δ, we find all roots $\alpha \in \Delta$, such that $\alpha + \varrho$ is regular, i.e., does not vanish on coroots. We do not introduce a special notation for the vector space containing Δ and for its dual. In the application we have in mind these spaces are the real forms of $\mathfrak{t}^*$ and $\mathfrak{t}$ respectively, where $\mathfrak{t}$ is a Cartan subalgebra of a semisimple complex Lie algebra.

For $\alpha, \beta \in \Delta$ put $n(\alpha,\beta) = \alpha(\mathrm{H}_\beta)$, where H_β is the coroot corresponding to β. For any $\mathcal{W}$-invariant scalar product on the ambient vector space one has

$$n(\alpha,\beta) = \frac{2(\alpha,\beta)}{(\beta,\beta)}.$$

Recall that

$$n(\alpha,\beta) \in \mathbb{Z}$$

by the definition of a root system. Thus

$$n(\alpha,\beta)\, n(\beta,\alpha) \;=\; 4\cos^2(\widehat{\alpha,\beta}) \in \{0,1,2,3,4\}$$

and, as a matter of fact,

$$n(\alpha,\beta) \in \{-3,-2,-1,0,1,2,3\}.$$

As usual, let s_α denote the reflection corresponding to $\alpha \in \Delta$. One has $s_\alpha(\beta) = \beta - n(\beta,\alpha)\alpha$ for $\beta \in \Delta$.

Let $\Pi = \{\alpha_1, \ldots, \alpha_r\}$ be a base of Δ. The elements of Π are called simple roots. For the sake of brevity we write H_i and s_i instead of H_{α_i} and s_{α_i} respectively.

The choice of Π gives rise to an *ordering* of Δ. Namely, for $\alpha, \beta \in \Delta$ the notation $\alpha \leq \beta$ means that $\beta - \alpha = \sum k_i\alpha_i, \ \ k_i \geq 0$. We also write $\alpha < \beta$ if and only if $\alpha \leq \beta$ and $\alpha \neq \beta$.

Proposition 1 *Let $\alpha, \beta \in \Delta^+$ and $\alpha < \beta$. Then there exists a number i, $1 \leq i \leq r$, such that $\alpha + \alpha_i \in \Delta$ and $\alpha + \alpha_i \leq \beta$.*

Proof Write $\beta - \alpha = \sum_i k_i\alpha_i, \ \ k_i \geq 0$, and put $k := \sum k_i$. Then $k > 0$ and the assertion is trivial if $k = 1$. Thus, by induction on k, it suffices to prove that either $\beta - \alpha_i \in \Delta$ or $\alpha + \alpha_i \in \Delta$ for some i, $\ 1 \leq i \leq r$, such that $k_i > 0$. Since $||\beta - \alpha||^2 = \sum_i k_i(\beta - \alpha, \alpha_i)$, there exists a number i with $k_i > 0$, such that $(\beta - \alpha, \alpha_i) > 0$. For such an i one has $(\beta, \alpha_i) > 0$ or $(\alpha, \alpha_i) < 0$.

Assume that $(\beta, \alpha_i) > 0$. Then $n(\beta, \alpha_i)$ and $n(\alpha_i, \beta)$ are both positive. If $n(\beta, \alpha_i) > 1$ and $n(\alpha_i, \beta) > 1$ then $n(\beta, \alpha_i) = n(\alpha_i, \beta) = 2$, and it follows that $\beta = \alpha_i$. Since $\beta > \alpha$, this is impossible. Thus at least one of the two numbers equals 1, and so we obtain

$$\beta - \alpha_i = \begin{cases} s_i(\beta), & \text{if } n(\beta, \alpha_i) = 1, \\ -s_\beta(\alpha_i), & \text{if } n(\alpha_i, \beta) = 1, \end{cases}$$

showing that $\beta - \alpha_i \in \Delta$.

Finally, if $(\alpha, \alpha_i) < 0$ then a similar argument shows that $\alpha + \alpha_i \in \Delta$. □

Proposition 2 *The following properties of a root α are equivalent:*

(i) $(\alpha, \alpha_i) \geq 0$ *for all* i, $\ i = 1, \ldots, r$;

(ii) *if* $\beta \in \Delta$ *and* $\alpha < \beta$ *then* $|| \alpha || < || \beta ||$.

Proof (i)⇒(ii). By Proposition 1 there exists a number i, $1 \leq i \leq r$, such that

$$\beta = \alpha + \alpha_i + \gamma,$$

where $\alpha + \alpha_i \in \Delta$, $\gamma = \sum_j l_j\alpha_j, \ \ l_j \geq 0$. Hence

$$|| \beta ||^2 = || \alpha + \alpha_i ||^2 + 2(\alpha, \gamma) + 2(\alpha_i, \gamma) + || \gamma ||^2 =$$

$$= || \alpha + \alpha_i ||^2 - || \alpha_i ||^2 + 2 \sum_j l_j(\alpha, \alpha_j) + || \alpha_i + \gamma ||^2 >$$

$$> || \alpha + \alpha_i ||^2 - || \alpha_i ||^2 = || \alpha ||^2 + 2(\alpha, \alpha_i) \geq || \alpha ||^2.$$

(ii)⇒(i). Assume $(\alpha, \alpha_i) < 0$ for some i so that

$$m := -n(\alpha, \alpha_i) - 1 \geq 0.$$

Then $\beta := \alpha + \alpha_i \in \Delta$ and

$$|| \alpha ||^2 = || s_i(\alpha) ||^2 = || \alpha + \alpha_i + m\alpha_i ||^2 = || \beta ||^2 + (2(\alpha + \alpha_i, \alpha_i) + m || \alpha_i ||^2)\, m =$$

$$= || \beta ||^2 + \ (n(\alpha,\alpha_i) + 2 + m\ ) \ || \alpha_i ||^2 \ m = || \beta ||^2 + || \alpha_i ||^2 \ m \geq || \beta ||^2,$$

contradictory to (ii). □

Proposition 3 *Assume that $\alpha \in \Delta$ satisfies* (i), (ii). *Then $\alpha \in \Delta^+$. Moreover, if Δ is irreducible then $\alpha = \sum_{i=1}^r \ k_i\alpha_i$, where $k_i > 0$ for all i, $i = 1, \ldots, r$.*

Proof If $\alpha \in \Delta^-$ then $\alpha < -\alpha$ and we obtain a contradiction with (ii). Thus $k_i \geq 0$ for all i. Let $I^+ := \{i \in \{1, \ldots, r\} \mid k_i > 0\}$ and $I^0 := \{i \in \{1, \ldots, r\} \mid k_i = 0\}$. Since $(\alpha_i, \alpha_j) \leq 0$ if $i \neq j$, it follows that

$$\sum_{i \in I^+} k_i(\alpha_i, \alpha_j) \ = \ 0 \qquad \text{for} \ \ j \in I^0$$

or, equivalently,

$$(\alpha_i, \alpha_j) = 0 \qquad \text{for} \quad i \in I^+, \ j \in I^0.$$

Since Δ is irreducible, we see that $I^0 = \emptyset$. □

Proposition 4 *Let α, α' be two roots satisfying* (i), (ii). *If $||\alpha|| = ||\alpha'||$ then either $(\alpha, \alpha') = 0$ or $\alpha = \alpha'$.*

Proof We may assume that Δ is irreducible. By Proposition 3 we have

$$\alpha = \sum_i \ k_i\alpha_i, \quad \alpha' = \sum_i \ k'_i\alpha_i, \quad k_i, k'_i > 0.$$

Choose j, $\ 1 \leq j \leq r$, so that $(\alpha', \alpha_j) > 0$. Then

$$(\alpha, \alpha') = \sum_i \ k_i(\alpha_i, \alpha') \geq k_j(\alpha_j, \alpha') > 0.$$

Since $||\alpha|| = ||\alpha'||$, we have two possible cases: $\alpha = \alpha'$ or $\widehat{(\alpha, \alpha')} = \frac{\pi}{3}$. However, in the second case $\alpha - \alpha' \in \Delta$, showing that either $\alpha < \alpha'$ or $\alpha' < \alpha$. Since this is impossible by (ii), the proof is complete. □

Proposition 5 *Assume that Δ is irreducible. Then a maximal element in the subset of all roots of given length is unique. If all elements of Δ have the same length then* (i), (ii) *are equivalent to*

(iii′) *α is the maximal root.*

If there are roots of different lengths, then (i), (ii) *are equivalent to*

(iii″) *α is either the maximal root or the maximal short root.*

Proof Let $h = ||\alpha||$. Then (ii) means that α is maximal in the set of all roots of length h. The uniqueness of such an element follows from Proposition 4. □

Let $\tilde{\alpha}$ denote the maximal root of an irreducible root system Δ. If Δ has roots of different lengths then we denote by α^+ the maximal short root.

Proposition 6 *There exists exactly one number k, $\ 1 \leq k \leq r$, such that*

$$\alpha^+ + \alpha_k \in \Delta.$$

Proof The existence of α_k follows from Proposition 1 and from the fact that $\alpha \le \tilde{\alpha}$ for all $\alpha \in \Delta$. Assume that $\beta := \alpha^+ + \alpha_k \in \Delta$ and $\gamma := \alpha^+ + \alpha_l \in \Delta$. It is then clear that β and γ are long roots. Since

$$\| \beta \|^2 = \| \alpha^+ \|^2 + \| \alpha_k \|^2 + 2(\alpha^+, \alpha_k) \ge \| \alpha^+ \|^2 + \| \alpha_k \|^2 ,$$

it follows that α_k is a short root. By the same reason α_l is short. Thus α_k, α_l and α^+ have the same length and, consequently,

$$(\beta, \gamma) = (\alpha^+ + \alpha_k\ ,\ \alpha^+ + \alpha_l) = \| \alpha^+ \|^2 \ +\ (\alpha^+, \alpha_k)\ +\ (\alpha^+, \alpha_l)\ +\ (\alpha_k, \alpha_l) \ge$$

$$\ge \| \alpha^+ \|^2 + (\alpha_k, \alpha_l) = \| \alpha^+ \|^2 \ (1 + \cos\ (\widehat{\alpha_k, \alpha_l})) > 0.$$

Since β and γ have the same length and $(\beta, \gamma) > 0$, we obtain that $\beta - \gamma \in \Delta \cup \{0\}$. But

$$\beta - \gamma = \alpha_k - \alpha_l.$$

Therefore $k = l$. □

Recall that $\mathcal{P}$ is the additive group of all linear forms $\lambda = \sum k_i \alpha_i \in \mathfrak{t}^*$, $k_i \in \mathbb{Q}$, such that $\lambda(\mathrm{H}_i) \in \mathbb{Z}$, $i = 1, \ldots, r$. The subset $\mathcal{P}^+$ (resp. $\mathcal{P}^{++}$) of $\mathcal{P}$ is defined by the inequalities $(\lambda, \alpha_i) \ge 0$ (resp. $(\lambda, \alpha_i) > 0$), $i = 1, \ldots, r$.

Proposition 7 (A.Borel, see [B]) *For each $\alpha \in \Delta$ there exists at most one simple root α_i, such that $(\alpha + \varrho, \alpha_i) < 0$. If $\alpha + \varrho$ is regular and $(\alpha + \varrho, \alpha_i) < 0$ then $s_i(\alpha + \varrho) \in \mathcal{P}^{++}$.*

Proof Observe that $(\alpha + \varrho, \alpha_i) < 0$ is equivalent to $(\alpha, \alpha_i) \le -\| \alpha_i \|^2$. Let α_i and α_j be two different simple roots with this property. Then α, α_i, and α_j are contained in the same irreducible component of Δ and

$$\| \alpha + \alpha_i + \alpha_j \|^2 = \| \alpha \|^2 + \| \alpha_i \|^2 + \| \alpha_j \|^2 + 2(\alpha, \alpha_i) + 2(\alpha, \alpha_j) + 2(\alpha_i, \alpha_j) \le$$

$$\le \| \alpha \|^2 + \| \alpha_i \|^2 + \| \alpha_j \|^2 - 2\| \alpha_i \|^2 - 2\| \alpha_j \|^2 = \| \alpha \|^2 - \| \alpha_i \|^2 - \| \alpha_j\|^2 \le 0.$$

We use here the following observation. If an irreducible root system is not of type G_2, then the ratio of lengths does not exceed $\sqrt{2}$ for any two roots. In the case of G_2 one of the roots α_i, α_j is long.

As a result we obtain

$$\| \alpha + \alpha_i + \alpha_j \| = 0.$$

Thus $-\alpha_i - \alpha_j = \alpha \in \Delta$. However, the above inequalities also show α_i and α_j are orthogonal. Since the sum of two orthogonal simple roots is never a root, we get a contradiction.

Assume now that $\alpha + \varrho$ is regular and that $(\alpha + \varrho, \alpha_i) < 0$. We know already that $(\alpha + \varrho, \alpha_j) > 0$ if $j \ne i$. Our goal is to prove that

$$(s_i(\alpha + \varrho), \alpha_j) > 0 \qquad (+)$$

for all j. This is clear if $j = i$ or if α_j is orthogonal to α_i. Thus we assume further that Δ is irreducible and that $j \ne i$. Observe that $n(\alpha, \alpha_i) \le -2$ and consider two possible cases: $n(\alpha, \alpha_i) = -2$ and $n(\alpha, \alpha_i) = -3$.

1) $n(\alpha, \alpha_i) = -2$. Suppose for a moment that $n(\alpha_i, \alpha_j) \le -2$. Then either $\alpha = -\alpha_i$, so that (+) is obvious, or one is led to the inequalities

$$\| \alpha \| > \| \alpha_i \| > \| \alpha_j \|,$$

which contradict the irreducibility of Δ. Therefore we may assume that

$$n(\alpha_i, \alpha_j) \ge -1.$$

But then

$$(s_i(\alpha + \varrho), \alpha_j) = (\alpha + \alpha_i + \varrho, \alpha_j) =$$
$$= (\alpha, \alpha_j) + \frac{\| \alpha_j \|^2}{2}(n(\alpha_i, \alpha_j) + 1) \ge (\alpha, \alpha_j) \ge 0,$$

and (+) follows from the regularity of $\alpha + \varrho$.

2) Let $n(\alpha, \alpha_i) = -3$. Then Δ has type G_2, $\alpha_i = \alpha_1$ is the short simple root, and α is a long root, which can be equal α_2 or $-3\alpha_1 - \alpha_2$. Since $-3\alpha_1 - \alpha_2 + \varrho = 2\alpha_1 + 2\alpha_2$ is singular, the only case which remains is $\alpha = \alpha_2$. But then $s_1(\alpha + \varrho) = 7\alpha_1 + 4\alpha_2 \in \mathcal{P}^{++}$. □

We summarize the results of this section in the following theorem.

Let Δ be an irreducible reduced root system, $\alpha_1, \dots, \alpha_r$ the simple roots of Δ, and $\tilde{\alpha}$ the maximal root. If Δ contains roots of different lengths then α^+ denotes the maximal short root. Define the number k, $1 \le k \le r$, as in Proposition 6 and put $\alpha^{++} := \alpha^+ + \alpha_k$.

Theorem *Assume that $\alpha \in \Delta$ has the property that $\alpha + \varrho$ is regular. Denote by w the (unique) element of W, such that $w(\alpha + \varrho) \in \mathcal{P}^{++}$, and let l be the length of w. Then all possibilities for α, l, w, and $w(\alpha + \varrho)$ are as follows:*

1) $\alpha = \tilde{\alpha}, \quad w = \mathrm{id}, \quad l = 0;$

2) $\alpha = \alpha^+, \quad w = \mathrm{id}, \quad l = 0;$

3) $\alpha = s_k(\alpha^{++}), \quad w = s_k, \quad l = 1, \quad w(\alpha + \varrho) = \alpha^+ + \varrho;$

4) $\alpha = -\alpha_i, \quad w = s_i, \quad l = 1, \quad w(\alpha + \varrho) = \varrho \quad (i = 1, \dots, r).$

The roots α^+ and $s_k(\alpha^{++})$ are listed in the following table:

type of Δ	α^+	k	$s_k(\alpha^{++})$
B_r	$\alpha_1 + \alpha_2 + \ldots + \alpha_{r-1} + \alpha_r$	r	$\alpha_1 + \alpha_2 + \ldots + \alpha_{r-1}$
C_r	$\alpha_1 + 2\alpha_2 + \ldots + 2\alpha_{r-1} + \alpha_r$	1	$2\alpha_2 + \ldots + 2\alpha_{r-1} + \alpha_r$
F_4	$\alpha_1 + 2\alpha_2 + 3\alpha_3 + 2\alpha_4$	3	$\alpha_1 + 2\alpha_2 + 2\alpha_3 + 2\alpha_4$
G_2	$2\alpha_1 + \alpha_2$	1	α_2

Proof If $(\alpha, \alpha_i) \ge 0$ for all i, $i = 1, 2, \dots, r$, then α is of the form 1) or 2) by Proposition 2. Otherwise, by Proposition 7, there is exactly one simple root α_i, such that $(\alpha + \varrho, \alpha_i) < 0$.

In particular, $\alpha + \alpha_i \in \Delta \cup \{0\}$ and, again by Proposition 7, $\beta := s_i(\alpha) - \alpha_i = s_i(\alpha + \varrho) - \varrho \in \mathcal{P}^+$. If $\alpha + \alpha_i \in \Delta$ then $\beta = s_i(\alpha + \alpha_i) \in \Delta$. This root is contained in $\mathcal{P}^+$, but is not maximal, because $\beta + \alpha_i \in \Delta$. Thus $\beta = \alpha^+$. Since $\beta + \alpha_i \in \Delta$, Proposition 6 yields $i = k$. Hence $\alpha = s_k(\alpha^{++})$, i.e., α is of the form 3). Finally, if $\alpha + \alpha_i = 0$, then we obtain 4). □

4.8 Cohomology of the tangent sheaf

Let H be a group, V a finite-dimensional complex vector space, $\varphi : H \to \mathrm{GL}(V)$ a linear representation. There exists a sequence of H-invariant subspaces

$$0 = V_0 \subset V_1 \subset V_2 \subset \ldots \subset V_n = V,$$

such that the representations $H \to \mathrm{GL}(V_{i+1}/V_i)$, induced by φ, are irreducible for all i, $i = 0, \ldots, n-1$. A sequence $\{V_i\}$ with these properties is called a *Jordan - Hölder series* of the representation $\varphi : H \to \mathrm{GL}(V)$.

If G is a complex Lie group, $H \subset G$ a closed complex Lie subgroup, and $\varphi : H \to \mathrm{GL}(V)$ a holomorphic representation, then a Jordan - Hölder series determines a sequence of homogeneous holomorphic vector bundles $\{\mathbb{V}_i\}$ on G/H. If $X = G/H$ is a flag manifold then Bott's theorem applies to each quotient $\mathbb{V}_{i+1}/\mathbb{V}_i$. This observation makes it possible to compute the cohomology of the tangent sheaf $\mathcal{T} = \mathcal{T}_X$. We start by the following vanishing theorem, which is also due to R.Bott [B].

Theorem 1 *Let G be a semisimple complex Lie group, $P \subset G$ a parabolic subgroup, and $X = G/P$ the corresponding flag manifold. Then*

$$H^p(X, \mathcal{T}) = 0 \qquad \textit{for} \;\; p \geq 1.$$

Proof Consider the exact sequence of P-modules

$$0 \to \mathfrak{p} \to \mathfrak{g} \to \mathfrak{g}/\mathfrak{p} \to 0$$

with the P-action induced by the adjoint representation. Let

$$0 \to \mathbb{V} \to \mathbb{E} \to \mathbb{T} \to 0$$

be the corresponding sequence of homogeneous vector bundles on X. The representation $P \to \mathrm{GL}(\mathfrak{g})$ extends in a natural way to a holomorphic representation $G \to \mathrm{GL}(\mathfrak{g})$. It follows that $\mathbb{E}$ is the inverse image of a bundle over a point under the mapping $G/P \to G/G$. In particular, $\mathbb{E}$ is holomorphically trivial. Observe that $\mathbb{T}$ is the tangent bundle of X and denote by $\mathcal{V}$, $\mathcal{E}$, and $\mathcal{T}$ the G-sheaves of germs of holomorphic sections of $\mathbb{V}$, $\mathbb{E}$, and $\mathbb{T}$ respectively. Then

$$H^p(X, \mathcal{E}) = H^p(X, \mathcal{O}) \otimes \mathfrak{g} = 0$$

for $p \geq 1$ by Bott's theorem, and the exact cohomology sequence yields

$$H^p(X, \mathcal{T}) \simeq H^{p+1}(X, \mathcal{V}), \qquad p \geq 1.$$

Thus we have to show that $H^p(X, \mathcal{V}) = 0$ for $p \geq 2$. With this purpose consider a Jordan-Hölder series $0 = V_0 \subset V_1 \subset \ldots \subset V_i \subset \ldots \subset V_n = V$ of the P-module $V = \mathfrak{p}$. Then we have the exact sequences

$$0 \to V_i \to V_{i+1} \to V_{i+1}/V_i \to 0, \qquad i = 0, \ldots, n-1,$$

of P-modules and the corresponding exact sequences

$$0 \to \mathcal{V}_i \to \mathcal{V}_{i+1} \to \mathcal{V}_{i+1}/\mathcal{V}_i \to 0, \qquad i = 0, \dots, n-1,$$

of locally free G-sheaves on X. Each quotient V_{i+1}/V_i, $i = 0, \dots, n-1$, is an irreducible P-module whose highest weight is a weight of the adjoint representation. Such a weight is either a root or zero. Therefore

$$H^p(X, \mathcal{V}_{i+1}/\mathcal{V}_i) = 0 \,, \qquad p \geq 2,$$

for all i, $i = 0, \dots, n-1$, by Bott's theorem and by the result of the preceding section. From the exact cohomology sequences

$$\dots \to H^p(X, \mathcal{V}_i) \to H^p(X, \mathcal{V}_{i+1}) \to H^p(X, \mathcal{V}_{i+1}/\mathcal{V}_i) \to H^{p+1}(X, \mathcal{V}_i) \to \dots$$

one obtains by induction on i that $H^p(X, \mathcal{V}_i) = 0$ for $p \geq 2$ and $i = 0, \dots, n-1$. In particular, $H^p(X, \mathcal{V}) = 0$ for $p \geq 2$. This completes the proof. □

Remark In the above proof we used the equality $h^{p,0}(X) = 0 \quad (p \geq 1)$. More generally, one has $h^{p,q}(X) = 0$ if $p \neq q$, see [BoHi].

Corollary The complex structure of a flag manifold X is locally rigid.

Proof Since $H^1(X, \mathcal{T}_X) = 0$, this results from the Kodaira-Spencer theory, see [KoSp]. □

We now proceed to the computation of $H^0(X, \mathcal{T})$. Assume without loss of generality that G is simple and let $X = G/P$, where P is a parabolic subgroup of G containing B^-. The subgroup P is determined by a subset $\Phi \subset \Pi$ so that $\Delta_{\mathfrak{p}}^{s} = (\Phi)$, where (Φ) is the set of all roots which are linear combinations of the elements of Φ (see §3.1). It turns out that, as a rule, the Lie algebra $H^0(X, \mathcal{T})$ coincides with $\mathfrak{g}$. We find all exceptions and so obtain the proof of Theorem 2 of §3.3. It should be noted that the number of the maximal parabolic subgroup in Theorem 2 of §3.3 is equal to the number of the unique simple root in the complement to $(-w_0)(\Phi)$. As usual, we denote by V^λ an irreducible G-module with highest weight $\lambda \in \mathcal{P}^+$.

Theorem 2 *Let G be a simple complex Lie group, Δ the root system of $\mathfrak{g}$ with respect to some Cartan subalgebra, Π a base of Δ, and $\Phi \subset \Pi$ a subset. Suppose that a parabolic subgroup $P \subset G$ is defined as above and let $X = G/P$ be the corresponding flag manifold. If all roots of Δ have the same length or $s_k(\alpha^{++}) \notin (\Phi)$ then*

$$H^0(X, \mathcal{T}) \simeq \mathfrak{g}.$$

If $s_k(\alpha^{++}) \in (\Phi)$ then

$$H^0(X, \mathcal{T}) \simeq \mathfrak{g} \oplus V^{\alpha^+}.$$

This happens in the following three cases:

1) *G has type B_r, $\Phi = \{\alpha_1, \dots, \alpha_{r-1}\}$, $H^0(X, \mathcal{T})$ is a simple algebra of type D_{r+1}, $r \geq 3$;*

2) *G has type C_r, $\Phi = \{\alpha_2, \dots, \alpha_r\}$, $H^0(X, \mathcal{T})$ is a simple algebra of type A_{2r-1}, $r \geq 2$;*

3) *G has type G_2, $\Phi = \{\alpha_2\}$, $H^0(X, \mathcal{T})$ is a simple algebra of type B_3.*

Proof Put
$$h(\gamma) := \sum_{\alpha \in \Pi - \Phi} m_\alpha,$$
where
$$\gamma = \sum_{\alpha \in \Pi} m_\alpha \alpha \in \Delta^+,$$

and consider a sequence of vector subspaces of $\mathfrak{g}$ defined by

$$\mathfrak{g}^{(0)} := \mathfrak{p}, \quad \mathfrak{g}^{(d)} := \mathfrak{p} \oplus \{\oplus_{1 \le h(\gamma) \le d} \mathfrak{g}_\gamma\}, \quad 1 \le d \le m := h(\tilde{\alpha}).$$

Then $\mathfrak{g}^{(d)} \subset \mathfrak{g}^{(d+1)}$ for all d, $0 \le d \le m-1$, and each $\mathfrak{g}^{(d)}$ is P-invariant. Observe that $\mathfrak{g}^{(m)} = \mathfrak{g}$ and let
$$T^{(d)} := \mathfrak{g}^{(d)}/\mathfrak{p}, \quad 1 \le d \le m.$$

The sequence

$$0 = T^{(0)} \subset T^{(1)} \subset \ldots \subset T^{(d)} \subset \ldots \subset T^{(m)} = \mathfrak{g}/\mathfrak{p}$$

is a P-invariant filtration of $\mathfrak{g}/\mathfrak{p}$. We have the associated filtration of the tangent sheaf
$$0 = \mathcal{T}^{(0)} \subset \mathcal{T}^{(1)} \subset \ldots \subset \mathcal{T}^{(d)} \subset \ldots \subset \mathcal{T}^{(m)} = \mathcal{T},$$

where all $\mathcal{T}^{(d)}$ are locally free G-sheaves. Each P-module

$$T^{(d)}/T^{(d-1)}, \quad 1 \le d \le m,$$

can be decomposed into irreducible P-modules whose highest weights are some positive roots which do not belong to (Φ). We denote by Υ be the set of all such roots. Then $\gamma \in \Delta^+$ belongs to Υ if and only if $\gamma + \alpha_i \notin \Delta$ for all simple roots $\alpha_i \in \Phi$. In what follows we distinguish three cases.

(a) *All roots of Δ have the same length.* Then, by what we have seen in §4.7, all forms $\gamma + \varrho$, where $\gamma \in \Upsilon$, $\gamma \neq \tilde{\alpha}$, are singular. Since $\tilde{\alpha}$ is the highest weight of $T^{(m)}/T^{(m-1)}$, we have

$$H^p(X, \mathcal{T}^{(d)}/\mathcal{T}^{(d-1)}) = 0 \qquad \text{if } p > 0 \text{ or } d < m$$

and
$$H^0(X, \mathcal{T}^{(m)}/\mathcal{T}^{(m-1)}) \simeq V^{\tilde{\alpha}} \simeq \mathfrak{g}.$$

The exact sequence

$$0 \to H^0(X, \mathcal{T}^{(d-1)}) \to H^0(X, \mathcal{T}^{(d)}) \to H^0(X, \mathcal{T}^{(d)}/\mathcal{T}^{(d-1)}) \to$$
$$\to H^1(X, \mathcal{T}^{(d-1)}) \to H^1(X, \mathcal{T}^{(d)}) \to H^1(X, \mathcal{T}^{(d)}/\mathcal{T}^{(d-1)}) \to \ldots$$

gives us (by induction on d) that

$$H^0(X, \mathcal{T}^{(d)}) = H^1(X, \mathcal{T}^{(d)}) = 0$$

for all $d, 0 \le d \le m-1$. The same exact sequence for $d = m$ shows that

$$H^0(X, \mathcal{T}^{(m)}) \simeq H^0(X, \mathcal{T}^{(m)}/\mathcal{T}^{(m-1)}) \simeq \mathfrak{g}.$$

(b) $s_k(\alpha^{++}) \notin (\Phi)$. Recall that $\alpha^+ + \alpha_i \in \Delta$ if and only if $i = k$. We claim that also $s_k(\alpha^{++}) + \alpha_i \in \Delta$ if and only if $i = k$. Indeed, if $i = k$ then $s_k(\alpha^{++}) + \alpha_k = s_k(\alpha^+) \in \Delta$. On the other hand, if $i \neq k$ then

$$(s_k(\alpha^{++}), \alpha_i) = (\alpha^+, \alpha_i) - n(\alpha_i, \alpha_k)(\alpha^+, \alpha_k) - (\alpha_i, \alpha_k) \ge 0,$$

showing that

$$\| s_k(\alpha^{++}) + \alpha_i \| > \| s_k(\alpha^{++}) \| = \| \alpha^{++} \|.$$

Since α^{++} is a long root, it follows that $s_k(\alpha^{++}) + \alpha_i$ is not a root. If $\alpha_k \in \Phi$ then $\alpha^+ \notin \Upsilon$ and $s_k(\alpha^{++}) \notin \Upsilon$, and so we obtain $H^0(X, \mathcal{T}) \simeq \mathfrak{g}$ as in (a). Assume now that $\alpha_k \notin \Phi$. Then $\tilde{\alpha}$, α^+, and $s_k(\alpha^{++})$ are in Υ. Let $T^{(p)}/T^{(p-1)}$ and, respectively, $T^{(q)}/T^{(q-1)}$ be the quotients where α^+ and $s_k(\alpha^{++})$ occur as highest weights. We claim that $p > q$. For, if $p \le q$ then

$$H^1(X, \mathcal{T}^{(j)}) = H^2(X, \mathcal{T}^{(j)}) = 0, \quad j \le q-1,$$

and

$$H^1(X, \mathcal{T}^{(q)}) \simeq H^1(X, \mathcal{T}^{(q)}/\mathcal{T}^{(q-1)}) \simeq V^{\alpha^+}.$$

In the exact sequence

$$H^0(X, \mathcal{T}^{(d)}/\mathcal{T}^{(d-1)}) \to H^1(X, \mathcal{T}^{(d-1)}) \to$$

$$\to H^1(X, \mathcal{T}^{(d)}) \to H^1(X, \mathcal{T}^{(d)}/\mathcal{T}^{(d-1)}) = 0 \quad (d \ge q+1)$$

the first G-module either equals 0 or is isomorphic to the adjoint module (if $d = m-1$). Therefore

$$H^1(X, \mathcal{T}^{(d)}) \simeq H^1(X, \mathcal{T}^{(d-1)})$$

for all d, $d \ge q+1$, and $H^1(X, \mathcal{T}) \neq 0$. Since this contradicts Theorem 1, we have $p > q$. Thus

$$H^1(X, \mathcal{T}^{(p)}/\mathcal{T}^{(p-1)}) = 0$$

and

$$H^0(X, \mathcal{T}^{(0)}) = \ldots = H^0(X, \mathcal{T}^{(p-1)}) = 0.$$

Therefore the sequence

$$0 \to H^0(X, \mathcal{T}^{(p)}) \to H^0(X, \mathcal{T}^{(p)}/\mathcal{T}^{(p-1)}) \to H^1(X, \mathcal{T}^{(p-1)}) \to H^1(X, \mathcal{T}^{(p)}) \to 0 \quad (*)$$

is exact. Since

$$H^1(X, \mathcal{T}^{(p-1)}) \simeq H^1(X, \mathcal{T}^{(p-2)}) \simeq \ldots \simeq H^1(X, \mathcal{T}^{(q)}) \simeq H^1(X, \mathcal{T}^{(q)}/\mathcal{T}^{(q-1)}),$$

the middle terms in (*) are isomorphic to V^{α^+}. By the same reason as above they cancel, so that

$$H^0(X, \mathcal{T}^{(p)}) = 0.$$

From this we obtain

$$H^0(X, \mathcal{T}^{(d)}) = 0, \quad d = p+1, \dots, m-1.$$

Thus the homomorphism of G-modules

$$H^0(X, \mathcal{T}^{(m)}) \to H^0(X, \mathcal{T}^{(m)}/\mathcal{T}^{(m-1)}) \simeq \mathfrak{g}$$

is injective. Since $H^0(X, \mathcal{T}^{(m)}) \neq 0$ and $\mathfrak{g}$ is an irreducible G-module, this homomorphism is also surjective.

(c) $s_k(\alpha^{++}) \in (\Phi)$. In § 4.7 we have seen that $\alpha_i < s_k(\alpha^{++})$ for all $i \neq k$ (and also for $i = k$ if G has type F_4). Assume without loss of generality that $P \neq G$. Then $\Phi = \Pi - \{\alpha_k\}$. In particular, the case F_4 cannot occur and we obtain 1), 2), or 3). Observe that $X = \mathbb{IG}_{r+1}(2r+2)$, $\mathbb{P}_{2r-1}$, or $\mathbb{Q}(7)$ and $\mathrm{Aut}^\circ(X)$ is a simple group of type D_{r+1}, A_{2r-1}, or B_3 respectively. Note that 1) reduces to 2) if $r = 2$. Since $\tilde{\alpha}$, $\alpha^+ \in \Upsilon$ and $s_k(\alpha^{++}) \notin \Upsilon$, it follows that

$$H^1(X, \mathcal{T}^{(d)}/\mathcal{T}^{(d-1)}) = 0$$

and

$$H^0(X, \mathcal{T}^{(d)}) \simeq H^0(X, \mathcal{T}^{(d-1)}) \oplus H^0(X, \mathcal{T}^{(d)})/\mathcal{T}^{(d-1)})$$

for all $d, 1 \leq d \leq m$. Consequently,

$$H^0(X, \mathcal{T}^{(m)}) \simeq V^{\tilde{\alpha}} \oplus V^{\alpha^+} \simeq \mathfrak{g} \oplus V^{\alpha^+}. \qquad \square$$

Remark Using the same technique, one can study the Dolbeault cohomology of homogeneous line bundles over flag manifolds. For Hermitian symmetric spaces the decomposition of $\wedge^p(\mathfrak{g}/\mathfrak{p})^*$ into irreducible P-modules leads to explicit algorithms for computing $H^p(G/P, \Omega^p \otimes \mathcal{O}_{\mathbb{L}})$, where $\mathbb{L}$ is an ample line bundle, see [Sn3], [Sn4].

5 Function Theory on Homogeneous Manifolds

Let K be a connected compact Lie group, $G = K_{\mathbb{C}}$ the reductive linear algebraic group obtained by complexification, and $H \subset G$ a closed complex Lie subgroup. In this chapter we study holomorphic functions in K-invariant domains $\Omega \subset G/H$. For any such domain there is a representation of K on the Fréchet vector space $\mathcal{O}(\Omega)$. Therefore our starting point is a theorem of Harish-Chandra, which extends the classical Fourier expansion to the representation theory of compact Lie groups on Fréchet spaces. As an application, we prove that for G/H holomorphically separable the subgroup H is closed in the Zariski topology of G. Furthermore, under this assumption G/H is a quasi-affine algebraic variety. Algebraic subgroups of (not necessarily reductive) linear algebraic groups having this property are called observable. An algebraic subgroup $H \subset G$ is observable if and only if G/H is an orbit in a finite-dimensional rational G-module. Using the methods of the geometric invariant theory, we obtain a description of the class of observable subgroups. Namely, an algebraic subgroup H of a connected linear algebraic group G is observable if and only if there exist an irreducible rational G-module V and a vector $v \in V$ with $G[v]$ closed in $\mathbb{P}(V)$, such that $H \subset G_v$ and the unipotent radical of H is contained in the unipotent radical of G_v. If G is reductive, then this algebraic condition is necessary and sufficient for G/H to be holomorphically separable.

Returning to the case $G = K_{\mathbb{C}}$, we prove that if G/H contains a K-invariant Stein domain then the connected component H° of the isotropy subgroup H is reductive. This yields a characterization of Stein homogeneous manifolds, due to Y.Matsushima and A.L.Onishchik. Namely, a homogeneous complex manifold G/H of a complex reductive Lie group G is Stein if and only if the isotropy subgroup H is also reductive, and in this case G/H is an affine algebraic variety. In the last section we consider K-invariant functions in K-invariant domains in G/H. We prove that if such a function is plurisubharmonic then the corresponding function on $K\backslash G$ is convex along geodesics in a G-invariant Riemannian metric. We also give some conditions on H sufficient for the converse to be true.

5.1 Representations of compact Lie groups on Fréchet spaces

In this section we prove the fundamental theorem of non-commutative harmonic analysis, due to Harish-Chandra [Ha-Ch]. Another exposition can be found in [War]. Let G be a Lie group countable at infinity, F a Fréchet vector space over $\mathbb{C}$, and $g \mapsto \rho(g)$ a *continuous* representation of G on F. This means that ρ is a homomorphism of G into the group of invertible continuous linear operators on F such that the mapping $f \mapsto \rho(g)f$ of G into F is continuous for every $f \in F$. A vector $f \in F$ is said to be *differentiable* if the mapping $g \mapsto \rho(g)f$ is of class C^∞. A vector $f \in F$ is said to be *G-finite* if the linear span of its orbit is a finite-dimensional vector subspace. Let F^∞ (resp. F^0) denote the set of all differentiable (resp. G-finite) vectors. Then F^0 and F^∞ are G-invariant linear subspaces in F.

Since any finite-dimensional continuous representation of G is of class C^∞, one has $F^0 \subset F^\infty$. Our first objective is the proof of the following two theorems.

Theorem 1 (Gårding) *F^∞ is dense in F.*

Theorem 2 (Harish-Chandra) *If G is compact then F^0 is dense in F.*

For the proof we need three lemmas. Fix a left invariant Haar measure μ on G. For G compact we assume that μ is normalized by the condition $\mu(G) = 1$. As usual, $C_0(G)$ denotes the space of continuous functions with compact support on G. Put $C_0^k(G) := C^k(G) \cap C_0(G)$, where $k = 1, 2, \dots, \infty$. For any $\varphi \in C_0(G)$ define a linear operator $\rho(\varphi) : F \to F$ by

$$\rho(\varphi)f = \int_G \varphi(x)\rho(x)f\,d\mu(x) \qquad (f \in F).$$

Lemma 1 *If $\varphi \in C_0^\infty(G)$ then $\rho(\varphi)F \subset F^\infty$.*

Proof For any $g \in G$ we have

$$\rho(g)\rho(\varphi)f = \int_G \varphi(x)\rho(gx)f\,d\mu(x) = \int_G \varphi(g^{-1}x)\rho(x)f\,d\mu(x).$$

Fix $X \in \mathfrak{g}$ and let $g_t = \exp tX \ \ (t \in \mathbb{R})$. Then

$$\varphi(g_t^{-1}x) - \varphi(x) = \int_0^1 \frac{d}{ds}\varphi(g_{st}^{-1}x)\,ds = -t\int_0^1 (X\varphi)(g_{st}^{-1}x)\,ds,$$

where X is the (right invariant) vector field on G, corresponding to the one-parameter transformation group $x \mapsto g_t x$. It follows that

$$\frac{\rho(g_t) - \mathrm{id}}{t}\,\rho(\varphi)f = -\int_0^1 \rho(g_{st})\rho(X\varphi)f\,ds$$

and

$$\lim_{t\to 0}\frac{\rho(g_t) - \mathrm{id}}{t}\,\rho(\varphi)f = -\rho(X\varphi)f. \qquad (*)$$

This shows that the mapping $\Phi : G \to F$, $\Phi(g) = \rho(g)\rho(\varphi)f$, is of class C^1. More precisely, let X^* be the (left invariant) vector field on G, corresponding to the one-parameter transformation group $x \mapsto xg_t$. Consider the Lie derivative

$$(X^*\Phi)(g) = \lim_{t\to 0}\frac{\Phi(gg_t) - \Phi(g)}{t}.$$

By $(*)$ the limit exists and

$$(X^*\Phi)(g) = -\rho(g)\rho(X\varphi)f,$$

the last expression being of the same type as $\Phi(g)$. Repeating this argument we conclude that Φ is of class C^∞. □

For any $X \in \mathfrak{g}$ we have the linear operator $\rho^\infty(X) : F^\infty \to F^\infty$, defined by

$$\rho^\infty(X) \cdot f := \lim_{t \to 0} \frac{\rho(g_t)f - f}{t}.$$

Corollary *With the notation of Lemma 1 one has*

$$\rho^\infty(X) \cdot \rho(\varphi)f = -\rho(X\varphi)f \qquad (f \in F)$$

and

$$\rho(\varphi) \cdot \rho^\infty(X)f = -\rho(X^*\varphi)f \qquad (f \in F^\infty).$$

Proof The first equality is $(*)$. The proof of the second one is similar. □

Lemma 2 *Let φ_j $(j \geq 1)$ be a sequence in $C_0^\infty(G)$ with the following properties:*

(i) *$\varphi_j \geq 0$ for all j;*

(ii) *$\int_G \varphi_j(x) d\mu(x) = 1$ for all j;*

(iii) *for any neighborhood U of $e \in G$ one has*

$$\text{supp } \varphi_j \subset U$$

for all j except a finite number.

Then

$$\lim_{j \to \infty} \rho(\varphi_j)f = f$$

for any $f \in F$.

Proof Let ν be a continuous seminorm on F. Fix $f \in F$ and $\epsilon > 0$. Then one can choose a neighborhood U of $e \in G$ so that $\nu(\rho(x)f - f) \leq \epsilon$ for $x \in U$. Since

$$\rho(\varphi_j)f - f = \int_G \varphi_j(x)(\rho(x)f - f)d\mu(x),$$

we have $\nu(\rho(\varphi_j)f - f) \leq \epsilon$ if supp $\varphi_j \subset U$. □

In what follows we assume that the Lie group G is compact and denote it by K. Consider the representation of K on $C(K)$, given by the left translations. Let A_K be the $\mathbb{C}$-subalgebra of $C(K)$ consisting of all K-finite vectors. Note that these vectors are also K-finite with respect to the right translations.

Lemma 3 *For any $f \in F$ there exists a sequence $\psi_j \in A_K$ such that*

$$\lim_{j \to \infty} \rho(\psi_j)f = f.$$

Proof By the Peter-Weyl theorem A_K is dense in $C(K)$. This implies that $\rho(A_K)f$ is dense in $\rho(C(K))f$. The existence of a sequence ψ_j with the required property follows from Lemma 2. □

Proof of Theorems 1 and 2 Theorem 1 is a direct consequence of Lemma 1 and Lemma 2. Now, for $\varphi \in A_K$ one has

$$\varphi(k^{-1}x) = \sum_{i=1}^{d} a_i(k)\varphi_i(x) \quad (k, x \in G),$$

where $\varphi_1, ..., \varphi_d \in A_K$, $a_i(k) \in \mathbb{C}$. This yields

$$\rho(k)\rho(\varphi)f = \sum_{i=1}^{d} a_i(k)\rho(\varphi_i)f,$$

hence

$$\rho(\varphi)f \in F^0.$$

The statement of Theorem 2 follows now from Lemma 3. □

For K a compact Lie group let $\hat{K}$ denote the set of all equivalence classes of finite-dimensional irreducible linear representations of K. For each $\delta \in \hat{K}$ let ξ_δ denote the character of the representation δ, $d(\delta)$ the degree of δ, i.e., the dimension of the representation space of δ, and $\chi_\delta = d(\delta)\xi_\delta$.

Let ρ be a continuous linear representation of K on a Fréchet space F. For each $\delta \in \hat{K}$ there is a continuous linear operator $E_\delta : F \to F$ defined by

$$E_\delta(f) := \rho(\overline{\chi_\delta})f = \int_K \overline{\chi_\delta(x)} \cdot \rho(x)f \; d\mu(x).$$

We shall need some properties of the operator family $\{E_\delta\}_{\delta \in \hat{K}}$. These are a consequence of the Schur orthogonality relations, which we now recall.

Denote by V_δ the irreducible K-module corresponding to $\delta \in \hat{K}$. Choose a basis of V_δ. Then to each $k \in K$ there corresponds a matrix, whose entries are certain (real analytic) functions on K. These functions are called the *matrix elements* of the representation δ. Their linear span is independent of the choice of the basis of V_δ. Note that A_K is spanned as a vector space by the matrix elements of all finite-dimensional representations of K. Fix a K-invariant Hermitian inner product on V_δ. Let $a_{ij}(k)$ be the matrix elements of the representation δ defined with respect to a unitary basis of V_δ.

Lemma 4 (Schur orthogonality relations for matrix elements)

1) $\int_K a_{ij}(x)\overline{a_{pq}(x)}d\mu(x) = \frac{1}{d(\delta)}\delta_{ip}\delta_{jq}$;

2) *For $\delta, \epsilon \in \hat{K}$, $\delta \neq \epsilon$, the matrix elements of δ and ϵ are mutually orthogonal.*

Proof See [Ki], part II, §9. □

The following corollaries are often useful.

Corollary 1 (Schur orthogonality relations for characters) *For $\delta, \epsilon \in \hat{K}$ one has:*

1) $(\xi_\delta, \xi_\delta) = 1$;

2) $(\xi_\delta, \xi_\epsilon) = 0$ *if* $\delta \neq \epsilon$. □

Recall that the convolution of two functions on K is defined by

$$\varphi * \psi\ (x) := \int_K \varphi(y)\varphi(y^{-1}x)d\mu(y).$$

Corollary 2 *Let a_{ij} be the matrix elements of δ and let $\epsilon \neq \delta$. Then:*

1) $a_{ij} * \chi_\delta = a_{ij}$ *for all* i, j;

2) $a_{ij} * \chi_\epsilon = 0$ *for all* i, j.

Proof By 1) of Lemma 4 we have

$$(a_{ij} * \chi_\delta)(x) = \int_K a_{ij}(y)\chi_\delta(y^{-1}x)d\mu(y) =$$

$$= d(\delta) \int_K a_{ij}(y)\{\sum_{p,q} a_{pq}(y^{-1})a_{qp}(x)\}d\mu(y) =$$

$$= d(\delta) \sum_{p,q} a_{qp}(x) \int_K a_{ij}(y)\overline{a_{qp}(y)}d\mu(y) = a_{ij}(x).$$

This proves 1). A similar calculation using 2) of Lemma 4 proves 2). □

Corollary 3 *For $\delta, \epsilon \in \hat{K}$ one has :*

1) $\chi_\delta * \chi_\delta = \chi_\delta$;

2) $\chi_\delta * \chi_\epsilon = 0$ *if* $\delta \neq \epsilon$. □

As a consequence, we have the following properties of the operator family $\{E_\delta\}_{\delta\in\hat{K}}$.

Theorem 3 1) *Each E_δ is a continuous projection, i.e., $E_\delta^2 = E_\delta$;*
2) $E_\delta E_\epsilon = 0$ *if* $\delta \neq \epsilon$;
3) $\rho(k)E_\delta = E_\delta\rho(k)$ *for all* $k \in K$.

Proof Observe that $\rho(\varphi)\rho(\psi) = \rho(\varphi * \psi)$ for any two functions $\varphi, \psi \in C(K)$. Therefore

$$E_\delta E_\epsilon = \rho(\overline{\chi_\delta})\rho(\overline{\chi_\epsilon}) = \rho(\overline{\chi_\delta * \chi_\epsilon}).$$

Thus 1) and 2) follow from 1) and 2) of Corollary 3 respectively. Observe also that $\chi_\delta(k^{-1}xk) = \chi_\delta(x)$ for all $x, k \in K$. Therefore

$$\rho(k)E_\delta f = \int_K \overline{\chi_\delta(x)}\rho(kx)fd\mu(x) = \int_K \overline{\chi_\delta(k^{-1}x)}\rho(x)fd\mu(x) =$$

$$= \int_K \overline{\chi_\delta(xk^{-1})}\rho(x)fd\mu(x) = \int_K \overline{\chi_\delta(x)}\rho(xk)fd\mu(x) = E_\delta\rho(k)f,$$

showing 3). □

Theorem 4 *Let $F_\delta := E_\delta(F)$. Then:*

(i) *F_δ is a closed K-invariant subspace of F for all $\delta \in \hat{K}$;*

(ii) *$F_\delta \subset F^0$ for all $\delta \in \hat{K}$;*

(iii) *if $f \in F^0$ then for all $\delta \in \hat{K}$ except a finite number one has $E_\delta f = 0$;*

(iv) *for $f \in F^0$ one has $f = \sum_{\delta \in \hat{K}} E_\delta f$;*

(v) *$F^0 = \bigoplus_{\delta \in \hat{K}} F_\delta$ (each $f \in F^0$ is uniquely written as a finite sum of $f_\delta \in F_\delta$).*

Proof (i) In view of 1) of Theorem 3 the subspace F_δ can be defined by

$$F_\delta = \{f \in F \mid E_\delta f = f\}.$$

Therefore F_δ is closed. By 3) of Theorem 3 F_δ is K-invariant.

(ii) The orbit $\rho(K)E_\delta f$ is contained in the linear span of

$$\rho(\overline{a_{ij}})f,\ i,j = 1,2,...,d,$$

where $d := d(\delta)$.

(iii) Assume first that the orbit $\rho(K)f$ is contained in an irreducible finite-dimensional K-submodule of F which we can identify with some V_δ. Take a unitary basis $v_1, ..., v_d \in V_\delta$ and define the matrix elements $a_{ij}(k)$ as above. Let $f = \sum_{j=1}^{d} b_j v_j$. Then

$$E_\delta f = \sum_j b_j E_\delta v_j = \sum_j b_j \int_K \overline{\chi_\delta(x)}\rho(x)v_j d\mu(x) =$$

$$= \sum_{i,j} b_j \{\int_K \overline{\chi_\delta(x)} a_{ij}(x) d\mu(x)\} v_i =$$

$$= \sum_i b_i v_i \int_K \overline{\chi_\delta(x)} a_{ii}(x) d\mu(x) = \sum_i b_i v_i = f,$$

where we used 1) of Lemma 4.

Note that the equality $E_\delta f = f$ remains valid if f is contained in a finite-dimensional K-submodule isomorphic to $mV_\delta := V_\delta \oplus ... \oplus V_\delta$ (m times). By 2) of Theorem 3 we have $E_\epsilon f = 0$ if $\epsilon \neq \delta$. Now, any vector $f \in F^0$ can be written as a sum

$$f = f^{(1)} + ... + f^{(r)}, \qquad (**)$$

where each $f^{(p)}$ is contained in a K-submodule isomorphic to $m_p V_{\delta_p}$. Denote by $\delta_1, ..., \delta_r$ the corresponding classes of irreducible representations. Without loss of generality we may assume that $\delta_p \neq \delta_q$ if $p \neq q$. By 2) of Theorem 3

$$E_\delta f = \sum_{p=1}^{r} E_\delta f^{(p)} = \sum_{p=1}^{r} E_\delta E_{\delta_p} f^{(p)} = 0$$

for any $\delta \in \hat{K} \setminus \{\delta_1, ..., \delta_r\}$.

(iv) From (**) we obtain

$$f = \sum_{p=1}^{r} f^{(p)} = \sum_{p=1}^{r} E_{\delta_p} f^{(p)} = \sum_{p=1}^{r} E_{\delta_p}(f - \sum_{p\neq q} f^{(q)}) =$$

$$= \sum_{p=1}^{r} E_{\delta_p} f \ - \ \sum_{q\neq p} E_{\delta_p} f^{(q)} = \sum_{p=1}^{r} E_{\delta_p} f - \sum_{p\neq q} E_{\delta_p} E_{\delta_q} f^{(q)} = \sum_{p=1}^{r} E_{\delta} f.$$

(v) By (iv) the decomposition exists. What we have to prove is uniqueness. Assume that $\sum_{p=1}^{r} g_p = 0$, where $g_p = E_{\delta_p} h_p$, $h_p \in F$, and $\delta_p \neq \delta_q$ for $p \neq q$. Using 1) and 2) of Theorem 3 we conclude that

$$g_q = E_{\delta_q} h_q = E_{\delta_q}(\sum_{p=1}^{r} E_{\delta_p} h_p) = E_{\delta_q}(\sum_{p=1}^{r} g_p) = 0 \qquad \square$$

If $F = V_\delta$ then Lemma 4 shows that $E_\delta = \mathrm{id}$, $E_\epsilon = 0$ $(\epsilon \neq \delta)$. In this special case we have $F_\delta = F$, $F_\epsilon = \{0\}$ $(\epsilon \neq \delta)$. Using (ii) of Theorem 4 one can easily describe F_δ in the general case. Namely, F_δ is the *isotypic component* of F of type δ, i.e., F_δ consists of those vectors in F, whose K-orbit is contained in a finite-dimensional K-submodule isomorphic to mV_δ. For any vector $f \in F$ we call $E_\delta f$ the *δ-th Fourier component* of f.

Lemma 5 *If $E_\delta f = 0$ for all $\delta \in \hat{K}$ then $f = 0$.*

Proof Observe that $\overline{\chi_\delta} = \chi_{\delta'}$, where δ' is the equivalence class of the contragredient representation. By 1) of Corollary 2

$$\rho(a_{ij})f = \rho(a_{ij} * \chi_d)f = \rho(a_{ij})\rho(\chi_\delta)f = \rho(a_{ij})E_{\delta'} f = 0$$

for all matrix elements. Therefore $\rho(\psi)f = 0$ for all $\psi \in A_K$.

By Lemma 3 f is the limit of $\rho(\psi_j)f$, where $\psi_j \in A_K$. Therefore $f = 0$. $\square$

The formal series $\sum_{\delta\in\hat{K}} E_\delta f$ is called the *Fourier series* of f.

Theorem 5 (Harish-Chandra) *For any $f \in F^\infty$*

$$f = \sum_{\delta\in\hat{K}} E_\delta f,$$

where the convergence is absolute with respect to any continuous seminorm on F.

Proof Let $\mathfrak{K}$ be the universal enveloping algebra of $\mathfrak{k}_{\mathbb{C}} = \mathfrak{k} \otimes \mathbb{C}$. We regard the elements of $\mathfrak{K}$ as left invariant linear differential operators on K. Put

$$\Omega = 1 - (X_1^2 + \ldots + X_r^2),$$

where $\{X_1, \ldots, X_r\}$ is an orthogonal basis of $\mathfrak{k}$ with respect to an invariant positive definite inner product. Then Ω belongs to the center of $\mathfrak{K}$, so that the operator Ω is also right invariant.

Recall that in any class $\delta \in \hat{K}$ we have chosen a unitary K-module V_δ. Fix a unitary basis of V_δ and denote by $a_{kl}(x)$, $x \in K$, the matrix elements of δ with respect to this basis. By Schur's Lemma Ω acts on V_δ as a scalar operator. Let $c(\delta)$ be the corresponding number. Since any element of $\mathfrak{k}$ acts on V_δ as a skew-adjoint operator, it follows that $c(\delta) \in \mathbb{R}$ and $c(\delta) \geq 1$.

For each k, $k = 1, 2, ..., d(= d(\delta))$, the linear span of $a_{k1}, ..., a_{kd}$ in $C(K)$ is invariant under the right translations. Moreover, this K-module is isomorphic to V_δ. Therefore $\Omega a_{kl} = c(\delta) \cdot a_{kl}$ and $\Omega \xi_\delta = c(\delta) \cdot \xi_\delta$. Since $c(\delta)$ is real, it follows that $\Omega \overline{\xi_\delta} = c(\delta) \cdot \overline{\xi_\delta}$.

For any $X \in \mathfrak{k}$ we have defined the linear operator $\rho^\infty(X) : F^\infty \to F^\infty$. It is clear that ρ^∞ is a linear representation of the Lie algebra $\mathfrak{k}$. This representation extends to a linear representation of $\mathfrak{K}$, which will be denoted by the same letter. By Corollary to Lemma 1

$$E_\delta \rho^\infty(\Omega) f = \rho(\overline{\chi}_\delta)\rho^\infty(\Omega) f = \rho(\Omega\overline{\chi}_\delta) f = c(\delta)\rho(\overline{\chi}_\delta) f = c(\delta) E_\delta f$$

for any $f \in F^\infty$. Thus

$$E_\delta f = c(\delta)^{-1} E_\delta \rho^\infty(\Omega) f$$

and, consequently,

$$E_\delta f = c(\delta)^{-m} E_\delta \rho^\infty(\Omega^m) f$$

for all $f \in F^\infty$ and all $m \in \mathbb{Z}^+$.

Because K is compact, we can apply to the operator family $\{\rho(k) \mid k \in K\}$ the Banach-Steinhaus theorem. Namely, for any continuous seminorm ν on F there exists a continuous seminorm ν_0 on F such that $\nu(\rho(k)f) \leq \nu_0(f)$ for all $k \in K$ and $f \in F$. Therefore

$$\nu(E_\delta f) = \nu(\rho(\overline{\chi}_\delta) f) = \nu\Big(\int_K \overline{\chi}_\delta(x)\rho(x) f \, d\mu(x)\Big) \leq$$

$$\leq \max_K |\chi_\delta(x)| \cdot \int_K \nu(\rho(x) f) \, d\mu(x) \leq d(\delta)^2 \nu_0(f).$$

Since, as we have seen before,

$$E_\delta f = c(\delta)^{-m} E_\delta \rho^\infty(\Omega^m) f,$$

it follows that

$$\nu(E_\delta f) \leq c(\delta)^{-m} d(\delta)^2 \nu_0(\rho^\infty(\Omega^m) f).$$

We shall prove that

$$\sum_{\delta \in \hat{K}} d(\delta)^2 c(\delta)^{-m} < \infty \qquad (***)$$

if m is sufficiently large. Assuming this for a moment, we see that the series $\sum_{\delta \in \hat{K}} E_\delta f$ converges absolutely with respect to any continuous seminorm on F. Let f_0 denote its sum. Since E_δ is continuous,

$$E_\delta f_0 = E_\delta\Big(\sum_{\epsilon \in \hat{K}} E_\epsilon f\Big) = \sum_{\epsilon \in \hat{K}} E_\delta E_\epsilon f = E_\delta f$$

by 1) and 2) of Theorem 3. Thus $E_\delta(f - f_0) = 0$ for all $\delta \in \hat{K}$. By Lemma 5 it follows that $f = f_0$.

We still have to prove $(***)$. The proof can be reduced to the case when K is connected. Namely, let K_0 be the connected component of $e \in K$ and let $\hat{K}_0$ be the set of all equivalence classes of finite-dimensional irreducible linear representations of K_0. Since K is compact, the index $N := [K : K_0]$ is finite. For $\delta \in \hat{K}$, $\delta_0 \in \hat{K}_0$ let $[\delta : \delta_0]$ denote the number of times δ_0 occurs in the restriction $\delta|_{K_0}$. For each $\delta_0 \in \hat{K}_0$ put $\hat{K}(\delta_0) := \{\delta \in \hat{K} \mid [\delta : \delta_0] \geq 1\}$.

Let $I(\delta_0)$ be the representation of K induced by δ_0. By the Frobenius reciprocity theorem each $\delta \in \hat{K}$ occurs in $I(\delta_0)$ with multiplicity $[\delta : \delta_0]$ (see e.g. [Ki], part II, §13). Since the degree of $I(\delta_0)$ is equal to $Nd(\delta_0)$, we obtain the equality

$$\sum_{\delta \in \hat{K}} [\delta : \delta_0]\, d(\delta) = Nd(\delta_0).$$

In particular, it follows that

$$\sum_{\delta \in \hat{K}(\delta_0)} d(\delta)^2 \leq N^2 d(\delta_0)^2.$$

By the definition of Ω we have $\Omega\xi_{\delta_0} = c(\delta_0) \cdot \xi_{\delta_0}$, where $c(\delta_0) = c(\delta)$ for any $\delta \in \hat{K}(\delta_0)$. Therefore

$$\sum_{\delta \in \hat{K}} c(\delta)^{-m} d(\delta)^2 \leq \sum_{\delta_0 \in \hat{K}_0} c(\delta_0)^{-m} \sum_{\delta \in \hat{K}(\delta_0)} d(\delta)^2 \leq N^2 \sum_{\delta_0 \in \hat{K}_0} c(\delta_0)^{-m} d(\delta_0)^2.$$

Hence we may assume that K is connected.

Let $T^c \subset K$ be a maximal torus, $\mathfrak{t}$ the Cartan subalgebra in $\mathfrak{k}_{\mathbb{C}}$ obtained by complexification of $\mathfrak{t}^c$, and Δ the root system of $\mathfrak{k}_{\mathbb{C}}$ with respect to $\mathfrak{t}$. We introduce an ordering in Δ and denote, as usual, by ϱ the half-sum of positive roots. For each character $T^c \to \mathbb{C}^*$ its differential at e extends to $\mathfrak{t}$ as a complex linear form. Let $\mathfrak{t}^*_{\mathbb{Z}}$ be the additive subgroup of $\mathfrak{t}^*$ consisting of all such forms, $\mathfrak{t}^*_{\mathbb{R}} \subset \mathfrak{t}^*$ the real vector subspace generated by $\mathfrak{t}^*_{\mathbb{Z}}$, and $\lambda(\delta)$ the highest weight of $\delta \in \hat{K}$. Then $\lambda(\delta) \in \mathfrak{t}^*_{\mathbb{Z}}$.

The Weyl formula for the dimension of an irreducible representation tells us that $d(\delta) = p(\lambda(\delta))$, where p is some polynomial function on $\mathfrak{t}^*_{\mathbb{R}}$. It is also known that

$$c(\delta) = 1 - \langle \lambda(\delta), \lambda(\delta) + 2\varrho \rangle,$$

where $\langle .\, , .\rangle$ is a complex bilinear form on $\mathfrak{t}^*_{\mathbb{C}}$ such that the associated quadratic form $(-q)$ is negative definite on $\mathfrak{t}^*_{\mathbb{R}}$ (see [Bou], Chap. 8, § 6, n.4). In other words,

$$c(\delta) = 1 + q(\lambda(\delta) + \varrho) - q(\varrho),$$

where q is a positive definite quadratic form. Obviously, one can find a compact set C in $\mathfrak{t}^*_{\mathbb{R}}$, such that

$$q(\lambda + \varrho) - q(\varrho) \geq \frac{q(\lambda)}{2}$$

for all $\lambda \in \mathfrak{t}^*_{\mathbb{R}}$ outside C. The set

$$\Gamma := \{\delta \in \hat{K} \mid \lambda(\delta) \in C\}$$

is finite since $\mathfrak{t}_{\mathbb{Z}}^* \cap C$ is finite. On the other hand,

$$\sum_{\delta \in \hat{K}-\Gamma} c(\delta)^{-m} d(\delta)^2 = \sum_{\delta \in \hat{K}-\Gamma} (1 + q(\lambda(\delta) + \varrho) - q(\varrho))^{-m} p(\lambda(\delta))^2 \leq$$

$$\leq \sum_{\delta \in \hat{K}-\Gamma} (1 + q(\lambda(\delta))/2)^{-m} |p(\lambda(\delta))|^2 \leq 2^m \sum_{\lambda \in \mathfrak{t}_{\mathbb{Z}}^*} (1 + q(\lambda))^{-m} |p(\lambda)|^2 < \infty$$

for m sufficiently large. This completes the proof of $(***)$. □

5.2 Differentiable vectors and Fourier series in $\mathcal{O}(X)$

Let G be a transformation group of a complex space X. Then G has a linear representation ρ on the Fréchet vector space $F := \mathcal{O}(X)$, defined by

$$\rho(g)f := f \circ g^{-1} \qquad (g \in G,\ f \in F).$$

We want to apply the results of the preceding section to the representation ρ in the case when G is a compact Lie transformation group of X. For this we need the following simple general fact.

Proposition *Let G be a topological transformation group of a complex space X. Then the representation ρ is continuous. If G is a Lie transformation group then any vector $f \in F$ is differentiable.*

Proof The canonical Fréchet topology on $\mathcal{O}(X)$ has the following property (see [GR2], Ch. V, § 6, First Lemma).

Let $\{X_\nu\}_{\nu=1,2,\dots}$ be an open covering of X, $\varrho_\nu : \mathcal{O}(X) \to \mathcal{O}(X_\nu)$ the restriction mappings, and W a topological space. A mapping $\alpha : W \to \mathcal{O}(X)$ is continuous if and only if all composition mappings $\varrho_\nu \circ \alpha : W \to \mathcal{O}(X_\nu)$ are continuous.

According to the definition in §1.2, the mapping

$$G \to \mathcal{O}(U),\ g \mapsto \rho(g)f|_U,$$

is continuous for $f \in F$ and for any open relatively compact subset $U \subset\subset X$. It follows that the mapping $G \to \mathcal{O}(X)$, $g \mapsto \rho(g)f$, is also continuous.

Assume now that G is a Lie transformation group of X. By the result of §1.6 the G-action on X is real analytic. In particular, for every one-parameter subgroup $g_t = \exp t\mathrm{X}$, where $\mathrm{X} \in \mathfrak{g}$, $t \in \mathbb{R}$, and for every open $U \subset\subset X$ there exists the limit

$$h_U := \lim_{t \to 0} \frac{\rho(g_t)f - f}{t}\Big|_U \in \mathcal{O}(U).$$

Obviously, $h_U|_{U \cap V} = h_V|_{U \cap V}$ if $V \subset\subset X$ is another open subset. Thus h_U is the restriction to U of a global section $h \in \mathcal{O}(X)$. The above property of the canonical Fréchet topology implies that

$$\lim_{t \to 0} \frac{\rho(g_t)f - f}{t} = h$$

in $\mathcal{O}(X)$. This shows that the mapping $G \to F$, $g \mapsto \rho(g)f$, is of class C^1. Repeating the same argument we see that this mapping is of class C^∞. □

Remark Although the G-action is real analytic, it is not true in general that the mapping $g \mapsto \rho(g)f$ is real analytic (see the remark in §4.1).

Let K be a compact Lie transformation group of a complex space X.

Theorem 1 *For any $f \in \mathcal{O}(X)$ there exists a sequence $\psi_j \in A_K$ such that*

$$\lim_{j\to\infty}\rho(\psi_j)f = f.$$

Each vector $\rho(\psi_j)f$ is K-finite.

Proof The first part is the statement of Lemma 3 in §5.1. The second one follows from the proof of Theorem 2 in §5.1. □

Theorem 2 *For any $f \in \mathcal{O}(X)$*

$$f = \sum_{\delta\in\hat{K}} E_\delta f,$$

where the convergence is absolute with respect to any continuous seminorm on $\mathcal{O}(X)$.

Proof The above proposition shows that Theorem 5 of §5.1 holds for any $f \in \mathcal{O}(X)$. □

In the sequel we shall need a more precise statement. Namely, we shall construct a convergent majorant for the above series, having certain distinguished properties. As in §5.1, denote by V_δ an irreducible K-module of class $\delta \in \hat{K}$ with a fixed K-invariant Hermitian inner product. Let $d(\delta) := \dim V_\delta$ and let $a_{ij}(k)$ ($k \in K$, $i, j = 1, \ldots, d(\delta)$) be the matrix elements with respect to a unitary basis of V_δ. Given a continuous linear representation of K on a Fréchet vector space F, define linear operators $M_{\delta,ij} : F \to F$ by

$$M_{\delta,ij}f := d(\delta)\rho(\overline{a_{ij}})f = d(\delta)\int_K \overline{a_{ij}(k)} \cdot \rho(k)f\, d\mu(k).$$

Since $a_{ij} \in A_K$, we have

$$M_{\delta,ij}(F) \subset F^0.$$

Observe also that

$$E_\delta = \sum_{i=1}^{d(\delta)} M_{\delta,ii}.$$

Theorem 3 *Let X be a reduced complex space, K a compact Lie transformation group of X, and $f \in F := \mathcal{O}(X)$. Then the real C^∞ function*

$$p_\delta(f) := \sum_{i,j=1}^{d(\delta)} |M_{\delta,ij}f|^2$$

is K-invariant. One has

$$|E_\delta f\ (x)| \ \leq\ d(\delta)^{1/2}\ p_\delta(f)(x)^{1/2}$$

for all $x \in X$. The series

$$\sum_{\delta\in\hat{K}} d(\delta)^{1/2}\ p_\delta(f)^{1/2}$$

converges uniformly on compact sets in X.

Proof Let $k_0 \in K$ and $i, j \in \{1, \ldots, d(\delta)\}$ be fixed. Then

$$\rho(k_0)M_{\delta,ij}f = d(\delta)\int_K \overline{a_{ij}(k)}\cdot\rho(k_0k)f\ d\mu(k) = d(\delta)\int_K \overline{a_{ij}(k_0^{-1}k)}\cdot\rho(k)f\ d\mu(k) =$$

$$= d(\delta)\sum_{q=1}^{d(\delta)}\overline{a_{iq}(k_0^{-1})}\int_K \overline{a_{qj}(k)}\cdot\rho(k)f\ d\mu(k) = \sum_{q=1}^{d(\delta)} a_{qi}(k_0)\cdot M_{\delta,qj}f.$$

Therefore

$$\sum_i |M_{\delta,ij}f|^2(k_0^{-1}x) = \sum_i |M_{\delta,ij}f|^2(x)$$

for all j. This shows that $p_\delta(f)$ is K-invariant. The first assertion is proved. The second one follows from the inequality

$$\sum_i |c_{ii}|\ \leq \sqrt{d}\cdot\big(\sum_{i,j}|c_{ij}|^2\big)^{1/2},$$

which is valid for any complex $d\times d$-matrix (c_{ij}).

We still have to prove the last one. For this we apply the argument from the proof of Theorem 5, §5.1. Let ν be any continuous seminorm on F. Then there exists a continuous seminorm ν_0 on F such that

$$\nu(M_{\delta,ij}f) = d(\delta)\ \nu\Big(\int_K \overline{a_{ij}(k)}\cdot\rho(k)f\ d\mu(k)\Big) \leq d(\delta)\nu_0(f).$$

It follows that

$$\nu(M_{\delta,ij}f) \leq c(\delta)^{-m}d(\delta)\nu_0(\rho^\infty(\Omega^m)f)$$

for all $\delta \in \hat{K}$ and $m \in \mathbb{Z}$. Consequently,

$$|M_{\delta,ij}f(x)| \leq L_m c(\delta)^{-m}d(\delta)$$

and

$$\sqrt{p_\delta(f)(x)} = \Big(\sum_{i,j=1}^{d(\delta)}|M_{\delta,ij}f(x)|^2\Big)^{1/2} \leq L_m c(\delta)^{-m}d(\delta)^2$$

for all x in a compact set C, where L_m does not depend on δ and x. We have seen in §5.1 that

$$\sum_{\delta\in\hat{K}} d(\delta)^2 c(\delta)^{-m} < \infty,$$

if m is sufficiently large. It is then clear that for any $l \geq 2$ there exists $m = m(l) > 0$ such that

$$\sum_{\delta \in \hat{K}} d(\delta)^l c(\delta)^{-m} < \infty.$$

The last assertion of the theorem follows from the inequality

$$d(\delta)^{1/2} p_\delta(f)^{1/2}(x) \leq L_m d(\delta)^{5/2} c(\delta)^{-m},$$

where $x \in C$ and $m = m(5/2)$. □

Example 1 Let $K = \mathrm{U}(1)^n$. A *Reinhardt domain* in $\mathbb{C}^n$ is a domain which is invariant under the action of K on $\mathbb{C}^n$ given by

$$(z_1, \ldots, z_n) \mapsto (e^{i\varphi_1} z_1, \ldots, e^{i\varphi_n} z_n).$$

Let X be a Reinhardt domain in $\mathbb{C}^n$. The Fourier series of $f \in \mathcal{O}(X)$ coincides with the usual Laurent series

$$f = \sum_{m \in \mathbb{Z}^n} f_m,$$

where

$$f_m(z) = \frac{1}{(2\pi i)^n} \int_{|\zeta_1| = \ldots = |\zeta_n| = 1} \zeta_1^{-m_1 - 1} \ldots \zeta_n^{-m_n - 1} \cdot f(\zeta_1 z_1, \ldots, \zeta_n z_n) d\zeta_1 \ldots d\zeta_n$$

$$(m = (m_1, \ldots, m_n) \in \mathbb{Z}^n).$$

In this notation $f_m = E_{\delta_m} f$, where $\delta_m \in \hat{K}$ is the equivalence class of the one-dimensional representation $(e^{i\varphi_1}, \ldots, e^{i\varphi_n}) \mapsto e^{-i(m_1\varphi_1 + \ldots + m_n\varphi_n)}$. All non-zero isotypic components of $F = \mathcal{O}(X)$ are one-dimensional.

Example 2 For the action of $K = \mathrm{U}(n)$ on $X = \mathbb{C}^n$ the Fourier decomposition of $f \in \mathcal{O}(X)$ has the form

$$f = \sum_{m=0}^{\infty} f_m,$$

where f_m is the m-th Taylor polynomial of f (with center 0). Denote by $\delta_m \in \hat{K}$ the equivalence class of the representation dual to the m-th symmetric power of the identity representation (it is known that all these symmetric powers are irreducible). The isotypic component $F_{\delta_m} \subset F = \mathcal{O}(X)$ is the subspace of homogeneous polynomials of degree m. We have $f_m = E_{\delta_m} f$. All other isotypic components are trivial, i.e., $F_\delta = \{0\}$ if $\delta \neq \delta_m$.

5.3 Reductive complex Lie groups

A complex Lie group G is called *reductive* if: (a) G has finitely many connected components; (b) the connected component G° of $e \in G$ has a compact real form (see §1.8).

This definition is closely related to the corresponding definition for linear algebraic groups. We recall that a linear algebraic group is called reductive if its unipotent radical is trivial. A reductive linear algebraic group over $\mathbb{C}$ satisfies (a), (b). Conversely, each reductive complex Lie group G has a unique structure of a reductive linear algebraic group. Thus one can consider the subalgebra of regular functions $\mathbb{C}[G] \subset \mathcal{O}(G)$.

Since a reductive complex Lie group G has finitely many connected components, any two maximal compact subgroups of G are conjugate under some $g \in G^\circ$. Moreover, the intersection of a maximal compact subgroup $K \subset G$ with every connected component of G is non-empty and connected (see [Mo2], Theorem 3.1). Fix a maximal compact subgroup K of a reductive complex Lie group G. Then K° is a real form of G°, i.e., $\mathfrak{g} = \mathfrak{k} \bigoplus i\mathfrak{k}$. We shall use the same notation as in §1.8 (even if G is disconnected) and write $G = K_{\mathbb{C}}$.

Recall that a C^1 submanifold M of a complex manifold X is said to be *totally real* if for any $p \in M$ the tangent space at p does not contain complex lines. If M is such a submanifold of (real) dimension d and $A \subset X$ a closed analytic subset containing M, then $\dim_p A \geq d$ for all $p \in M$, where $\dim_p A$ is the complex dimansion of A at p. In particular, if X is connected and $\dim_{\mathbb{C}} X = \dim_{\mathbb{R}} M$ then

$$f \in \mathcal{O}(X),\ f|_M = 0 \implies f = 0.$$

Since K is a (real analytic) totally real submanifold of $K_{\mathbb{C}}$, we obtain the following result.

(*) *Let D be a domain in G and let $z_0 \in D$. If $f \in \mathcal{O}(D)$ vanishes on $D \cap Kz_0$ then $f = 0$.*

Now, each representation of K on a finite-dimensional complex vector space extends to a holomorphic representation of G on the same vector space. The latter one is in fact a rational representation of G viewed as an algebraic group. (This is the so-called 'unitary trick' of H. Weyl, which is well-known if G is connected. If G is disconnected, the same assertion is true because $G = G^\circ K$ and $G^\circ \cap K = K^\circ$.) The following fact is an immediate consequence.

(**) *The restriction mapping $\mathcal{O}(G) \to C(K)$ gives rise to an isomorphism of algebras*

$$\mathbb{C}[G] \simeq A_K.$$

Let $H \subset G$ be an arbitrary subgroup, $\tau : H \to \mathbb{C}^*$ a homomorphism, and $D \subset G$ a non-empty open subset which is K-invariant on the left and H-invariant on the right. We have the representation ρ of K on $\mathcal{O}(D)$, defined by

$$(\rho(k)f)(z) = f(k^{-1}z), \quad \text{where } k \in K, z \in D.$$

The subspace

$$\mathcal{O}_\tau(D) := \{f \in \mathcal{O}(D) \mid f(zh) = f(z)\tau(h) \quad \text{for all} \quad z \in D,\ h \in H\}$$

is closed and K-invariant. Let $\hat{H}$ denote the Zariski closure of H in G and let $\hat{D} := D \cdot \hat{H}$.

Theorem 1 *Assume that $D \cap G^\circ$ is connected. Then the subspace $\mathbb{C}[G] \cap \mathcal{O}_\tau(D)$ is dense in $\mathcal{O}_\tau(D)$. If $\mathcal{O}_\tau(D) \neq 0$ then there exists a unique rational character*

$\hat{\tau} : \hat{H} \to \mathbb{C}^*$ such that $\hat{\tau}|_H = \tau$. The restriction mapping $\mathcal{O}(\hat{D}) \to \mathcal{O}(D)$ gives rise to a topological isomorphism of Fréchet spaces

$$\mathcal{O}_\tau(D) \simeq \mathcal{O}_{\hat{\tau}}(\hat{D}).$$

Proof Let $F := \mathcal{O}_\tau(D)$. For $\varphi \in C(K)$ we have defined the operator $\rho(\varphi) : F \to F$. Namely,

$$(\rho(\varphi)f)(z) = \int_K \varphi(x) f(x^{-1}z) d\mu(x) \qquad (f \in F, z \in D).$$

If $\varphi \in A_K$ then the proof of Theorem 2 of §5.1 shows that $\rho(\varphi)f \in F^0$. More precisely, there exist $f_1, \ldots, f_d \in F$ such that

$$(\rho(\varphi)f)(k^{-1}z) = \sum_{i=1}^{d} a_i(k) f_i(z)$$

for all $k \in K, z \in D$. It is clear that $a_i \in A_K$. By (**) one can extend a_i to regular functions on G. Denote the extensions again by a_i. Let $z_0 \in D$ be an arbitrary point. Replacing in the above equality z by kz we obtain

$$(\rho(\varphi)f)(z) = \sum_{i=1}^{d} a_i(z_0 z^{-1}) f_i(z)$$

for all $z \in Kz_0$. Applying (*) to each connected component of D we conclude that this equality is valid everywhere in D. Thus $\rho(\varphi)f \in \mathcal{O}(D)$ extends to a regular function on G. The fact that $\mathbb{C}[G] \cap \mathcal{O}_\tau(D)$ is dense in $\mathcal{O}_\tau(D)$ follows now from Theorem 1 of §5.2.

Assume that a function $f \in \mathcal{O}(D)$ vanishes on some connected components of D, but does not vanish everywhere. Then one can find $\alpha, \beta \in \mathbb{C}$ and $k \in K$ so that $f' := \alpha f + \beta \rho(k) f$ vanishes on a smaller number of connected components. This shows that if $F \neq \{0\}$ then there exists a function $f \in F$, which does not vanish on every connected component of D. But f can be approximated by the functions of the form $\rho(\varphi)f$, where $\varphi \in A_K$. Thus one can find $\varphi \in A_K$ such that $\rho(\varphi)f$ does not vanish on every connected component of D.

We already know that $\rho(\varphi)f$ extends to a regular function $f_0 \in \mathbb{C}[G]$. Since $f_0(zh) = f_0(z)\tau(h)$ for all $h \in H$ when $z \in D$, the same is true when z is any point in G. In view of our choice of φ the ratio $f_0(zw)/f_0(z)$ is a well-defined rational function on each connected component of $G \times G$. For $w \in H$ this ratio does not depend on z. Therefore the same is true for $w \in \hat{H}$. Put $\hat{\tau}(h) := f_0(zh)/f_0(z)$, where $h \in \hat{H}$ and $z \in G$ is any point such that $f_0(z) \neq 0$.

Since one can find z so that $f_0(zh) \neq 0$, it follows that $\hat{\tau}(h) \in \mathbb{C}^*$. In order to show that $\hat{\tau} : \hat{H} \to \mathbb{C}^*$ is a character, take two elements $h_1, h_2 \in \hat{H}$ and choose $z \in G$ so that $f_0(z) f_0(zh_1) \neq 0$. Then

$$\hat{\tau}(h_1 h_2) = \frac{f_0(zh_1h_2)}{f_0(z)} = \frac{f_0((zh_1)h_2)}{f_0(zh_1)} \cdot \frac{f_0(zh_1)}{f_0(z)} = \hat{\tau}(h_1)\hat{\tau}(h_2).$$

Thus $\hat{\tau}$ is a rational character of $\hat{H}$, which is an extension of τ. Since such an extension is obviously unique, the first statement is proved.

Given $f \in F = \mathcal{O}_\tau(D)$ define $\hat{f} \in \mathcal{O}(\hat{D})$ by $\hat{f}(zh) = f(z)\hat{\tau}(h)$, where $z \in D$, $h \in \hat{H}$. We certainly have to check that

$$f(zh) = f(z)\hat{\tau}(h) \qquad \text{if} \quad z, zh \in D. \tag{+}$$

Assume that z and zh are fixed. Then it suffices to verify (+) for $\rho(\varphi)f$ instead of f, where $\varphi \in A_K$. (Recall that such elements are dense in F.) But we have seen above that $\rho(\varphi)f$ is a regular function. In this case (+) follows from the definition of $\mathcal{O}_\tau(D)$ and from the fact that H is Zariski dense in $\hat{H}$.

After it is shown that $\hat{f}$ is well-defined, the last statement of the theorem is clear. Namely, $\hat{f} \in \mathcal{O}_{\hat{\tau}}(\hat{D})$ and

$$\mathcal{O}_\tau(D) \to \mathcal{O}_{\hat{\tau}}(\hat{D}), \quad f \mapsto \hat{f},$$

is the inverse mapping for the restriction operator $\mathcal{O}_{\hat{\tau}}(\hat{D}) \to \mathcal{O}_\tau(D)$. □

We retain the above notation. For any $f \in F = \mathcal{O}_\tau(D)$ we have the Fourier decomposition

$$f = \sum_{\delta \in \hat{K}} E_\delta f.$$

Recall that the subspace $F_\delta = E_\delta F \subset F$ is the isotypic component of type δ.

Theorem 2 *Let D be as in Theorem 1. Then each isotypic component F_δ consists of regular functions. One has the estimate*

$$\dim F_\delta \leq d(\delta).$$

Proof Let $f \in F = \mathcal{O}_\tau(D)$ and $\varphi \in A_K$. We have seen in the proof of the preceding theorem that each function of the form $\rho(\varphi)f$ is regular (more precisely, extends to a regular function on G). In particular, this applies to $E_\delta f = \rho(\overline{\chi_\delta})f$.

In order to obtain the estimate consider first the case $D = G$, $H = \{e\}$. Then $\dim F_\delta = d(\delta)$ by the Frobenius reciprocity theorem. Since we have a K-equivariant embedding $F_\delta \subset \mathbb{C}[G]$, the estimate follows. □

This result shows that F_δ can be regarded as a linear subspace of $\mathbb{C}[G]$, which does not depend on D if H and τ are fixed. We denote this subspace by $F_\delta(H,\tau)$. A subgroup $H \subset G$ is called *multiplicity-free* if $\dim F_\delta(H,\tau) \leq 1$ for each τ.

Corollary 1 *If $F_\delta(H,\tau) \neq \{0\}$ then τ extends to a rational character of $\hat{H}$. Moreover, $F_\delta(H,\tau) = F_\delta(\hat{H},\hat{\tau})$.*

Proof Take $D = G$. Then $\hat{D} = G$, and the assertion follows from Theorem 1. □

Corollary 2 *H is multiplicity-free if and only if $\hat{H}$ is multiplicity-free.* □

Let G be a connected reductive linear algebraic group. A multiplicity-free algebraic subgroup $H \subset G$ is also called a *spherical* subgroup.

Theorem 3 (B.N.Kimelfeld - E.B.Vinberg [KimVi]) *Let G be a connected reductive linear algebraic group. An algebraic subgroup $H \subset G$ is spherical if and only if a Borel subgroup $B \subset G$ has an open orbit on G/H.*

Proof Assume first that

$$\dim F_\delta(H,\tau) > 1$$

for some rational character $\tau : G \to \mathbb{C}^*$ and for some $\delta \in \hat{K}$. Recall that an irreducible representation of K is uniquely determined by its highest weight. Let $\chi \in \mathrm{X}(B)$ be the highest weight corresponding to δ. Then our assumption means that there exist two linearly independent functions $f_1, f_2 \in \mathbb{C}[G]$, such that

$$f_i(gh) = f_i(g)\tau(h)$$

and

$$f_i(bg) = \chi(b) f_i(g),$$

where $g \in G$, $h \in H$, $b \in B, \chi \in \mathrm{X}(B)$, $i = 1,2$. The ratio $f := f_1/f_2$ can be regarded as a B-invariant function on G/H. Since $f \neq$ const, it follows that B has no open orbit on G/H.

In order to prove the converse statement, we need the following facts from the theory of algebraic groups. The proofs can be found in [KSS].

(a) (M.Rosenlicht) Let G be an algebraic group acting regularly on an irreducible variety X. Then there exist a G-stable Zariski open subset $U \subset X$, an irreducible variety Y, and a surjective morphism $\pi : U \to Y$, such that:

(i) the fibers $\pi^{-1}(\pi(x))$ $(x \in U)$ are exactly the G-orbits;

(ii) π induces an isomorphism of $\mathbb{C}(Y)$ onto the field of G-invariant rational functions $\mathbb{C}(U)^G$.

(b) (M.Rosenlicht) Let G be a connected algebraic group. If a function $f \in \mathbb{C}[G]$ has no zeros, then f is a character multiplied by a constant.

(c) (cf. A.Grothendieck, sém. Chevalley, 2e année, 1958, exposé 5, p. 21) The Picard group of a connected linear algebraic group is finite.

Assume now that each orbit of B on G/H has codimension ≥ 1. By (a) there exists a non-constant rational function $f \in \mathbb{C}(G)$ such that $f(bgh) = f(g)$ for all $g \in G, b \in B, h \in H$. Denote by D_0 (resp. D_∞) the divisor of zeros (resp. poles) of f. By (c) one can find a positive integer m such that $mD_0 = (f_1)$ and $mD_\infty = (f_2)$ for some $f_1, f_2 \in \mathbb{C}[G]$. Since $(f^m) = mD_0 - mD_\infty = (f_1/f_2)$, we may assume (after multiplying f_1 by a non-vanishing regular function) that $f^m = f_1/f_2$. The divisors (f_1) and (f_2) are left B-invariant and right H-invariant. Thus

$$f_i(bg) = s_b^{(i)}(g) f_i(g) \qquad \text{and} \qquad f_i(gh) = r_h^{(i)}(g) f_i(g)$$

for $i = 1,2$ and for all $g \in G, b \in B, h \in H$, where $s_b^{(i)}, r_h^{(i)} \in \mathbb{C}[G]$ are some non-vanishing functions. Since $f_1/f_2 = f^m$ is left B-invariant and right H-invariant, we have

$$s_b^{(1)}(g) = s_b^{(2)}(g) =: s_b(g) \qquad \text{and} \qquad r_h^{(1)}(g) = r_h^{(2)}(g) =: r_h(g).$$

In view of (b) we can write

$$s_b(g) = c_b \sigma_b(g) \qquad \text{and} \qquad r_h(g) = d_h \rho_h(g),$$

where $c_b, d_h \in \mathbb{C}^*, \sigma_b, \rho_h \in \mathrm{X}(G)$. It is easy to see that

$$c_{b_1 b_2} = c_{b_1} c_{b_2} \sigma_{b_1}(b_2) \qquad \text{and} \qquad \sigma_{b_1 b_2} = \sigma_{b_1}\sigma_{b_2}$$

for all $b_1, b_2 \in B$. Similarly,

$$d_{h_1 h_2} = d_{h_1} d_{h_2} \rho_{h_2}(h_1) \qquad \text{and} \qquad \rho_{h_1 h_2} = \rho_{h_1}\rho_{h_2}$$

for all $h_1, h_2 \in H$. Since $\mathrm{X}(G)$ is a free abelian group of finite rank with discrete topology, any continuous homomorphism from an algebraic group to $\mathrm{X}(G)$ is trivial. Thus $\sigma_b(g) = \rho_h(g) = 1$ for all $g \in G, b \in B, h \in H$. It follows that $b \mapsto c_b$ (resp. $h \mapsto d_h$) is a character of B (resp. of H). Letting $\chi(b) := c_b$ and $\tau(h) := d_h$, we obtain

$$f_i(bgh) = \chi(b)\tau(h) f_i(g)$$

for $i = 1, 2$. Let δ be an irreducible representation of K with highest weight χ. Then we obtain $\dim F_\delta(H, \tau) > 1$, so that H is not multiplicity-free. □

Example 1 A subgroup H containing a maximal unipotent subgroup of G is spherical. In particular, a parabolic subgroup of G is spherical. Indeed, a maximal unipotent subgroup $U \subset G$ is the commutator subgroup of some Borel subgroup $B = B^+ \subset G$. Let B^- be an opposite Borel subgroup, i.e., $B^+ \cap B^-$ is a maximal torus of G. Then $B^- \cdot U$ is open in G. If $H \supset U$, then $B^- \cdot H$ is also open in G, and so the orbit $B^-(e \cdot H)$ is open in G/H.

Example 2 Let $H = G_\theta$ be the fixed point set of an automorphism $\theta : G \to G$ of order two. Then H is a spherical subgroup. Indeed, denote again by θ the corresponding automorphism of the Lie algebra $\mathfrak{g}$. There exists a compact real form $\mathfrak{k} \subset \mathfrak{g}$ invariant under θ (see [He], Ch. X, Lemma 5.2). We have $\mathfrak{k} = \mathfrak{l} + \mathfrak{m}$, where $\mathfrak{l} = \{\mathrm{X} \in \mathfrak{k} \mid \theta(\mathrm{X}) = \mathrm{X}\}$ and $\mathfrak{m} = \{\mathrm{X} \in \mathfrak{k} \mid \theta(\mathrm{X}) = -\mathrm{X}\}$. Note that $\mathfrak{h} = \mathfrak{l} + i\mathfrak{l}$. Pick a maximal abelian subspace $\mathfrak{a} \subset \mathfrak{m}$, take a maximal abelian subalgebra of $\mathfrak{k}$ containing $\mathfrak{a}$, and denote by $\mathfrak{t}$ the complexification of this subalgebra in $\mathfrak{g}$. Then $\mathfrak{t}$ is a θ-invariant Cartan subalgebra of $\mathfrak{g}$. Therefore the root system Δ of $(\mathfrak{g}, \mathfrak{t})$ has the following property. If $\alpha \in \Delta$ then the linear form α^θ on $\mathfrak{t}$, defined by $\alpha^\theta(\mathrm{H}) = \alpha(\theta\mathrm{H})$, is also a root. As usual, denote by Δ^+ (resp. Δ^-) the set of all positive (resp. negative) roots relative to some ordering of Δ. The ordering can be chosen so that $\alpha \in \Delta^+$, $\alpha^\theta \neq \alpha$ implies $\alpha^\theta \in \Delta^-$. Then one checks easily that the Borel subalgebra $\mathfrak{b} = \mathfrak{t} \oplus \bigoplus_{\alpha \in \Delta^+} \mathfrak{g}^\alpha$ satisfies $\mathfrak{h} + \mathfrak{b} = \mathfrak{g}$. Thus B has an open orbit on G/H.

Remark Spherical subgroups play an important role in the theory of almost homogeneous algebraic varieties. Namely, let H be an algebraic subgroup of G and let X be an algebraic variety under a regular G-action. Assume that X is almost homogeneous with respect to G and that the open orbit $\Omega \subset X$ is isomorphic to G/H. Then H is spherical if and only if the number of G-orbits on any such X is finite, see [A10], [LuVu]. One can show that the number of B-orbits on X is also finite, see [Br1], [Vi]. One says that a homogeneous variety G/H is *spherical* if $H \subset G$ is a spherical subgroup. More generally, a normal almost homogeneous G-variety X is called *spherical* if the open G-orbit $\Omega \subset X$ is spherical. The mapping

$$G/H \xrightarrow{\sim} \Omega \hookrightarrow X$$

is then called a spherical embedding (of G/H). Spherical varieties can be viewed as a generalization of flag manifolds and, in the same time, of toric varieties. The reader is referred to [BrLuVu], [Br4] for the properties of spherical varieties, to [LuVu], [Kn] for the combinatorial theory of spherical embeddings, and to [Kr], [Mik], [Br2] for the classification of spherical subgroups.

5.4 Quasi-affine homogeneous varieties

In this section we consider algebraic varieties defined over $\mathbb{C}$. An algebraic variety X is called *quasi-affine* if X can be embedded into an affine variety as a Zariski open subset. An algebraic subgroup H of a linear algebraic group G is said to be *observable* if every rational finite-dimensional H-module is an H-submodule of a rational finite-dimensional G-module.

Lemma 1 *Let $H \subset I \subset G$ be linear algebraic groups. If $H \subset I$ is observable and $I \subset G$ is observable, then $H \subset G$ is observable.*

Lemma 2 *An algebraic subgroup H of a linear algebraic group G is observable if and only if for any rational finite-dimensional H-module V there exist a rational finite-dimensional G-module W and an epimorphism of H-modules $W \to V$.*

Proof Lemma 1 follows from the definition of an observable subgroup. Lemma 2 is the statement dual to the definition. □

Let $\mathrm{X}(H)$ denote the group of rational characters of H. For $\tau \in \mathrm{X}(H)$ put

$$\mathbb{C}[G]_\tau := \{f \in \mathbb{C}[G] \mid f(gh) = f(g)\tau(h) \quad \text{for all} \quad g \in G, h \in H\}.$$

It is easy to see that a rational one-dimensional H-module V is an H-submodule of a rational finite-dimensional G-module if and only if $\mathbb{C}[G]_\tau \neq \{0\}$, where τ is the character associated to V. Namely, let f be a non-zero function in $\mathbb{C}[G]_\tau$. Consider the linear representation of G on $\mathbb{C}[G]$ by right translations. Then the G-submodule of $\mathbb{C}[G]$, generated by f, contains V as an H-submodule. Conversely, assume that some G-module W has an H-invariant line $\mathbb{C}w_0$ such that $h \cdot w_0 = \tau(h)w_0$ for $h \in H$. Then for any $\lambda \in W^*$ the function $g \mapsto \lambda(g \cdot w_0)$, $g \in G$, belongs to $\mathbb{C}[G]_\tau$.

Lemma 3 *Assume that*

$$\mathbb{C}[G]_\tau \neq \{0\} \Longrightarrow \mathbb{C}[G]_{\tau^{-1}} \neq \{0\} \tag{+}$$

for any $\tau \in \mathrm{X}(H)$. Then H is an observable subgroup in G.

Proof Let V be a rational H-module and $v_1, \ldots, v_n$ a basis of V. Write

$$h \cdot v_j = \sum_{i=1}^{n} a_{ij}(h) v_i$$

and denote by $a_j(h)$ be the j-th column of the matrix $(a_{ij}(h))$. Since the restriction mapping $\pi : \mathbb{C}[G] \to \mathbb{C}[H]$ is onto, we can extend a_j to a regular function $G \to \mathbb{C}^n$, which will be denoted by the same letter. Consider the linear representation of G on $n \cdot \mathbb{C}[G] := \mathbb{C}[G] \oplus \ldots \oplus \mathbb{C}[G]$ (n summands) by right translations. Let M be the G-submodule of $n \cdot \mathbb{C}[G]$ generated by $a_1, \ldots, a_n$. It is easy to see that

$$a_j(xh) - \sum_i a_{ij}(h) a_i(x) \in \text{Ker } \pi \cap M \quad (x \in G, h \in H).$$

Consequently,

$$M_0 := \mathbb{C}a_1 + \ldots + \mathbb{C}a_n + (\text{Ker } \pi \cap M)$$

is an H-submodule of M such that $\pi(M_0) \subset n \cdot \mathbb{C}[H]$ is H-isomorphic to V.

We have an isomorphism of rational H-modules $V \simeq M_0/M_1$, where M_0 is contained in a rational G-module M. Let $m := \dim M_1$. The mapping

$$M_0 \bigotimes \bigwedge^m M_1 \to \bigwedge^{m+1} M_0, \quad x_0 \otimes (x_1 \wedge \ldots \wedge x_m) \mapsto x_0 \wedge x_1 \wedge \ldots \wedge x_m,$$

determines an embedding of H-modules

$$V \bigotimes \bigwedge^m M_1 \hookrightarrow \bigwedge^{m+1} M_0 \subset \bigwedge^{m+1} M.$$

Our goal is to show that V can be embedded as an H-module in some rational finite-dimensional G-module. Since

$$V = (V \bigotimes \bigwedge^m M_1) \bigotimes (\bigwedge^m M_1)^*,$$

it suffices to prove this for $(\bigwedge^m M_1)^*$. On the other hand, $\bigwedge^m M_1$ is an H-submodule of a rational finite-dimensional G-module. (Namely, one can take $\bigwedge^m M$ as a G-module containing $\bigwedge^m M_1$.) In view of ($^+$) the dual module has the same property.
□

Theorem 1 (A.Bialynicki-Birula - G.Hochschild - G.D.Mostow [BiHoMo]) *Let G be a connected linear algebraic group, H an algebraic subgroup of G, and $X = G/H$. Then the following conditions are equivalent:*

(i) *H is an observable subgroup of G;*

(ii) *the algebra of regular functions $\mathbb{C}[X]$ separates points of X;*

(iii) *the field of quotients of $\mathbb{C}[X]$ coincides with $\mathbb{C}(X)$;*

(iv) *X is equivariantly isomorphic to an orbit in a rational representation space of G;*

(v) *X is a quasi-affine variety.*

Proof (i)$\Longrightarrow$(iii). Let f be a non-zero rational function on X regarded as a rational right H-invariant function on G. We shall consider $\mathbb{C}[G]$ as an H-module with H acting by right translations. Since G is an affine variety, f can be written as a quotient of two regular functions. Therefore

$$S := \mathbb{C}[G]f \cap \mathbb{C}[G] \neq \{0\}.$$

Obviously, S is an ideal in $\mathbb{C}[G]$ and S is H-invariant. We want to find in S a non-zero H-invariant element.

Take any non-zero irreducible H-submodule $V \subset S$ and denote by V^* the dual H-module. By assumption, V^* is an H-submodule of a finite-dimensional rational G-module W. Since V^* is irreducible, we can choose W to be also irreducible. (If there is an exact sequence of G-modules $0 \to W_1 \to W \to W_2 \to 0$ with $W_1 \neq \{0\}$ and $W_2 \neq \{0\}$ then either $V^* \subset W_1$ or V^* is H-isomorphic to a submodule of W_2.) Then W is isomorphic to a G-submodule of $\mathbb{C}[G]$ with G acting on $\mathbb{C}[G]$ by right translations. Let $\theta : V^* \to \mathbb{C}[G]$ be the resulting H-monomorphism. Choose a basis $v_1, \ldots, v_n$ of V so that $v_1(g_0) \neq 0$ and $v_i(g_0) = 0$ for each $i > 1$, where $g_0 \in G$ is some fixed point. Denote by $\alpha_1, \ldots, \alpha_n$ the dual basis of V^* and put $p_i := \theta(\alpha_i)$. The regular function

$$p_x(y) := \sum_{i=1}^{n} p_i(xy)v_i(y) \in S$$

is H-invariant for any fixed $x \in G$. It remains to show that $p_x \neq 0$ for some $x \in G$. By our construction one can take $x := g_1 g_0^{-1}$, where $g_1 \in G$ is chosen so that $p_1(g_1) \neq 0$.

(ii) $\Longrightarrow$ (iii). We use the following elementary fact. If $k \subset K \subset L$ are fields and L/k is an extension of finite type, then K/k is also an extension of finite type. In particular, let Q be the field of quotients of $\mathbb{C}[X]$. Since $\mathbb{C}(X)$ is finitely generated over $\mathbb{C}$, the subfield $Q \subset \mathbb{C}(X)$ also has this property. Therefore there exists a finitely generated subalgebra $A \subset \mathbb{C}[X]$ whose field of quotients is Q. Let Y be an affine variety with $\mathbb{C}[Y] = A$. Then we have a regular map $\varphi : X \to Y$ corresponding to the embedding $A \hookrightarrow \mathbb{C}[X]$. Since $\mathbb{C}[X]$ separates points of X, the map φ is injective. Hence

$$\dim X \leq \dim Y = \text{tr.deg}\, Q \leq \text{tr.deg}\, \mathbb{C}(X) = \dim X,$$

and, consequently, $\mathbb{C}(X)$ is a finite algebraic extension of Q. In this situation $[\mathbb{C}(X) : Q] = \text{card}\, \varphi^{-1}(y)$ for y in a Zariski open subset of Y. Since φ is injective, we get $Q = \mathbb{C}(X)$.

(iii) $\Longrightarrow$ (iv). Let $\mathbb{C}(X) = \mathbb{C}(f_1, \ldots, f_n)$. By assumption, $f_i = p_i/q_i$, where $p_i, q_i \in \mathbb{C}[X]$, $i = 1, \ldots, n$. Let V be a finite-dimensional G-submodule in $\mathbb{C}[X]$ containing $p_1, \ldots, p_n, q_1, \ldots, q_n$. Further, let $A \subset \mathbb{C}[X]$ be a subalgebra generated by V and Y an affine variety with $\mathbb{C}[Y] = A$. Then G acts regularly on Y, and the embedding $A \hookrightarrow \mathbb{C}[X]$ gives rise to an equivariant regular map $X \to Y$. Since elements of A separate points of X, this map is injective. On the other hand, Y is equivariantly embedded in the dual space V^*. Since X is homogeneous, we obtain an equivariant isomorphism of X onto an orbit in V^*.

(iv) $\Longrightarrow$ (v). If X is an orbit in a rational representation space, then X is Zariski open in its Zariski closure.

(v) $\Longrightarrow$ (ii) is obvious.

(v) $\Longrightarrow$ (i). Let $f \in \mathbb{C}[G]_\tau$, $f \neq 0$. The set $Z := \{g \in G \,|\, f(g) = 0\}$ is invariant under the right translations by H. Let Z' denote the image of Z in G/H. Then Z' is a Zariski closed subset of G/H. Since G/H is quasi-affine, we can find a non-zero function $p' \in \mathbb{C}[G/H]$ such that $p'|_{Z'} = 0$. Let $p \in \mathbb{C}[G]$ be the pull-back of p' so that $p|_Z = 0$. By Hilbert's Nullstellensatz there exists $q \in \mathbb{C}[G]$ such that $p^m = qf$ for some $m \in \mathbb{Z}^+$. Since $p(gh) = p(g)$ if $h \in H$, it follows that $q \in \mathbb{C}[G]_{\tau^{-1}}$. Thus $(^+)$ is fulfilled, and H is observable by Lemma 3. □

For the future use we prove the following proposition, which is also contained in [BiHoMo]. As usual, G° denotes the connected component of $e \in G$.

Theorem 2 *Let H be an algebraic subgroup of a linear algebraic group G. Then the following conditions are equivalent:*
(a) *$H \subset G$ is observable;*
(b) *$H^\circ \subset G^\circ$ is observable;*
(c) *$H \cap G^\circ \subset G^\circ$ is observable.*

Proof (a) $\Longrightarrow$(b). In view of Lemma 1 it suffices to prove that H° is observable in H. Let $h_1 = e, h_2, \ldots, h_n$ be a system of representatives in H for the elements of H/H°. For a rational representation $\varrho : H^\circ \to \mathrm{GL}(V)$ denote by W the vector space of all regular mappings $f : H \to V$ satisfying $f(xh) = \varrho(h)^{-1} f(x)$, where $x \in H, h \in H^\circ$. Then W is a rational H-module with respect to the action $f(x) \overset{k}{\mapsto} f(k^{-1}x)$ $(k \in H)$, and the map $\mu : W \to V,\ \mu(f) := f(e)$, is H°-equivariant. Given $v \in V$, one can define $f \in W$ by

$$f(h_i h) := \varrho(h)^{-1} v_i, \qquad i = 1, \ldots, n,$$

where $v_1 = v$ and $v_2, \ldots, v_n \in V$ are arbitrary. It is then clear that $\mu(f) = v$, showing that μ is an epimorphism. Thus H° is observable in H by Lemma 2.

(b)$\Longrightarrow$(c). We may assume here that G is connected. Let $X := G/H,\ X^\circ := G/H^\circ$, and $\Gamma := H/H^\circ$. In view of Theorem 1, we suppose that X° is quasi-affine. Then we have to prove that X is quasi-affine. The group Γ acts on X° by

$$\gamma \cdot gH^\circ = gh^{-1}H^\circ, \quad \text{where } g \in G,\ \gamma = hH^\circ \in \Gamma.$$

For $p \in \mathbb{C}[X^\circ]$ we write

$$p^\gamma(gH^\circ) := p(\gamma^{-1} \cdot gH^\circ).$$

Observe that p comes from X if and only if $p^\gamma = p$ for all $\gamma \in \Gamma$. Let $f \in \mathbb{C}(X)$ and denote by f° the pull-back of f to X°. By (ii) of Theorem 1 there exists $p \in \mathbb{C}[X^\circ],\ p \neq 0$, such that $pf^\circ \in \mathbb{C}[X^\circ]$. Since f° is Γ-invariant, we also have $p^\gamma f^\circ \in \mathbb{C}[X^\circ]$. Let

$$q^\circ := \prod_{\gamma \in \Gamma} p^\gamma.$$

Then $q^\circ f^\circ \in \mathbb{C}[X_0]$. Since q° is Γ-invariant, q° is a pull-back of some $q \in \mathbb{C}[X]$ to X°. It follows that $qf \in \mathbb{C}[X]$, showing that (ii) is fulfilled for X. Therefore X is quasi-affine.

(c)$\Longrightarrow$(a). Let $\tau \in \mathrm{X}(H)$ and $\tau^\circ := \tau|_{H \cap G^\circ}$. Then there exists a non-zero function $f \in \mathbb{C}[G^\circ]_{\tau^\circ}$. Since $G^\circ H$ is an open subgroup in G, we can define $f \in \mathbb{C}[G]$ by

$$f(x) := \begin{cases} f^\circ(g)\tau(h) & \text{if } x = gh \in G^\circ H, \\ 0 & \text{if } x \notin G^\circ H. \end{cases}$$

Then $f \in \mathbb{C}[G]_\tau$, and H is observable by Lemma 3. □

In the preceding section we have defined the class of spherical subgroups of connected reductive linear algebraic groups. In closing, we want to show that for observable subgroups this definition can be simplified in the following way.

Theorem 3 *Let G be a connected reductive linear algebraic group. An observable subgroup $H \subset G$ is spherical if and only if each irreducible rational G-module occurs in $\mathbb{C}[G/H]$ at most once.*

Proof Suppose this condition is fulfilled and let $\tau \in X(H)$. We have to show that dim $F_\delta(H, \tau) \leq 1$ for every $\delta \in \hat{K}$. Assuming the contrary, we get two linearly independent functions $f_1, f_2 \in \mathbb{C}[G]$, such that

$$f_i(bgh) = \chi(b)\tau(h)f_i(g)$$

for all $g \in G, h \in H$, and $b \in B$, where B is a Borel subgroup of G and $\chi \in X(B)$ is the highest weight corresponding to δ. Now, the one-dimensional H-module with character τ^{-1} is an H-submodule of some rational finite-dimensional G-module. As we have seen above, this implies that there exists an $f \in \mathbb{C}[G]$, $f \neq 0$, such that $f(gh) = \tau^{-1}(h)f(g)$ for all $g \in G, h \in H$. Consider the linear representation of G on $\mathbb{C}[G]$ by left translations and let M be the G-submodule of $\mathbb{C}[G]$ generated by f. Replacing f by a highest weight vector in M, we obtain a non-zero regular function $f \in \mathbb{C}[G]$, such that

$$f(bgh) = \chi'(b)\tau(h)^{-1}f(g)$$

for all $g \in G, h \in H$, and $b \in B$, where $\chi' \in X(B)$ is some rational character. But then ff_1 and ff_2 are two linearly independent weight vectors of B in $\mathbb{C}[G]$, having the same character $\chi\chi'$ and satisfying $ff_i(gh) = ff_i(g)$ for all $h \in H$. Therefore the G-module $\mathbb{C}[G/H]$ has two distinct isomorphic irreducible G-submodules. This contradiction completes the proof. □

Example Let $H = U$ be a maximal unipotent subgroup. We know already that the subgroup U is spherical, see Example 1 of §5.4. In this case each irreducible G-module occurs in $\mathbb{C}[G/U]$ exactly once, see the proposition in §4.3.

5.5 Holomorphically separable homogeneous manifolds

For any point x of a complex space X let

$$A_x := \{y \in X \mid f(y) = f(x) \quad \text{for all } f \in \mathcal{O}(X)\}.$$

Then A_x is an analytic set in X. The space X is said to be *holomorphically separable* (resp. *weakly holomorphically separable*) if for every $x \in X$ we have $A_x = \{x\}$ (resp. x is isolated in A_x). Thus, X is holomorphically separable if and only if for any two points $x, y \in X, x \neq y$, there exists a holomorphic function $f \in \mathcal{O}(X)$ such that $f(x) \neq f(y)$. In this section we consider (weakly) holomorphically separable homogeneous manifolds of the form $X = G/H$, where G is a reductive complex Lie group, $H \subset G$ a closed complex Lie subgroup. It is convenient to start with a more general setting. Namely, let H be an arbitrary subgroup of G, $\hat{H}$ the Zariski closure of H in G, and $D \subset G$ an open set which is invariant under the left translations by elements of K.

Theorem 1 *Assume that $D \cap G_0$ is connected and $D \cdot \hat{H} = D$. Let*

$$\mathcal{O}(D)^H := \{f \in \mathcal{O}(D) \mid f(zh) = f(z) \quad \textit{for all } z \in D, h \in H\}.$$

Assume further that the following condition is fulfilled:

$$z, w \in D,\ f(z) = f(w) \quad \textit{for all}\ f \in \mathcal{O}(D)^H \quad \Longrightarrow z = wh \quad \textit{for some}\ h \in H. \quad (*)$$

Then $H = \hat{H}$.

Proof Taking the trivial character τ in Theorem 1 of §5.3, we obtain that any $f \in \mathcal{O}(D)^H$ is right-invariant with respect to $\hat{H}$. Let $h \in \hat{H}$ and let $z \in D$ be an arbitrary point. Since $f(zh) = f(z)$ for any $f \in \mathcal{O}(D)^H$, it follows that $h \in H$. □

Corollary *Let H be a closed complex Lie subgroup of a connected reductive complex Lie group G. Denote by $\alpha : G \to G/H$ and $\beta : G/H \to G/\hat{H}$ the canonical projection mappings. Let $\hat{\Omega}$ be a K-invariant domain in $G/\hat{H}$ such that:*

(a) *the homomorphism $\pi_1(\hat{\Omega}) \to \pi_1(G/\hat{H})$ is onto;*

(b) $\Omega := \beta^{-1}(\hat{\Omega})$ *is a holomorphically separable open submanifold of G/H.*

Then $H = \hat{H}$.

Proof Let $D := \alpha^{-1}(\Omega)$. Then $D \cdot \hat{H} = D$. All vertical arrows in the diagram

$$\begin{array}{ccc} D & \subset & G \\ \downarrow & & \downarrow \alpha \\ \Omega & \subset & G/H \\ \downarrow & & \downarrow \beta \\ \hat{\Omega} & \subset & G/\hat{H} \end{array}$$

are the projection maps of locally trivial fiber bundles. It follows from (a) that D and Ω are connected. Since $\mathcal{O}(D)^H = \mathcal{O}(\Omega)$, the condition (*) is exactly (b). Thus $\hat{H} = H$ by Theorem 1. □

Example The existence of a holomorphically separable K-invariant domain in G/H does not imply in general that the subgroup H is algebraic. For example, let $G = \mathbb{C}^*$ so that $K = \mathrm{U}(1)$ and let H be the cyclic subgroup of G generated by some $d \in \mathbb{C}^*$ with $|d| > 1$. Then α maps the ring $\{z \in \mathbb{C} \mid 1 < |z| < r\},\ r < |d|$, biholomorphically onto a domain in the elliptic curve G/H. On the other hand, H is Zariski dense in G.

Theorem 2 *Let H be an algebraic subgroup of a connected complex reductive Lie group G, K a maximal compact subgroup of G, and Ω a K-invariant domain in $X = G/H$. If, in addition, Ω is weakly holomorphically separable then X is a quasi-affine algebraic variety. In particular, X is a holomorphically separable manifold.*

Proof We first reduce the proof to the case of connected H. Namely, let $X^\circ := G/H^\circ$ and let $\pi : X^\circ \to X$ be the covering map. Denote by Ω° be any connected component of $\pi^{-1}(\Omega)$. Obviously, Ω° is weakly holomorphically separable. If we know that X° is quasi-affine then H° is observable by Theorem 1 of §5.4. But then H is also observable by Theorem 2 of §5.4, and so X is quasi-affine.

We now assume that H is connected. Let $\alpha : G \to G/H$ be the canonical projection map and observe that $D := \alpha^{-1}(\Omega)$ is connected. Taking $\tau(h) = 1$ ($h \in H$) in Theorem 1 of §5.3, we obtain that $\mathbb{C}[X]$ is dense in $\mathcal{O}(\Omega)$. Let U be a neighborhood of $x \in \Omega$, such that $A_x \cap U = \{x\}$, and let $y \in U$, $y \neq x$. There exists a holomorphic function $f \in \mathcal{O}(\Omega)$ with $f(x) \neq f(y)$. Since $\mathbb{C}[X]$ is dense in $\mathcal{O}(\Omega)$, one can find a regular function $f \in \mathbb{C}[X]$ with the same property. Consider the regular map $\varphi : X \to Y$ defined in the proof of Theorem 1, (ii) $\Longrightarrow$ (iii), in § 5.4. We have $\varphi^{-1}(\varphi(x)) \cap U = \{x\}$. If A is chosen G-stable, then φ is equivariant. But then $\varphi : X \to \varphi(X)$ is an equivariant finite covering over a quasi-affine variety. Hence X is also quasi-affine by Theorem 2 of § 5.4. □

Let $X = G/H$ be a homogeneous manifold, where G is a connected complex Lie group, $H \subset G$ a closed complex Lie subgroup. Let $x_0 = e \cdot H$ and let J be a closed subgroup in G defined by

$$J := \{g \in G \mid g \cdot A_{x_0} = A_{x_0}\}.$$

From Proposition 2 of §1.7 it follows that J is a complex Lie subgroup. Obviously, $J \supset H$, and so we have the holomorphic map

$$\sigma : X = G/H \to G/J =: Y, \quad \sigma(gH) = gJ.$$

By definition, σ induces an algebra isomorphism $\mathcal{O}(X) \simeq \mathcal{O}(Y)$ and Y is holomorphically separable. The map σ is called the *holomorphic reduction* of X (cf. [Gi2]). It is clear that X is holomorphically separable (resp. weakly holomorphically separable) if and only if σ is an isomorphism (resp. a covering map).

Theorem 3 (W.Barth - M.Otte [BaOt2]) *Let G be a connected reductive complex Lie group, $H \subset G$ a closed complex Lie subgroup, and $X = G/H$. Then the following conditions are equivalent:*

(i) *X is weakly holomorphically separable;*
(ii) *X is holomorphically separable;*
(iii) *H is an algebraic subgroup and X is a quasi-affine algebraic variety.*

Proof Assume first that X is holomorphically separable. Applying the above corollary to $\hat{\Omega} := G/\hat{H}$, we see that H is Zariski closed. If X is only weakly holomorphically separable, then we obtain in the same way that J is Zariski closed. Since J/H is discrete, it follows that H is Zariski closed as well. For an algebraic subgroup $H \subset G$ we have Theorem 2. Namely, for such a subgroup (i) implies that X is quasi-affine. □

5.6 Stein homogeneous manifolds

Let G be a transformation group of a complex space X. Any element $g \in G$ induces a continuous automorphism $f \mapsto f \circ g$ of the Fréchet algebra $\mathcal{O}(X)$. Denote by $\mathcal{O}(X)^G$ the closed subalgebra of $\mathcal{O}(X)$ consisting of all invariant functions, i.e., $f \in \mathcal{O}(X)^G$ if and only if $f \circ g = f$ for all $g \in G$. The following result goes back to Y.Matsushima and A.Morimoto (see [MatMor], Théorème 5).

Theorem 1 *Let $\pi : X \to Y$ be a surjective holomorphic mapping between reduced complex spaces. Suppose that X is a Stein space acted on by a compact Lie transformation group K. Assume that the following conditions are fulfilled:*

(i) *$\pi(kx) = \pi(x)$ for all $x \in X, k \in K$;*

(ii) *the image of the mapping $\mathcal{O}(Y) \to \mathcal{O}(X)$, $f \mapsto f \circ \pi$, coincides with $\mathcal{O}(X)^K$. Then Y is also a Stein space.*

Proof Let $y_1, y_2 \in Y$ be two distinct points. Then $A_1 := \pi^{-1}(y_1)$ and $A_2 := \pi^{-1}(y_2)$ are closed analytic subsets in X. As a consequence of Theorem B, the restriction mapping

$$\mathcal{O}(X) \to \mathcal{O}(A_1 \cup A_2)$$

is onto. Since $A_1 \cap A_2 = \emptyset$, there exists $f \in \mathcal{O}(X)$ such that $f|_{A_1} = 1$, $f|_{A_2} = 0$. Consider the function

$$\hat{f}(x) := \int_K f(kx)d\mu(k),$$

where μ is the normalized Haar measure on K. Then $\hat{f} \in \mathcal{O}(X)^K$ and, in view of (ii), $\hat{f}$ can be regarded as a holomorphic function on Y. It is clear that this function separates the points y_1, y_2. In order to show that Y is holomorphically convex, consider a discrete infinite sequence $\{y_k\}$ in Y. Again by Theorem B there exists $f \in \mathcal{O}(X)$ such that $f|_{\pi^{-1}(y_k)} = k$ for all k. Applying the same averaging procedure as above, we obtain $\hat{f} \in \mathcal{O}(Y)$ with $\lim_{k\to\infty} \hat{f}(y_k) = \infty$. □

Corollary 1 *Let (X, Y, π) be a locally trivial holomorphic fiber bundle. Assume that the total space X is a Stein manifold acted on by a compact Lie transformation group K so that each fiber $\pi^{-1}(y)$ is a K-stable submanifold of X with $\mathcal{O}(\pi^{-1}(y))^K = \mathbb{C}$. Then Y is also a Stein manifold.*

Proof If $f \in \mathcal{O}(X)^K$ then f is constant on each fiber and therefore comes from the base. This yields (ii). Since (i) is also fulfilled, one can apply the above result. □

Recall that a reductive complex Lie group G is biholomorphic to an affine algebraic variety. In particular, G is a Stein manifold.

Corollary 2 *Let G be a complex Lie group and $H \subset G$ a closed reductive complex Lie subgroup. If G is a Stein manifold, then G/H is also a Stein manifold. In particular, if G and H are both reductive then G/H is Stein.*

Proof. Let L be a maximal compact subgroup of H. Consider the action of L on G, on H, and on itself by right translations. Then L has the natural linear representations on $\mathcal{O}(H)$ and $C(L)$. Furthermore, the restriction mapping $\mathcal{O}(H) \to C(L)$ is a homomorphism of L-modules. Therefore $\mathcal{O}(H)^L = \mathbb{C}$ by (**) of §5.3. Now, G is the total space of a locally trivial holomorphic fiber bundle over G/H. All fibers are L-stable and each of them has no non-constant L-invariant holomorphic functions. Therefore G/H is a Stein manifold by Corollary 1. □

Lemma *Let $U \subset \mathbb{C}^n$ be an open set, φ a non-negative C^2 strictly plurisubharmanic function on U, and $M \subset U$ a C^1 submanifold such that $\varphi|_M = 0$. Then M is totally real.*

Proof Let $p \in M$. Choose holomorphic coordinates $z_1, \ldots, z_n$ in $\mathbb{C}^n$ satisfying

$$z_i(p) = 0, \quad \frac{\partial^2 \varphi}{\partial z_k \partial \bar{z}_l}(0) = \delta_{kl}.$$

Putting

$$S = (s_{kl}),\ s_{kl} = \frac{\partial^2 \varphi}{\partial z_k \partial z_l}(0),$$

we obtain

$$\varphi(z) = |z|^2 + \mathrm{Re}(zSz^t) + o(|z|^2),$$

where $z = (z_1, \ldots, z_n)$, $|z| := \sqrt{|z_1|^2 + \ldots + |z_n|^2} \to 0$. The real tangent space to U at p is identified with $\mathbb{C}^n$. Assume that the tangent space to M contains a non-zero complex subspace. Since the expression $\mathrm{Re}(zSz^t)$ changes its sign under the transform $z \mapsto iz$, we conclude that

$$\mathrm{Re}(zSz^t) \geq 0$$

on some non-zero vector tangent to M. But then φ grows in the corresponding direction. Since M is contained in the minimum set of φ, we get a contradiction. □

Remark One can show that the zero set of φ is locally contained in a totally real C^1 submanifold, see [HaWe].

Theorem 2 (see [A5]) *A compact Lie transformation group K of a Stein manifold X has a totally real orbit.*

Proof Let $\varphi : X \to \mathbb{R}^+$ be a strictly plurisubharmanic exhaustion function. Then

$$\hat{\varphi}(x) = \int_K \varphi(kx) d\mu(k)$$

is also an exhaustion function and $\hat{\varphi}$ is also strictly plurisubharmonic. The minimum set of $\hat{\varphi}$ is K-stable. By the above lemma any K-orbit in this set is totally real. □

Proposition *Let G be a connected reductive complex Lie group, $H \subset G$ a closed complex Lie subgroup, K a maximal compact subgroup of G, and $X = G/H$. If X contains a K-invariant Stein domain Ω, then H° is reductive. Furthermore, let $\hat{H}$ be the Zariski closure of H, $\beta : G/H \to G/\hat{H}$ the canonical projection mapping, and $\hat{\Omega} = \beta(\Omega)$. If $\Omega = \beta^{-1}(\hat{\Omega})$ and the homomorphism $\pi_1(\hat{\Omega}) \to \pi_1(G/\hat{H})$ is onto, then $H = \hat{H}$ and H is reductive.*

Proof The second assertion follows from the first one and from the corollary in §5.5. We now prove the first assertion. Let Kx_0 be a totally real orbit in Ω and let $L := K_{x_0}$ be the corresponding isotropy subgroup. Replacing H by a conjugate subgroup, we may assume that $H = G_{x_0}$. Then L is a compact subgroup of H and $Kx_0 = K/L$ is a totally real submanifold of $X = G/H$. Obviously,

$$\dim_{\mathbb{R}} K/L \leq \dim_{\mathbb{C}} G/H.$$

Therefore

$$\dim_{\mathbb{R}} L \geq \dim_{\mathbb{C}} H.$$

Since $\mathfrak{l} \cap i\mathfrak{l} = \{0\}$, the last inequality implies that $\mathfrak{h} = \mathfrak{l} \oplus i\mathfrak{l}$. This means that H° has a compact real form. □

Theorem 3 (Y. Matsushima [Mat2], A.L.Onishchik [On1]) *Let G be a connected reductive complex Lie group, $H \subset G$ a closed complex Lie subgroup, and $X = G/H$. The following conditions are equivalent:*
(a) *H is reductive;*
(b) *X is a Stein manifold;*
(c) *H is an algebraic subgroup and X is an affine variety.*

Proof If H is reductive then X is Stein by Corollary 2. If X is Stein then the above proposition applies to $\Omega := X$, showing that H is reductive. This proves (a)$\Longleftrightarrow$(b). The implication (c)$\Longrightarrow$(b) is evident, and (a)$\Longrightarrow$(c) is contained in the next theorem. □

Let $\mathrm{R}_u(G)$ denote the unipotent radical of a linear algebraic group G.

Theorem 4 (see [Bi]) *Let G be a connected linear algebraic group and $H \subset G$ an algebraic subgroup such that $\mathrm{R}_u(H) \subset \mathrm{R}_u(G)$. Then $X = G/H$ is affine. In particular, if G is solvable then X is always affine.*

Proof (the ground field is $\mathbb{C}$) Assume first that G is unipotent. We claim that there exists a Zariski closed subset $A \subset G$, such that $G = A \cdot H$ and for every $g \in G$ the decomposition $g = a \cdot h$ with $a \in A, h \in H$ is unique. From this it follows that $G/H \simeq A$ is affine.

We use the induction on $d := \dim G/H$. For $d = 0$ there is nothing to prove. If $d > 0$ then by Engel's theorem there exists a vector $\mathrm{E} \in \mathfrak{g}$ such that $\mathrm{E} \not\in \mathfrak{h}$ and $[\mathfrak{h}, \mathrm{E}] \subset \mathfrak{h}$. Hence $\mathfrak{h}_1 := \mathbb{C} \cdot \mathrm{E} + \mathfrak{h}$ is a Lie algebra, and we denote by H_1 the corresponding connected Lie subgroup in G. Since G is unipotent, H_1 is automatically Zariski closed. By the induction hypothesis there is a decomposition $G = A_1 \cdot H_1$ with the above properties. Let $\varphi : G \to H_1$ be the associated projection map. Note that φ is a regular map between algebraic varieties, but not a group homomorphism. Furthermore, $\mathfrak{h}$ is an ideal of codimension 1 in $\mathfrak{h}_1$ and $H_1 = A_2 \ltimes H$, where $A_2 = \exp(\mathbb{C} \cdot \mathrm{E})$. Let $A := A_1 \cdot A_2$. Since $A = \varphi^{-1}(A_2)$, the set A is Zariski closed. Obviously, $G = A \cdot H$ and for every $g \in G$ the decomposition is unique.

Consider now the general case and write

$$G = L_G \ltimes \mathrm{R}_u(G), \quad H = L_H \ltimes \mathrm{R}_u(H),$$

where $L_G \subset G$ and $L_H \subset H$ are maximal reductive subgroups. Let $Z := G/\mathrm{R}_u(H)$. Then we have an isomorphism of algebraic varieties

$$Z \simeq L_G \times (\mathrm{R}_u(G)/\mathrm{R}_u(H)),$$

and it follows that Z is affine. There is a regular action of $L_H = H/\mathrm{R}_u(H)$ on Z whose orbits are exactly the fibers of the canonical mapping $Z = G/\mathrm{R}_u(H) \to G/H$. In this situation G/H is isomorphic to the categorical quotient of Z with respect to L_H in the sense of the geometric invariant theory (see e.g. T.A.Springer, *Aktionen reduktiver Gruppen auf Varietäten*, [KSS], p. 3 - 39). In particular, G/H is affine. □

Remark 1 Let H be a closed complex Lie subgroup of a connected complex Lie group G. Under certain restrictions on the pair (G, H) the holomorphic separability of G/H is already sufficient for G/H to be Stein. We list here several cases when this is known to be true:

1) $H = \{e\}$ [MatMor];
2) G is nilpotent [GiHu1];
3) G is solvable and H has finitely many connected components [HuOe3];
4) G is reductive and $H = \Gamma$ is discrete.

In the latter case Γ is in fact finite by Theorem 3 of §5.5. This certainly implies that G/Γ is Stein. In general, the quotient G/Γ with Γ discrete can be holomorphically separable but not Stein (see [Oel2], where $G = \mathrm{SL}(2,\mathbb{C}) \ltimes \mathbb{C}^2$).

Remark 2 Let $H \subset G$ be as above and let $\sigma : G/H \to G/J$ be the holomorphic reduction (see §5.5). Generally speaking, the fiber J/H can have non-constant holomorphic functions. For example, if G is reductive then J contains the Zariski closure of H. In particular, for

$$H = \left\{ \begin{pmatrix} 1 & n \\ 0 & 1 \end{pmatrix} \mid n \in \mathbb{Z} \right\} \subset G = \mathrm{SL}(2,\mathbb{C})$$

we have

$$J = \left\{ \begin{pmatrix} 1 & z \\ 0 & 1 \end{pmatrix} \mid z \in \mathbb{C} \right\}$$

so that the fiber $J/H \simeq \mathbb{C}^*$ is even a Stein manifold. However, under certain restrictions one can prove that the holomorphic reduction of G/H has the same properties as the corresponding mapping in the theory of holomorphically convex spaces. Namely, if G and H are subject to any of the assumptions 1), 2), 3) of Remark 1 then $\mathcal{O}(J/H) = \mathbb{C}$ and the base G/J is Stein. (In case 1) see [Mor]. The references for 2) and 3) are the same as above.)

Remark 3 If nothing is assumed about G, then no condition on $H \subset G$ is known which would be necessary and sufficient for G/H to be a Stein manifold. Some necessary and some sufficient conditions can be found in [Mat3], [Sn2].

Remark 4 Let X be a reduced Stein space with an action of a compact Lie group K. The *categorical quotient* $X//K$ is the quotient of X with respect to the equivalence relation

$$R := \{ (x_1, x_2) \in X \times X \mid f(x_1) = f(x_2) \quad \text{for all} \quad f \in \mathcal{O}(X)^K \}.$$

Let $\pi : X \to X//K$ be the canonical projection map. If $U \subset X//K$ is an open subset, then $\pi^{-1}(U)$ is open and K-invariant. Assigning to each open set $U \subset X//K$ the $\mathbb{C}$-algebra of K-invariant holomorphic functions on $\pi^{-1}(U)$, we get a presheaf on $X//K$. The corresponding sheaf is denoted by $\mathcal{O}_X^K$. The following theorem is due to P.Heinzner [Hei1]. The case when X is acted on by $K_{\mathbb{C}}$ was considered earlier by D.Snow [Sn1].

The $\mathbb{C}$-ringed space $(X//K, \mathcal{O}_X^K)$ is a complex space and π is a holomorphic mapping. The complex space $(X//K, \mathcal{O}_X^K)$ is Stein.

Note that the last assertion follows from Theorem 1. The result is an analytic version of the Finiteness Theorem in the theory of algebraic transformation groups (see [KSS], loc. cit.). Namely, the Fréchet algebra $\mathcal{O}(X)^K$ is isomorphic to the function algebra of the Stein space $X//K$ and therefore finitely generated as a topological algebra.

5.7 Observable subgroups

In this section we return to quasi-affine homogeneous varieties. Our goal is a characterization of observable subgroups, and so we start by defining a special class of such subgroups called quasiparabolic. It turns out that an algebraic subgroup H of a connected linear algebraic group G is observable if and only if H is embedded in a quasiparabolic subgroup Q in such a way that the unipotent radical of H is contained in the unipotent radical of Q (see Theorem 2).

Let G be a connected linear algebraic group, V an irreducible rational G-module. Then G has a unique closed orbit in $\mathbb{P}(V)$. Let $\pi : V \setminus \{0\} \to \mathbb{P}(V)$ be the canonical projection and let $v \in V$ be a non-zero vector, such that $\pi(v)$ is a point of the closed orbit. The isotropy subgroup $H := G_v$ is called a *quasiparabolic subgroup* of G. If V is non-trivial then H has codimension one in the parabolic subgroup $P := G_{\pi(v)}$. Since the orbit $Gv \simeq G/H$ is a quasi-affine variety, H is an observable subgroup of G.

We shall freely use the following properties of quasiparabolic subgroups. By definition, Q is the kernel of the character $\varphi \in \mathrm{X}(P)$ defined by $p{\cdot}v = \varphi(p)v$ $(p \in P)$. Therefore $Q \supset P'$ and $\mathrm{R}_u(Q) = \mathrm{R}_u(P)$. Also, P can be recovered from Q as the normalizer of Q in G. Indeed, if N is this normalizer then N is a parabolic subgroup in G containing P. Thus P/Q is a parabolic subgroup in N/Q. Since P/Q is abelian, it follows that $N/Q = P/Q$ and, consequently, $N = P$. Further, if Q is a quasiparabolic subgroup with normalizer P, then Q° is a quasiparabolic subgroup with the same normalizer. Namely, $Q^\circ = \mathrm{Ker}\, \varphi^\circ$, where $\varphi^\circ \in \mathrm{X}(P)$ and $(\varphi^\circ)^n = \varphi$ for some $n \in \mathbb{Z}, n \geq 1$. By definition, the kernel of a character of G is an example of a quasiparabolic subgroup. In particular, G is a quasiparabolic subgroup in G.

Observable subgroups are exactly the isotropy subgroups of vectors in rational G-modules. In the geometric invariant theory, there is a procedure reducing the study of arbitrary orbits to the study of orbits containing 0 in their closure. For such orbits we have the following theorem.

Hilbert-Mumford Theorem *Let G be a connected reductive algebraic subgroup in $\mathrm{GL}(V)$. Assume that the closure of an orbit Gv contains 0. Then there exists a one-parameter subgroup $\gamma : \mathbb{C}^* \to G$, such that*

$$\lim_{z \to 0} \gamma(z)v = 0.$$

We refer the reader to [Bir] for the beautiful proof due to R.Richardson. We need another formulation using the notion of the support of a vector $v \in V$. Let $T \subset G$ be a fixed algebraic torus, $r = \dim T$, and Δ the root system of $\mathfrak{g}$ with respect to $\mathfrak{t}$. We embed the character group $\mathrm{X}(T)$ in $\mathfrak{t}^*$ by taking the differential of a character at $e \in T$. Thus $\mathrm{X}(T)$ is identified with a subgroup of $\mathfrak{t}^*$ which is denoted by $\mathfrak{t}^*_{\mathbb{Z}}$. Let $\mathfrak{t}^*_{\mathbb{R}}$ (resp. $\mathfrak{t}^*_{\mathbb{Q}}$) be the real (resp. rational) vector subspace in $\mathfrak{t}^*$ generated by $\mathfrak{t}^*_{\mathbb{Z}}$. Then $\mathfrak{t}^*_{\mathbb{Z}} \subset \mathfrak{t}^*_{\mathbb{Q}} \subset \mathfrak{t}^*_{\mathbb{R}}$, each of these sets is stable under the action of the Weyl group $\mathcal{W}$, and $\mathfrak{t}^*_{\mathbb{Z}}$ is a lattice of rank r in $\mathfrak{t}^*_{\mathbb{R}}$. We introduce a positive $\mathcal{W}$-invariant inner product (λ, μ), $\lambda, \mu \in \mathfrak{t}^*_{\mathbb{R}}$, in such a way that $(\lambda, \mu) \in \mathbb{Q}$ for all $\lambda, \mu \in \mathfrak{t}^*_{\mathbb{Q}}$. The same notation is used for the complex extension of this form to $\mathfrak{t}^*$. The bilinear form on $\mathfrak{t}$ obtained by duality is the orthogonal sum of an arbitrary form on the center of $\mathfrak{g}$ and the Killing forms multiplied by some numbers on the intersections of $\mathfrak{t}$ with the simple ideals of $\mathfrak{g}$. In particular, $\lambda(\mathrm{H}_\alpha) = 2(\lambda, \alpha)/(\alpha, \alpha)$ for all $\lambda \in \mathfrak{t}^*$ and for each coroot H_α. As usual, we put $||\lambda|| := \sqrt{(\lambda, \lambda)}$, $d(\lambda, \mu) := ||\lambda - \mu||$, and $d(\lambda, C) := \inf_{\mu \in C} d(\lambda, \mu)$, where C is a subset in $\mathfrak{t}^*_{\mathbb{R}}$.

For $\lambda \in \mathfrak{t}^*_{\mathbb{Z}}$ define a subspace $V_\lambda \subset V$ by

$$V_\lambda := \{v \in V \mid \mathrm{H}v = \lambda(\mathrm{H})v \quad \text{for all} \quad \mathrm{H} \in \mathfrak{t}\}.$$

Then $\mathcal{P}(V) := \{\lambda \in \mathfrak{t}^*_{\mathbb{Z}} \mid V_\lambda \neq 0\}$ is a finite set, called the it set of weights of V with respect to T, and we have the *weight decomposition*

$$V = \oplus_{\lambda \in \mathcal{P}(V)} V_\lambda.$$

The projection of $v \in V$ onto V_λ is denoted by v_λ. Thus

$$v = \sum_{\lambda \in \mathcal{P}(v)} v_\lambda, \quad \text{where} \quad \mathcal{P}(v) := \{\lambda \in \mathcal{P}(V) \mid v_\lambda \neq 0\}.$$

The *support* of a vector $v \in V$ is, by definition, the convex hull of $\mathcal{P}(v)$ in $\mathfrak{t}^*_{\mathbb{R}}$. The support of $v \in V$ is denoted by $\operatorname{supp}(v)$.

Proposition 1 *The closure $\overline{Gv}$ contains 0 if and only if there exists an element $g \in G$, such that $0 \notin \operatorname{supp}(gv)$.*

Proof If $0 \notin \operatorname{supp}(gv)$, then there exists a real linear form on $\mathfrak{t}^*_{\mathbb{R}}$ having positive values on $\mathcal{P}(gv)$. This form is given by $\lambda \mapsto \lambda(\mathrm{H})$, where $\mathrm{H} \in \mathfrak{t}$. Therefore

$$\exp(t\mathrm{H})gv = \sum_{\lambda \in \mathcal{P}(gv)} e^{t\lambda(\mathrm{H})}(gv)_\lambda \to 0 \qquad \text{as } t \to -\infty. \tag{1}$$

Conversely, let $0 \in \overline{Gv}$. By the Hilbert-Mumford theorem there exist $g \in G$ and $\mathrm{H} \in \mathfrak{t}$, such that (1) holds and, in addition, $\lambda(\mathrm{H}) \in \mathbb{Z}$ for all $\lambda \in \mathfrak{t}^*_{\mathbb{Z}}$. This implies $\lambda(\mathrm{H}) > 0$ for all $\lambda \in \mathcal{P}(v)$, showing that $0 \notin \operatorname{supp}(gv)$. □

In what follows we use the notation

$$C_1 \pm C_2 := \{\lambda_1 \pm \lambda_2 \mid \lambda_1 \in C_1, \lambda_2 \in C_2\}, \quad a \cdot C := \{a\lambda \mid \lambda \in C\},$$

where $C, C_1, C_2 \subset \mathfrak{t}^*_{\mathbb{R}}$, $a \in \mathbb{R}$.

Lemma 1 *Let $\mathrm{X}_\alpha \in \mathfrak{g}_\alpha$ be a root vector. Then*

$$\operatorname{supp}(\exp(t\mathrm{X}_\alpha)v) \subset \operatorname{supp}(v) + \mathbb{R}^+\alpha.$$

Proof Since $\operatorname{supp}(v_1 + \ldots + v_m)$ is contained in the convex hull of $\cup_{i=1}^m \operatorname{supp}(v_i)$ for any $v_1, \ldots, v_m \in V$, it is enough to prove the assertion for each weight vector v_λ, where $\lambda \in \mathcal{P}(v)$. But

$$\mathcal{P}(\exp(t\mathrm{X}_\alpha)v_\lambda) \subset \lambda + \mathbb{Z}^+\alpha,$$

hence

$$\operatorname{supp}(\exp(tX_\alpha)v_\lambda) \subset \lambda + \mathbb{R}^+\alpha. \qquad \square$$

Lemma 2 *Consider the action of G on $V \otimes \ldots \otimes V$ (m times). Then*

$$\operatorname{supp}(v \otimes \ldots \otimes v) = m \cdot \operatorname{supp}(v).$$

Proof Let $\mathcal{P}(v) = \{\lambda_1, \ldots, \lambda_r\}$, $\lambda_i \neq \lambda_j$. Then

$$\mathcal{P}(v \otimes \ldots \otimes v) = \{\lambda_{i_1} + \ldots + \lambda_{i_m} \mid 1 \leq i_1 \leq r, \ldots, 1 \leq i_m \leq r\}.$$

Since

$$\lambda_{i_1} + \ldots + \lambda_{i_m} = m(\frac{1}{m}\lambda_{i_1} + \ldots + \frac{1}{m}\lambda_{i_m}) \in m \cdot \operatorname{supp}(v),$$

it follows that

$$\operatorname{supp}(v \otimes \ldots \otimes v) \subset m \cdot \operatorname{supp}(v).$$

On the other hand, if $\lambda = \sum t_i\lambda_i$, $\sum t_i = 1$, $t_i \geq 0$, then

$$m\lambda = \sum t_i(m\lambda_i) \in \operatorname{supp}(v \otimes \ldots \otimes v). \qquad \square$$

Theorem 1 (F.A.Bogomolov [Bog]) *Suppose that $0 \in \overline{Gv}$. Then the isotropy subgroup G_v is contained in a proper quasiparabolic subgroup.*

Proof Replacing v by another vector in the same G-orbit, we may assume that

$$d(0, \operatorname{supp}(v)) \geq d(0, \operatorname{supp}(gv)) \tag{2}$$

for all $g \in G$. Then $0 \notin \operatorname{supp}(v)$ by Proposition 1. Define Λ to be a vector of minimal length in $\operatorname{supp}(v)$, so that $d := ||\Lambda|| = d(0, \operatorname{supp}(v))$. Let $F \subset \operatorname{supp}(v)$ be the face of minimal dimension containing Λ and let $k = \dim F$. Then F is contained in an affine subspace of the form $\lambda_0 + \mathbb{R}\lambda_1 + \ldots + \mathbb{R}\lambda_k$, where $\lambda_1, \ldots, \lambda_k \in \mathfrak{t}^*_{\mathbb{Q}}$ are linearly independent. Since Λ is an interior point of F, it follows that $(\Lambda, \lambda_i) = 0$ for all i, $i = 1, \ldots, k$. Letting $\Lambda = \lambda_0 + \sum_{i=1}^k t_i\lambda_i$, we get the system of linear equations

$$\sum_{i=1}^k (\lambda_i, \lambda_j)t_i = -(\lambda_0, \lambda_j), \qquad j = 1, \ldots, k,$$

whose coefficients are rational numbers. Therefore $\Lambda \in \mathfrak{t}^*_{\mathbb{Q}}$.

There exists a positive integer m such that $m\Lambda \in \mathfrak{t}^*_{\mathbb{Z}}$. Consider the action of G in the tensor product $V \otimes \ldots \otimes V$ (m times). Then

$$\overline{G(v \otimes \ldots \otimes v)} \ni 0$$

and

$$G_v \subset G_{v \otimes \ldots \otimes v}.$$

Thus it suffuces to prove the theorem for $v \otimes \ldots \otimes v$ instead of v. On the other hand, Lemma 2 shows that $v \otimes \ldots \otimes v$ satisfies the condition similar to (2) and that $m\Lambda$ is a vector of minimal length in $\mathrm{supp}(v \otimes \ldots \otimes v)$. Replacing $v \in V$ by $v \otimes \ldots \otimes v \in V \otimes \ldots \otimes V$, we reduce the proof to the case $\Lambda \in \mathfrak{t}^*_{\mathbb{Z}}$.

Choose an ordering in Δ satisfying

$$(\Lambda, \alpha) > 0 \Longrightarrow \alpha \in \Delta^+.$$

Let

$$\mathfrak{p} := \mathfrak{t} \oplus \bigoplus_{(\Lambda,\alpha)\geq 0} \mathfrak{g}_\alpha, \qquad \mathfrak{l} := \mathfrak{t} \oplus \bigoplus_{(\Lambda,\alpha)=0} \mathfrak{g}_\alpha.$$

Then $\mathfrak{p}$ is a parabolic subalgebra in $\mathfrak{g}$ and $\mathfrak{l}$ is a maximal reductive subalgebra of $\mathfrak{p}$. The corresponding groups are denoted by P and L. Since $\Lambda(\mathrm{H}_\alpha) \geq 0$ for all $\alpha \in \Delta^+$, there is an irreducible G-module V^Λ with highest weight Λ. Let Q be the isotropy subgroup of the highest weight vector in V^Λ. Then Q is a normal subgroup of codimension 1 in P, more precisely, $Q = \mathrm{Ker}\,\varphi$, where $\varphi \in \mathrm{X}(P)$ satisfies $\dot{\varphi}|_{\mathfrak{t}} = \Lambda$.

We show first that $G_v \subset P$. From the Bruhat decomposition it follows that $G = PN(T)P$, where $N(T)$ is the normalizer of T in G. Let $g = (p_1)^{-1} n p_2 \in G_v$, where $p_1, p_2 \in P$, $n \in N(T)$. Then

$$p_1 v = n p_2 v. \tag{3}$$

Since P is generated by T and the roots subgroups $\{\exp(t\mathrm{X}_\alpha)\}$ with $(\Lambda, \alpha) \geq 0$, it follows from Lemma 1 that $(\Lambda, \lambda) \geq d^2$ for all $\lambda \in \mathrm{supp}(pv)$, $p \in P$. Thus $d(0, \mathrm{supp}(pv)) \geq d(0, \mathrm{supp}(v))$, and (2) yields

$$d(0, \mathrm{supp}(pv)) = d(0, \mathrm{supp}(v)) = d$$

for all $p \in P$. Therefore the (unique) vector of minimal length in $\mathrm{supp}(pv)$ is Λ. From (3) and from the definition of a support we obtain $\mathrm{supp}(p_1 v) = w(\mathrm{supp}(p_2 v))$, where $w = nT \in \mathcal{W}$. Hence $w(\Lambda) = \Lambda$, showing that n normalizes P. Since a parabolic subgroup coincides with its normalizer, it follows that $n \in P$ and $g \in P$.

We show next that the connected component $(G_v)^\circ$ of G_v is contained in Q. Assuming the contrary we obtain $\varphi((G_v)^\circ) = \mathbb{C}^*$. Let E, E_+, and E_+° be an affine hyperplane, a closed half-space, and an open half-space in $\mathfrak{t}^*_{\mathbb{R}}$ defined by the equality $(\Lambda, \lambda) = d^2$ and by the inequalities $(\Lambda, \lambda) \geq d^2$ and $(\Lambda, \lambda) > d^2$ respectively. Put

$$V_+ := \bigoplus_{\lambda \in \mathcal{P}(V) \cap E_+} V_\lambda, \quad V_+^\circ := \bigoplus_{\lambda \in \mathcal{P}(V) \cap E_+^\circ} V_\lambda,$$

and observe that $V_+^\circ \subset V_+$ are P-invariant subspaces. The action of the unipotent radical of P on $V' := V_+/V_+^\circ$ is trivial. Denote by $\pi : V_+ \to V'$ the canonical epimorphism and let $v' := \pi(v)$. Since π is L-equivariant and, in particular, T-equivariant, v' is the sum of all $\pi(v_\lambda)$, where $\lambda \in \mathcal{P}(v) \cap E$. Define a new action of L in V' by

$$x \overset{l}{\mapsto} \varphi(l)^{-1} l x, \qquad x \in V', l \in L.$$

Then

$$\mathrm{supp}(lv') = \mathrm{supp}(lv) \cap E - \{\Lambda\},$$

where the support on the left hand side is taken with respect to the new L-action in V'. It follows that $\mathrm{supp}(lv')$ contains 0 for any $l \in L$. However, according to our assumption, there is a one-parameter subgroup $\gamma : \mathbb{C}^* \to G_v$, such that $\varphi(\gamma(z)) = z$ for all $z \in \mathbb{C}^*$. Furthermore, we can choose γ so that $\gamma(\mathbb{C}^*) \subset L \cap G_v$. Hence $\overline{Lv'}$ contains 0, and we obtain a contradiction with Proposition 1.

Since $(G_v)^\circ \subset Q$, it follows that $\varphi(G_v)$ is a finite subgroup in $\mathbb{C}^*$. Therefore $G_v \subset \operatorname{Ker} \varphi^k$ for some $k \geq 1$. But $\operatorname{Ker} \varphi^k$ is another quasiparabolic subgroup in P. This completes the proof. □

Remark F.A.Bogomolov proved in fact that $\overline{Gv}$ admits an equivariant regular map onto the closure of the orbit of a highest weight vector in an irreducible G-module. Another proof can be found in [Br3].

So far we considered the observable subgroups which correspond to the orbits containing 0 in their closure. We now proceed to the general case. Let G be a (not necessarily connected) linear algebraic group. As usual, $\mathrm{R}_u(G)$ denotes the unipotent radical of G. Obviously, $\mathrm{R}_u(G) = \mathrm{R}_u(G^\circ)$. We say that an algebraic subgroup $H \subset G$ is *canonically embedded* in G if $\mathrm{R}_u(H) \subset \mathrm{R}_u(G)$. A canonical embedding is denoted by $H \sqsubset G$. Observe that

$$\mathrm{R}_u(H\mathrm{R}_u(G)) = \mathrm{R}_u(H)\mathrm{R}_u(G),$$

hence $H \sqsubset H\mathrm{R}_u(G)$ for any subgroup $H \subset G$.

Proposition 2 *If $H \subset G$ is observable then $H\mathrm{R}_u(G)$ is also observable.*

Proof Assume we know this for G connected. By Theorem 2 of §5.4 the intersection $H \cap G^\circ$ is an observable subgroup in G°. Thus

$$H\mathrm{R}_u(G) \cap G^\circ = (H \cap G^\circ)\mathrm{R}_u(G)$$

is also observable in G°. But then by the same theorem $H\mathrm{R}_u(G)$ is observable in G.

We now prove the assertion for G connected. As usual, a rational (resp. a regular) function on G/H is regarded as a rational (resp. a regular) function on G invariant under H acting by right translations. We put $(l_g f)(x) := f(g^{-1}x),\ x, g \in G$. Let $f \in \mathbb{C}(G/H\mathrm{R}_u(G)) \subset \mathbb{C}(G/H)$. By (iii) of §5.4 we can write f in the form $f = f_1/f_2$, where $f_i \in \mathbb{C}[G/H] = \mathbb{C}[G]^H$. Then

$$(l_g f_1)(x) = f(g^{-1}x) \cdot (l_g f_2)(x) = f(x(x^{-1}g^{-1}x)) \cdot (l_g f_2)(x) = f(x) \cdot (l_g f_2)(x)$$

for any $g \in \mathrm{R}_u(G)$. Thus $f = l_g f_1 / l_g f_2,\ \ g \in \mathrm{R}_u(G)$, and $l_g f_1, l_g f_2 \in \mathbb{C}[G/H]$. A unipotent algebraic group has a non-zero fixed vector in any rational representation space. In particular, there is a function $\tilde{f}_1 := \sum c_g l_g f_1 \neq 0$ such that $l_g \tilde{f}_1 = \tilde{f}_1$ for all $g \in \mathrm{R}_u(G)$. Letting $\tilde{f}_2 := \sum c_g l_g f_2$ with the same c_g, we obtain $f = \tilde{f}_1/\tilde{f}_2$, where the numerator and the denominator are invariant under right translations by elements of H and left translations by elements of $\mathrm{R}_u(G)$. But then $\tilde{f}_i(xg) = \tilde{f}_i((xgx^{-1})x) = \tilde{f}_1(x),\ \ x \in G,\ g \in \mathrm{R}_u(G),\ i = 1, 2$, showing that $\tilde{f}_1, \tilde{f}_2 \in \mathbb{C}[G]^{H\mathrm{R}_u(G)}$. Therefore $H\mathrm{R}_u(G) \subset G$ is observable by (iii) of §5.4. □

Proposition 3 *Let G be a connected reductive linear algebraic group, $P \subset G$ a parabolic subgroup, and $H \subset G$ an observable subgroup such that $P' \subset H \subset P$. Then $H \sqsubset Q$, where Q is a quasiparabolic subgroup of G contained in P.*

Proof Since G/H is quasi-affine, there exists a rational G-module V and a vector $v \in V$, such that $G_v = H$. Let W be a vector subspace in V generated by Pv.

Then W is P-stable and H acts in W trivially. Thus W is acted on by the algebraic torus P/H, and we have the weight decomposition

$$W = W_1 \oplus \ldots \oplus W_m, \quad W_i = \{w \in W \mid pw = \varphi_i(p)w \quad \text{for all} \quad p \in P\} \neq \{0\},$$

where $\varphi_i \in \mathrm{X}(P)$, $\varphi_i \neq \varphi_j$. Note that $v = \sum_{i=1}^m c_i v_i$, where $c_i \neq 0$ for all i, $i = 1, \ldots, m$. Hence $\varphi_i|_H = 1$ for every i. Let

$$P_i := \{g \in G \mid gW_i = W_i \text{ and } g|_{W_i} \text{ is a scalar operator}\}, \; i = 1, \ldots, m.$$

Then P_i is a parabolic subgroup containing P, and φ_i extends to P_i by $gw = \varphi_i(g)w, \quad g \in P_i, w \in W_i$. We claim that

$$P = P_1 \cap \ldots \cap P_m. \tag{4}$$

In order to show this, fix a maximal torus $T \subset G$ in P, choose an ordering of the root system Δ so that $B^+ \subset P$, and denote by Π the set of simple roots. A parabolic subgroup containing B^+ is determined by a subset of Π (see §3.1). Let Φ, $\Phi_1, \ldots, \Phi_m$ be these subsets for P, $P_1, \ldots, P_m$ respectively. For $\alpha \in \Pi$ we have $\alpha \in \Phi$ (resp. $\alpha \in \Phi_i$) if and only if $\mathfrak{g}_{-\alpha} \subset \mathfrak{p}$ (resp. $\mathfrak{g}_{-\alpha} \subset \mathfrak{p}_i$). From the definition of P_i it is clear that

$$\Phi_i = \{\alpha \in \Pi \mid \lambda_i(\mathrm{H}_\alpha) = 0\},$$

where $\lambda_i \in \mathcal{P}^+$ is the differential of φ_i on $\mathfrak{t}$. Assuming $\alpha \in \Phi_1 \cap \ldots \cap \Phi_m$, consider the orbit $S^{(\alpha)}v$, where $S^{(\alpha)}$ is the connected three-dimensional subgroup with Lie algebra $\mathbb{C}\cdot \mathrm{E}_\alpha \oplus \mathbb{C}\cdot \mathrm{H}_\alpha \oplus \mathbb{C}\cdot \mathrm{F}_\alpha$, $\mathrm{E}_\alpha \in \mathfrak{g}_\alpha, \mathrm{F}_\alpha \in \mathfrak{g}_{-\alpha}$. Since $\mathrm{H}_\alpha \in \mathfrak{h}$ by assumption, it follows that $(S^{(\alpha)} \cap B^+)v = \{v\}$, and so $S^{(\alpha)}v$ is compact. Therefore $S^{(\alpha)}v = \{v\}$, hence $\mathrm{F}_\alpha \in \mathfrak{h} \subset \mathfrak{p}$ and $\alpha \in \Phi$, showing that

$$P_1 \cap \ldots \cap P_m \subset P.$$

Since the opposite inclusion is clear, we obtain (4).

Now, let $\lambda := \lambda_1 + \ldots + \lambda_m$. Then $\lambda \in \mathcal{P}^+$ and $\lambda(\mathrm{H}_\alpha) = 0$ if and only if $\lambda_i(\mathrm{H}_\alpha) = 0$ for all i, $i = 1, \ldots, m$. It follows from (4) that $\varphi := \varphi_1 \cdot \ldots \cdot \varphi_m \in \mathrm{X}(P)$ does not extend to a larger parabolic subgroup. Thus $Q := \mathrm{Ker}\, \varphi \subset P$ is quasiparabolic. Since $\varphi_i|_H = 1$ for all i, we have $H \subset Q$. Finally, $\mathrm{R}_u(H) = \mathrm{R}_u(P) = \mathrm{R}_u(Q)$, hence $H \sqsubset Q$. □

Proposition 4 *Let G be a (not necessarily connected) reductive algebraic subgroup in* $\mathrm{GL}(V)$ *and let $v \in V$. If the isotropy subgroup G_v is not contained in a proper reductive subgroup of G, then G has a fixed point in $\overline{Gv}$.*

Proof This is a well-known consequence of Luna's slice theorem, see [Lu], Cor.2. □

The following characterization of observable subgroups is due to A.A.Sukhanov (for connected subgroups).

Theorem 2 (cf. [Su]) *Let G be a connected linear algebraic group. An algebraic subgroup $H \subset G$ is observable if and only if $H \sqsubset Q$, where Q is a quasiparabolic subgroup of G.*

Proof If H is canonically embedded in a quasiparabolic subgroup Q then $Q^\circ/H\cap Q^\circ$ is affine by Theorem 4 of § 5.6. Thus $H \cap Q^\circ$ is observable in Q°. Hence H is observable in Q by Theorem 2 of §5.4. But then Lemma 1 of §5.4 tells us that H is observable in G.

The converse will be proved by induction on dim G and, for dim G fixed, by induction on dim G/H. The subgroup $H\mathrm{R}_u(G)$ is observable by Proposition 2, and the isomorphism

$$(G/\mathrm{R}_u(G))/(H\mathrm{R}_u(G)/\mathrm{R}_u(G)) \simeq G/H\mathrm{R}_u(G)$$

shows the variety on the left hand side is quasi-affine. If $\mathrm{R}_u(G) \neq \{e\}$ then, by the induction hypothesis, $H\mathrm{R}_u(G)/\mathrm{R}_u(G)$ is canonically embedded in a quasiparabolic subgroup $\tilde{Q} \subset G/\mathrm{R}_u(G)$. Let Q be the preimage of $\tilde{Q}$ under the canonical epimorphism $G \to G/\mathrm{R}_u(G)$. Then Q is a quasiparabolic subgroup in G, and $H \sqsubset H\mathrm{R}_u(G) \sqsubset Q$. Thus we may assume that G is reductive. If H is also reductive then $H \sqsubset G$, and the proof is complete.

For H non-reductive we show first that there is a *proper* quasiparabolic subgroup $Q \subset G$ containing H. Let M be a minimal reductive subgroup of G containing H. Take an embedding $G/H \simeq Gv \hookrightarrow V$, where V is a rational G-module, and put $Z := \overline{Mv}$. By Proposition 4 the group M has a fixed point $v_0 \in Z$. Let $V^M \subset V$ be the subspace of M-stable vectors, $V' \subset V$ an M-invariant complement, $\pi : V \to V'$ the projection map defined by the decomposition $V = V^M \oplus V'$, and $v' := \pi(v)$. Since $\overline{Mv} \ni v_0$ and $v_0 \in V^M$, it follows that $\overline{Mv'} \ni 0$. Since $M \subset G$ is observable, there exists an M-equivariant embedding $V' \hookrightarrow W$, where W is a rational G-module. Obviously, $\overline{Gv'} \ni 0$ in W, and Theorem 1 implies that $G_{v'} \subset Q$, where Q is a proper quasiparabolic subgroup of G. But

$$H = M_v \subset M_{v'} \subset G_{v'},$$

and it follows that $H \subset Q$. This embedding is not necessarily canonical, and so we have to change Q. Let $P = \mathrm{Norm}_G(Q)$. For the sake of brevity we put $U := \mathrm{R}_u(P) = \mathrm{R}_u(Q)$. Since $H \sqsubset HU$, we may assume that $H \supset U$. Then

$$Q/H \simeq (Q/U)/(H/U),$$

showing that H/U is an observable subgroup in Q/U. From Theorem 2 of §5.4 we obtain that $(H \cap Q^\circ)/U$ is an observable subgroup in Q°/U. By the induction hypothesis there exists a quasiparabolic subgroup $\tilde{I} \subset Q^\circ/U$ such that $(H \cap Q^\circ)/U \sqsubset \tilde{I}$. Denote by I the preimage of $\tilde{I}$ in Q° under the canonical epimorphism $Q^\circ \to Q^\circ/U$. Then $H \cap Q^\circ \sqsubset I \subset Q^\circ$, and I is an observable subgroup in G by Lemma 1 of §5.4. Let $\tau : Q \to Q/Q^\circ$ be the canonical map. Since $P \supset Q \supset Q^\circ \supset P'$, the group Q/Q° is abelian. Therefore $H^* := \tau^{-1}(\tau(H))$ is a normal subgroup in Q. Hence $J := IH^* \subset Q$ is a group containing I as a subgroup of finite index. In particular, J is an observable algebraic subgroup in Q and, consequently, in G. Note that $H \sqsubset J$. If dim $H <$ dim J, then we can apply the induction hypothesis to J and thus finish the proof. On the other hand, if dim $H =$ dim J then $H^\circ = J^\circ = I^\circ$. Thus, H° is a quasiparabolic subgroup in Q° and $F := \mathrm{Norm}_{Q^\circ}(H^\circ)$ is a parabolic subgroup in Q°. Note that $\mathrm{R}_u(F) \subset F'U$. For any $p \in P$ the subgroup $pFp^{-1} \subset Q^\circ$ is also parabolic, and so there exists $q \in Q^\circ$ such that $pFp^{-1} = qFq^{-1}$. Thus $P = Q^\circ P^*$, where $P^* := \mathrm{Norm}_P(F)$.

Since $\mathrm{Norm}_{Q^\circ}(F) = F$, we have $Q^\circ \cap P^* = F$. Therefore $P/P^* \simeq Q^\circ/F$ is a flag manifold, showing that P^* is a parabolic subgroup in P and in G. Furthermore, $P^*/F \simeq P/Q^\circ \simeq \mathbb{C}^*$ as groups, hence

$$(P^*)' \subset F'\mathrm{R}_u(F) \subset F'U \subset H \subset P^*,$$

and Proposition 3 applies. This completes the proof. □

5.8 Invariant plurisubharmonic functions and geodesic convexity

Let G be a connected reductive complex Lie group and $K \subset G$ a maximal compact subgroup. The orbits of the left K-action on G form the coset space $K\backslash G$. Denote by τ the canonical projection mapping $G \longrightarrow K\backslash G$ and let f be a function, defined in a domain $D \subset G$. If D and f are left K-invariant, then f determines a function in $\tau(D)$, which will be denoted by $\tilde{f}$.

The homogeneous manifold $K\backslash G$ can be equipped with a G-invariant Riemannian metric. Considered with any such metric, $K\backslash G$ is a symmetric space. More precisely, $K\backslash G$ is the product of a Euclidean space and a symmetric space of non-compact type. In particular, $K\backslash G$ has non-positive curvature. Since $K\backslash G$ is simply connected, any two points of this manifold can be joined by a unique geodesic (see [He], Ch.1, Theorem 13.3).

In this section we prove that if f is plurisubharmonic in D then $\tilde{f}$ is convex along geodesic segments in $\tau(D)$. We also discuss the conditions under which the plurisubharmonicity of f follows from the geodesic convexity of $\tilde{f}$. As an application we show that if D is a left K-invariant Stein domain in G then $\tau(D)$ is convex with respect to the family of all geodesics on $K\backslash G$.

In order to reduce the problem to smooth plurisubharmonic functions, we shall need the following fact.

Lemma 1 *Let G be a connected complex Lie group, D a domain in G, and f a plurisubharmonic function in D. Then for each $\varrho \in (0,1]$ there exist an open set $D_\varrho \subset D$ and a C^∞ plurisubharmonic function $f_\varrho : D_\varrho \to \mathbb{R}$ such that $D_\varrho \subset D_{\varrho'}$ if $\varrho' < \varrho$,*

$$\bigcup_{\varrho\in(0,1]} D_\varrho = D,$$

and

$$f(x) = \lim_{\varrho\to 0} f_\varrho(x)$$

for all $x \in D$. Moreover, if D and f are left invariant with respect to some subgroup of G, then D_ϱ and f_ϱ have the same property.

Proof Fix a coordinate neighborhood of $e \in G$, which is biholomorphically equivalent to the polydisk

$$\Delta_a^n = \{z = (z_1, \ldots, z_n) \in \mathbb{C}^n \,|\, |z_j| < a \quad \text{for all} \;\; j, \;\; j = 1, \ldots, n\},$$

where $a > 1$. The coordinate functions are also denoted by $z_1, \ldots, z_n$. For $\varrho \le 1$ let U_ϱ be the open neighborhood of e which is mapped onto Δ_ϱ^n by $(z_1, \ldots, z_n)$. In particular, $U := U_1$ is identified with Δ_1^n and we have the equality

$$d\mu(z) = c(z)d\lambda(z) \qquad (z \in U),$$

where μ is a left invariant Haar measure on G and λ is the Lebesgue measure on Δ_1^n. Choose μ so that $c(z) \le 1$. Let $\alpha(z) = \alpha(x+iy)$ be a non-negative C^∞ function on $\mathbb{C}$ with support in the unit disk, which depends only on $|z|$ and satisfies

$$\int_{\mathbb{C}} \alpha(x+iy)dxdy = 1.$$

Put

$$\sigma(z) := \alpha(z_1)\dots\alpha(z_n) \quad (z = (z_1,\dots,z_n) \in \Delta_1^n)$$

and

$$\sigma_\varrho(z) := \frac{1}{\varrho^{2n}}\sigma\left(\frac{z}{\varrho}\right) \qquad (z \in \Delta_\varrho^n,\ \ 0 < \varrho \le 1).$$

Then σ_ϱ can be considered as a C^∞ function on G with support in U_ϱ. Due to our choice of μ we have

$$\int_G \sigma_\varrho(z)\cdot d\mu(z) \le 1.$$

Let

$$D_\varrho := \{x \in D \mid x\cdot \overline{U}_\varrho \subset D\}.$$

It is clear that $D_\varrho \subset D_{\varrho'}$ if $\varrho' < \varrho$ and $\bigcup D_\varrho = D$. We now define f_ϱ by

$$f_\varrho(x) = \int_G f(xz)\sigma_\varrho(z)\cdot d\mu(z) = \int_G f(z)\sigma_\varrho(x^{-1}z)\cdot d\mu(z).$$

Then f_ϱ is a plurisubharmonic function of class C^∞. Since f is upper semicontinuous, it follows that for any $\epsilon > 0$ and for any $x \in D$ there exists $\varrho_0 = \varrho_0(x,\epsilon)$ such that $f_\varrho(x) < f(x) + \epsilon$ if $\varrho < \varrho_0$. (In case $f(x) = -\infty$ one has to make an obvious modification.)

An argument based on the mean value theorem (see e.g. [Gu2], p. 110) shows that

$$f(x) \le \int_U f(xz)\sigma_\varrho(z)\cdot d\lambda(z).$$

On the other hand, the integral on the right hand side does not exceed $f_\varrho(x)$ by our choice of the Haar measure. Thus $f(x) \le f_\varrho(x)$ for all $x \in D_\varrho$. It follows that $f_\varrho(x) \to f(x)$ as $\varrho \to 0$. The invariance of D_ϱ (resp. f_ϱ) is obvious from the definition. □

Denote by Vect(M) the Lie algebra of all vector fields on a C^∞ manifold M. For a complex manifold M=(M,J) let $\mathcal{V}(M)$ be the subalgebra of Vect(M), consisting of all infinitesimal automomorphisms of the complex structure J. By definition, a vector field X belongs to $\mathcal{V}(M)$ if and only if $L_X J = 0$, where L_X is the Lie derivative. A complex vector field Z on M is holomorphic if and only if $Z = X - iJX$, where $X \in \mathcal{V}(M)$. This shows that $\mathcal{V}(M)$ is J-invariant, i.e., $J\cdot\mathcal{V}(M) = \mathcal{V}(M)$.

Lemma 2 *Let M be a complex manifold, $X \in \mathcal{V}(M)$, and $f \in C^2(M)$ a real function such that $Xf = 0$ on M. If f is plurisubharmonic at $p \in M$ then*

$$(JX)^2 f(p) \ge 0. \tag{1}$$

Proof Let $Z := X - iJX$ be the holomorphic vector field with real part X. Since

$$[X, JX] = L_X(JX) = L_X(J) \cdot X + J \cdot L_X(X) = 0,$$

it follows that

$$Z\bar{Z}f = X^2 f + (JX)^2 f$$

for any $f \in C^2(M)$. The equality $Xf = 0$ implies that

$$(JX)^2 f = Z\bar{Z}f.$$

Since f is plurisubharmonic at p, we obtain (1). □

Corollary 1 *Let M be a complex manifold and let $X \in \mathcal{V}(M)$. Suppose that JX has a globally defined and relatively compact integral curve $\gamma : \mathbb{R} \longrightarrow M$. Let $f \in C^2(M)$ be a real function such that $Xf = 0$. Assume that f is plurisubharmonic at each point of this curve. Then $f(\gamma(t))$ is constant.*

Proof The inequality (1) shows that

$$\frac{d^2}{dt^2} f(\gamma(t)) \geq 0$$

for all $t \in \mathbb{R}$, so that $f(\gamma(t))$ is a convex function of $t \in \mathbb{R}$. On the other hand, this function is bounded. Therefore, $f(\gamma(t))$ is constant. □

Corollary 2 (J.-J. Loeb [Lo1]) *Let G be a connected semisimple complex Lie group acting holomorphically on a complex manifold M. Let S be a real form of G without compact normal subgroups. If a plurisubharmonic function f on M is S-invariant, then f is G-invariant. In particular, if M is a homogeneous manifold of G, then f is constant.*

Proof Restricting f to a G-orbit, we reduce the general case to the case of a homogeneous manifold. Lifting f to the group, we may assume that $M = G$. Then, by Lemma 1, it suffices to prove our statement for $f \in C^2(G)$. (Note that in our case $D_\varrho = D = G$ so that f_ϱ is defined on the whole group.)

We identify $\mathfrak{g}$ with the subalgebra of right invariant vector fields in $\mathcal{V}(G)$. Then $\mathfrak{g} = \mathfrak{s} + J \cdot \mathfrak{s}$. Choose a maximal compact subgroup $L \subset S$. Then we have the Cartan decomposition $\mathfrak{s} = \mathfrak{l} + \mathfrak{p}$. The dual algebra $\mathfrak{u} := \mathfrak{l} + J \cdot \mathfrak{p}$ is compact. In fact, the corresponding subgroup of G is a maximal compact subgroup.

By Corollary 1 the function $t \mapsto f(\exp tJX \cdot p)$ is constant for all $X \in \mathfrak{p}$ and $p \in M$. Therefore, f is annihilated by all vector fields from $J \cdot \mathfrak{p}$. Since $\mathfrak{l} = [\mathfrak{p}, \mathfrak{p}]$, it follows that $J \cdot \mathfrak{p}$ together with $\mathfrak{s}$ generates $\mathfrak{g}$ as a Lie algebra. Thus f is G-invariant. □

Theorem 1 (M.Lassalle [La2]) *Let $D \subset G$ be a left K-invariant domain and f a left K-invariant plurisubharmonic function in D. Then $\tilde{f}$ is geodesically convex, i.e., convex along any geodesic segment in $\tau(D)$.*

Proof Due to Lemma 1, we may assume that $f \in C^2(D)$. (A function which is the pointwise limit of a sequence of (geodesically) convex functions is (geodesically) convex.) A geodesic on $K\backslash G$ is of the form $t \mapsto \gamma(t) = \tau(\exp itX \cdot g_0)$, where $g_0 \in G$ and X is an element of the Lie algebra $\mathfrak{k}$ of K. Note that

$$\tilde{f}(\gamma(t)) = f(\exp itX \cdot g_0).$$

Since f is plurisubharmonic in D, it follows from Lemma 2 that

$$\frac{d^2}{dt^2}\tilde{f}(\gamma(t)) \geq 0$$

for all $t \in \mathbb{R}$. Thus $\tilde{f}$ is geodesically convex. □

Example In case G is abelian the converse is also true and the situation is well-understood. Namely, let f be an $(S^1)^n$-invariant function, defined in a Reinhardt domain $D \subset (\mathbb{C}^*)^n$. Then f is plurisubharmonic if and only if f is logarithmically convex, i.e.,

$$f(\exp x_1, ..., \exp x_n), \qquad x = (x_1, ..., x_n) \in \mathbb{R}^n,$$

is convex in the usual sense. Since $K\backslash G$ is flat, this condition is exactly the geodesic convexity.

Theorem 2 (O.S.Rothaus [Rot]) *Let $D \subset G$ be a left K-invariant Stein domain. Then $\tau(D)$ is geodesically convex, i.e., the geodesic segment joining two given points of $\tau(D)$ is contained in $\tau(D)$.*

Proof Let $f \in \mathcal{O}(D)$. In §5.2 we have defined the K-finite functions $M_{\delta,ij}f \in \mathcal{O}(D)$. Since D is connected, each $M_{\delta,ij}f$ extends to G as a regular function (see the proof of Theorem 1, §5.3). Consequently,

$$p_\delta(f) = \sum_{i,j=1}^{d(\delta)} |M_{\delta,ij}f|^2$$

extends to a K-invariant plurisubharmonic function on G. The extension will be denoted again by p_δ.

Now, let $X \in \mathfrak{k}$, $g_0 \in D$, and $g_\tau := \exp i\tau X \cdot g_0$. Assume that $g_1 \in D$. By Theorem 1

$$p_\delta(f)(g_\tau) \leq \max\,\{p_\delta(f)(g_0),\ p_\delta(f)(g_1)\}$$

for all $\tau \in [0, 1]$. On the other hand, Theorem 3 of §5.2 says that

$$f = \sum_{\delta \in \hat{K}} E_\delta f,$$

where

$$\sum_{\delta \in \hat{K}} |E_\delta f(x)| \leq \sum_{\delta \in \hat{K}} d(\delta)^{1/2} p_\delta(f)(x)^{1/2}$$

and the second series converges uniformly on compact sets in D. It follows that this series converges uniformly on the curve $C := \{g_\tau \mid \tau \in [0, 1]\}$. Therefore, f extends to a continuous function on $D \cup C$. Since $f \in \mathcal{O}(D)$ is arbitrary and since D is a Stein domain, we see that $C \subset D$. This shows that $\tau(D)$ is geodesically convex. □

Example (see [Lo3]) Let $G = \mathrm{SL}(2,\mathbb{C})$, $K = \mathrm{SU}(2)$, and

$$D = \{g = \begin{pmatrix} g_{11} & g_{12} \\ g_{21} & g_{22} \end{pmatrix} \in G \mid |g_{11}|^2 + |g_{21}|^2 > |g_{12}|^2 + |g_{22}|^2\}.$$

Then D is a left K-invariant domain in G. We claim that D is not Stein but, nevertheless, $\tau(D)$ is geodesically convex.

In order to prove that D is not Stein, it suffices to show that the intersection of D with the hypersurface $g_{22} = 1$ is not holomorphically convex. Let $z = g_{21}, w = g_{12}$. Then the intersection $D \cap \{g_{22} = 1\}$ is isomorphic to the domain $\Omega \subset \mathbb{C}^2$ defined by

$$\varphi(z,w) := 1 + |w|^2 - |z|^2 - |1 + zw|^2 < 0.$$

It is easy to check that $p = (1, -1/2)$ is a regular point of $\partial\Omega$ and that the Levi form of φ is negative definite on the complex tangent space to $\partial\Omega$ at p. Therefore Ω is not holomorphically convex.

Let $\mathrm{Herm}_2(\mathbb{C})$ denote the real vector space of all Hermitian matrices of order 2. Then $K\backslash G$ can be identified with the hypersurface

$$\Sigma = \{A \in \mathrm{Herm}_2(\mathbb{C}) \mid A > 0,\ \det A = 1\}.$$

The mapping $\tau : G \to \Sigma = K\backslash G$ is given by $\tau(g) = g^\star \cdot g$, where $g^\star$ is the complex conjugate of the transposed matrix. An easy calculation shows that $\tau(D) = \Sigma \cap U$, where

$$U = \{A = \begin{pmatrix} a & b \\ \bar{b} & c \end{pmatrix} \mid a > c\}.$$

Let

$$I = \begin{pmatrix} 1 & 0 \\ 0 & 1 \end{pmatrix}, \quad J = \begin{pmatrix} 1 & 0 \\ 0 & -1 \end{pmatrix}.$$

A geodesic through the origin $o = I \in \Sigma$ is of the form

$$t \mapsto \gamma(t) = k_0 \cdot \begin{pmatrix} e^{\lambda t} & 0 \\ 0 & e^{-\lambda t} \end{pmatrix} \cdot k_0^{-1},$$

where $\lambda, t \in \mathbb{R}$, $k_0 \in K$. It is clear that

$$\{\gamma(t) \mid t \in \mathbb{R}\} = \Sigma \cap (\mathbb{R} \cdot I + \mathbb{R} \cdot (k_0 J k_0^{-1})).$$

Therefore each geodesic in Σ is the intersection of Σ with a two-dimensional linear subspace of $\mathrm{Herm}_2(\mathbb{C})$.

We now use another model of the symmetric space $K\backslash G$. Namely, consider the linear subspace

$$\mathrm{Herm}_2^0(\mathbb{C}) := \{B \in \mathrm{Herm}_2(\mathbb{C}) \mid \mathrm{tr}\ B = 0\}$$

and let

$$\Delta = \{B \in \mathrm{Herm}_2^0(\mathbb{C}) \mid \det\,(I + B) > 0\}.$$

An element $B \in \mathrm{Herm}_2^0(\mathbb{C})$ has the form

$$B = \begin{pmatrix} a & b \\ \bar{b} & -a \end{pmatrix},$$

where $a \in \mathbb{R}, b \in \mathbb{C}$. Thus Δ is identified with the unit ball in $\mathbb{R}^3 = \mathbb{R} \times \mathbb{C}$:

$$\Delta = \{(a, b) \in \mathbb{R} \times \mathbb{C} \mid a^2 + |b|^2 < 1\}.$$

The mapping

$$\phi : \Sigma \to \mathrm{Herm}_2(\mathbb{C}), \quad \phi(A) := \frac{2A}{\mathrm{tr}\, A} - I,$$

defines a bijection of Σ onto Δ. The inverse mapping $\psi : \Delta \to \Sigma$ is given by

$$\psi(B) = \frac{(I + B)}{\sqrt{\det(I + B)}} \qquad (B \in \Delta).$$

These formulas have two consequences. Firstly, ϕ maps each geodesic of Σ onto a line segment in Δ, which is in fact the intersection of Δ with a line of $\mathrm{Herm}_2^0(\mathbb{C})$. Secondly, $\phi(\tau(D))$ is the intersection of Δ with a half-space. It follows that $\tau(D)$ is geodesically convex.

In the nonabelian case the converse to Theorem 1 is, in general, false. However, this can be true for functions which are right invariant with respect to a subgroup $H \subset G$. More precisely, let H be a closed complex subgroup of G and D a domain in G such that $K \cdot D \cdot H = D$. Denote by ${}^K C^2(D)^H$ the space of all left K-invariant and right H-invariant real C^2 functions in D. Let $f \in {}^K C^2(D)^H$. It is natural to ask when the geodesic convexity of $\tilde{f}$ in $\tau(D)$ implies the plurisubharmonicity of f in D. We shall give some conditions on H which guarantee the positive answer.

Lemma 3 *Let M be a complex manifold, $f \in C^2(M)$ and $U, V \in \mathcal{V}(M)$. Assume that $Uf = Vf = 0$. Put $X := U + JV$ and $Z := X - iJX$. Then*

$$\partial\bar{\partial} f(Z, \bar{Z}) = (JU)^2 f + (JV)^2 f + 2(J[U, V])f \tag{2}$$

at every point of M.

Proof Since $Uf = 0$, we have

$$X^2 f = (JV)^2 f + U(JV)f = (JV)^2 f + [U, JV]f = (JV)^2 f + J[U, V]f.$$

Similarly, since $Vf = 0$, we have

$$(JX)^2 f = (JU)^f - V(JU)f = (JU)^2 f - [V, JU]f = (JU)^2 f + J[U, V]f.$$

On the other hand,

$$\partial\bar{\partial} f(Z, \bar{Z}) = Z\bar{Z}f = X^2 f + (JX)^2 f.$$

Combining this with two previous equalities we obtain (2). □

Let M be a real manifold, $\mathrm{T}_p M$ the tangent space at $p \in M$, and

$$\varphi_p : \mathrm{Vect}(M) \to \mathrm{T}_p M$$

the mapping which assigns to each vector field its value at p. For a complex manifold M we denote by T_pM the holomorphic tangent space at $p \in M$, so that $\mathrm{T}_pM \otimes \mathbb{C} = T_pM \oplus \overline{T_pM}$.

Lemma 4 *Let M be a complex manifold, $\mathfrak{s}$ a linear subspace of $\mathcal{V}(M)$, and $f \in C^2(M)$ a real function, such that $Xf = 0$ for all $X \in \mathfrak{s}$. Assume that*

$$(JX)^2 f(p) \geq 0$$

for some $p \in M$ and for all $X \in \mathfrak{s}$. If there exist two linear subspaces $\mathfrak{s}_1, \mathfrak{s}_2 \subset \mathfrak{s}$ such that

$$\varphi_p(\mathfrak{s}_1 + J \cdot \mathfrak{s}_2) = \mathrm{T}_pM \tag{3}$$

and

$$[\, \mathfrak{s}_1 \,,\; J \cdot \mathfrak{s}_2 \,] \;\subset\; \mathfrak{s} + \operatorname{Ker} \varphi_p, \tag{4}$$

then f is plurisubharmonic at p.

Proof We have to prove that

$$(\partial\bar{\partial} f)_p(\zeta, \bar{\zeta}) \geq 0$$

for all $\zeta \in T_pM$. Write ζ in the form $\zeta = \xi - iJ\xi$, where $\xi \in \mathrm{T}_pM$. According to (3) there exist $U \in \mathfrak{s}_1$ and $V \in \mathfrak{s}_2$ such that

$$\varphi_p(U + JV) = \xi.$$

From (4) it follows that

$$[U, JV] = W + T,$$

where $W \in \mathfrak{s}$ and $\varphi_p(T) = 0$. Let $X = U + JV$ and $Z = X - iJX$. Then, using the equality (2), we obtain

$$(\partial\bar{\partial} f)_p(\zeta, \bar{\zeta}) = (\partial\bar{\partial} f)(Z, \bar{Z})(p) = (JU)^2 f(p) + (JV)^2 f(p) + 2(Wf)(p).$$

Since $Wf = 0$, our assertion follows from the assumption. □

Theorem 3 *Let G be a connected reductive complex Lie group, $K \subset G$ a maximal compact subgroup, $H \subset G$ a closed complex subgroup, and $D \subset G$ a domain in G such that $K \cdot D \cdot H = D$. Assume that one of the following conditions is satisfied:*

(i) *H contains a maximal unipotent subgroup of G;*

(ii) *H is conjugate to $L_{\mathbb{C}}$, where $L \subset K$ and (K, L) is a compact symmetric pair.*

Then a function $f \in {}^K C^2(D)^H$ is plurisubharmonic if and only if the corresponding function $\tilde{f} \in C^2(\tau(D))$ is geodesically convex in $\tau(D)$.

Proof Denote by $\mathfrak{g}$, $\mathfrak{k}$, and $\mathfrak{h}$ the Lie algebras of G,K, and H respectively. The proof is based on the following observation.

(*) *Let* $e \in D$. *Suppose that there exist two linear subspaces* $\mathfrak{k}_1 , \mathfrak{k}_2 \subset \mathfrak{k}$ *such that*

$$\mathfrak{k}_1 + i\mathfrak{k}_2 + \mathfrak{h} = \mathfrak{g} \tag{5}$$

and

$$[\, \mathfrak{k}_1 , i\mathfrak{k}_2 \,] \subset \mathfrak{k} + \mathfrak{h}. \tag{6}$$

Then $f \in {}^K C^2(D)^H$ *is plurisubharmonic at* $e \in D$ *if and only if the corresponding function* $\tilde{f} \in C^2(\tau(D))$ *is convex along any geodesic at the point* $o = K \cdot e \in \tau(D) \subset K\backslash G$.

Indeed, let M be the image of D under the natural projection mapping $G \longrightarrow G/H$. Then f defines a function on M, say $\check{f}$, which is plurisubharmonic at $p := e \cdot H \in M$ if and only if f is plurisubharmonic at $e \in G$. Take $\mathfrak{s} := \mathfrak{k}$ and apply Lemma 4 to $\check{f}$. Since (Ker $\varphi_p) \cap \mathfrak{g} = \mathfrak{h}$ and $T_pM = \mathfrak{g}/\mathfrak{h}$, we see that (5) (resp.(6)) is an exact equivalent of (3) (resp.(4)). On the other hand, the convexity of $\tilde{f}$ along any geodesic at o is precisely the inequality (1). Therefore (*) follows from Lemma 2 and Lemma 4.

(i) Let $g \in G$. We have to prove that f is plurisubharmonic at g if and only if $\tilde{f}$ is convex along any geodesic at $K \cdot g \in K\backslash G$. Consider the function $f_g(x) := f(xg)$, defined in $D \cdot g^{-1}$. Clearly, $f_g \in {}^K C^2(D \cdot g^{-1})^{gHg^{-1}}$. Replacing f by f_g, D by $D \cdot g^{-1}$, and H by gHg^{-1}, we may assume that $g = e$.

Let N be a maximal unipotent subgroup of G which is contained in H. Denote by B the normalizer of N. Then B is a Borel subgroup and $T := B \cap K$ is a maximal torus in K. Denote by $\mathfrak{n}$, $\mathfrak{b}$, and $\mathfrak{t}$ the corresponding Lie algebras.

Put $\mathfrak{k}_1 := \mathfrak{k}$, $\mathfrak{k}_2 := \mathfrak{t}$. Then we have

$$\mathfrak{g} = \mathfrak{k} + \mathfrak{b} = \mathfrak{k} + i\mathfrak{t} + \mathfrak{n} = \mathfrak{k}_1 + i\mathfrak{k}_2 + \mathfrak{n},$$

so that (5) is fulfilled. On the other hand, $\mathfrak{g} = i\mathfrak{k} + \mathfrak{b}$ and, consequently, $\mathfrak{k} \subset i\mathfrak{k} + \mathfrak{b}$. Therefore,

$$[\, \mathfrak{k}_1 , i\mathfrak{k}_2 \,] = [\, \mathfrak{k} , i\mathfrak{t} \,] \subset [\, i\mathfrak{k} + \mathfrak{b} , i\mathfrak{t} \,] =$$

$$= [\, \mathfrak{k} , \mathfrak{t} \,] + i[\, \mathfrak{b} , \mathfrak{t} \,] = [\, \mathfrak{k} , \mathfrak{t} \,] + [\, \mathfrak{b} , \mathfrak{t} \,] \subset \mathfrak{k} + \mathfrak{n} \subset \mathfrak{k} + \mathfrak{h} ,$$

and (6) is also fulfilled. Our assertion follows from (*).

(ii) First of all, we may assume, that $H = L_{\mathbb{C}}$. For, if $H = g_0 L_{\mathbb{C}} g_0^{-1}$, then we replace, as above, D and $f(x)$ by Dg_0 and $f(xg_0^{-1})$ respectively. The new domain and function are both right $L_{\mathbb{C}}$-invariant.

So let $H = L_{\mathbb{C}}$. Denote by $\mathfrak{p}$ the Cartan complement of the Lie subalgebra $\mathfrak{l} \subset \mathfrak{k}$, so that

$$\mathfrak{k} = \mathfrak{l} + \mathfrak{p}, \qquad [\, \mathfrak{p} , \mathfrak{p} \,] \subset \mathfrak{l}, \qquad \text{and} \qquad [\, \mathfrak{l} , \mathfrak{p} \,] \subset \mathfrak{p}.$$

It is known that $G = K \cdot \exp(i\mathfrak{p}) \cdot \exp(i\mathfrak{l})$. Moreover, the mapping

$$K \times \mathfrak{p} \times \mathfrak{l} \longrightarrow G, \qquad (\, k, \mathrm{A}, \mathrm{B} \,) \longmapsto k \exp(i\mathrm{A}) \exp(i\mathrm{B}),$$

is a diffeomorphism.

Take any point $g \in D$ and write $g = k\exp(i\mathrm{B})l$ with $k \in K$, $\mathrm{B} \in \mathfrak{p}$, $l \in \exp(i\mathfrak{l}) \subset L_{\mathbb{C}}$. To reduce to the case $g = e$, replace, as in (i), f by f_g. Then H is replaced by $gHg^{-1} = k\exp(i\mathrm{B})L_{\mathbb{C}}\exp(-i\mathrm{B})k^{-1}$. Define

$$\mathfrak{k}_1 := \mathfrak{k}_2 := \mathrm{Ad}k \cdot \mathfrak{p}.$$

For gHg^{-1} instead of H the equality (5) can be rewritten as

$$\mathfrak{p} + i\mathfrak{p} + \mathrm{Ad}\exp(i\mathrm{B}) \cdot (\mathfrak{l} + i\mathfrak{l}) = \mathfrak{g}. \tag{7}$$

Since (7) is true for $\mathrm{B} = 0$, it holds for all B outside a proper analytic subset in $\mathfrak{p}$. Now, again for gHg^{-1} instead of H, the inclusion (6) is equivalent to

$$[\,\mathfrak{p}\,,\,\mathfrak{p}\,] \subset i\mathfrak{k} + \mathrm{Ad}\exp(i\mathrm{B})(\,\mathfrak{l} + i\mathfrak{l}\,). \tag{8}$$

Since $[\mathfrak{p}, \mathfrak{p}] \subset \mathfrak{l}$, this would follow from

$$\mathfrak{l} + i\mathfrak{p} = \mathrm{Ad}\exp(i\mathrm{B}) \cdot \mathfrak{l} + i\mathfrak{p}. \tag{9}$$

But $\mathfrak{l} + i\mathfrak{p}$ is a Lie algebra and $i\mathrm{B} \in \mathfrak{l} + i\mathfrak{p}$. So the right hand side of (9) is always contained in the left hand side. Clearly, (9) is true for $\mathrm{B} = 0$. Therefore, (9) and, consequently, (8) hold for all B outside a proper analytic subset in $\mathfrak{p}$. Thus, there exists a dense open subset $\Omega \subset \mathfrak{p}$, such that both (7) and (8) are valid for $\mathrm{B} \in \Omega$. Together with (*) this shows that, for $g \in D \cap (K\exp(i\Omega)L_{\mathbb{C}})$, f is plurisubharmonic at g if and only if $\tilde{f}$ is convex along any geodesic at $K \cdot g \in K\backslash G$. By continuity the same is true for all $g \in D$. □

Remark Theorem 3, (ii) is due to J.-J. Loeb, see [Lo2]. His proof is based on the description of K-invariant plurisubharmonic functions on G/H in terms of usual convexity in an auxiliary Euclidean space. This approach was used earlier by M.Lassalle [La1] for a characterization of K-invariant Stein domains in G/H. Similar results in case (i) can be found in [Lo4].

In §5.3 we introduced the notion of a spherical subgroup. The subgroups satisfying (i) or (ii) of Theorem 3 are known to be spherical. However, Theorem 3 does not hold for general spherical subgroups.

Example (see [A13]) Let $G := \mathbb{C}^* \times \mathrm{SL}(2,\mathbb{C})$, $K := S^1 \times \mathrm{SU}(2)$, and

$$H := \left\{ \left(z\,, \begin{pmatrix} z & 0 \\ 0 & z^{-1} \end{pmatrix} \right) \;\middle|\; z \in \mathbb{C}^* \right\}.$$

Since a maximal torus in $\mathrm{SL}(2,\mathbb{C})$ is a spherical subgroup, the same is true for the subgroup $H \subset G$. Define a function f on G by

$$f(w, \begin{pmatrix} a & b \\ c & d \end{pmatrix}) = -\lambda |w|^2(|b|^2 + |d|^2) + \frac{|a|^2 + |c|^2}{|w|^2},$$

where $0 < \lambda < 1$. It is easily seen that f is left K-invariant and right H-invariant. A geodesic through $o = K \cdot e \in K\backslash G$, after lifting to G, becomes a one-parameter subgroup

$$t \mapsto (\exp pt\,, \exp t \begin{pmatrix} \alpha & \beta \\ \bar{\beta} & -\alpha \end{pmatrix}) \in G \qquad (p, \alpha \in \mathbb{R}, \beta \in \mathbb{C}).$$

Denote by $f(p,\alpha,\beta;t)$ the function on $\mathbb{R}$, obtained by restriction of f to this subgroup. A straightforward calculation shows that

$$f(p,\alpha,\beta;t) = (\exp(-2pt) - \lambda \exp 2pt) \cosh 2rt + \frac{\alpha}{r}(\exp(-2pt) + \lambda \exp 2pt) \sinh 2rt,$$

where $r = \sqrt{\alpha^2 + |\beta|^2}$, so that

$$\frac{d^2}{dt^2} f(p,\alpha,\beta;0) = 4(1-\lambda)\big((p-\alpha)^2 + |\beta|^2\big) \geq 0.$$

Therefore, $\tilde{f}$ is convex along each geodesic at o. On the other hand,

$$f|_{w=a=d=1,c=0} = 1 - \lambda(1 + |b|^2),$$

and we see that f is not plurisubharmonic at $e \in G$.

Concluding Remarks

1 Classification of homogeneous complex manifolds

The list of homogeneous complex manifolds in dimension 1 was certainly known to Poincaré. This list includes $\mathbb{C}$, $\mathbb{C}^*$, the unit disk $\Delta \subset \mathbb{C}$, and all elliptic curves. The automorphism group of a complex curve is a Lie group, and so there is no difference between the general homogeneity and the homogeneity under a Lie transformation group.

If dim $X > 1$ then it may happen that X is homogeneous under Aut(X), but does not admit a transitive Lie transformation group. One of the first examples of this type is due to W.Kaup [Ka2]. We reproduce this example in a slightly generalized form. Let $X_S := \mathbb{C}^n - \{(0,\ldots,0,\zeta)|\zeta \in S\}$, where $S \subset \mathbb{C}$ is a discrete set. There exists a function $f \in \mathcal{O}(\mathbb{C})$, such that $f(\zeta) = 0$ if $\zeta \in S$ and $f(\zeta) \neq 0$ is $\zeta \notin S$. The automorphism $\varphi : \mathbb{C}^n \to \mathbb{C}^n$, given by

$$z = (z_1,\ldots,z_n) \mapsto \varphi(z) = (z_1 + f(z_n), z_2,\ldots,z_n),$$

leaves the domain $X_S \subset \mathbb{C}^n$ invariant. We claim that for any point

$$z^0 = (z_1^0,\ldots,z_n^0) \in X_S$$

there is an automorphism of X_S sending z^0 to $(1,0,\ldots,0)$. Replacing z^0 by $\varphi(z^0)$ if necessary, we may assume that $z_1^0 \neq 0$. Then the linear mapping

$$(z_1, z_2,\ldots,z_{n-1},z_n) \mapsto (\frac{z_1}{z_1^0},\ z_2 - \frac{z_2^0 z_1}{z_1^0},\ \ldots,\ z_n - \frac{z_n^0 z_1}{z_1^0})$$

preserves X_S and has the desired property. This shows that Aut(X_S) acts transitively on X_S. On the other hand, if card $S \geq 2$ then X_S has more than two ends in the sense of H.Freudenthal. By a theorem of A.Borel [Bo1] a homogeneous space of a (real) Lie group with a connected isotropy subgroup has at most two ends. Since X_S is simply connected, it follows that X_S does not admit a transitive Lie transformation group (no matter whether this group preserves the complex structure or not).

We consider here only connected complex manifolds homogeneous under the action of a Lie group of holomorphic transformations. Such manifolds have been listed in dimension 2 (see [HuLiv], [OelRi1], and also [Hu2]) and in dimension 3 (see [Wi]). The list in dimension 3 gives little hope for a similar result in dimension 4. On the other hand, as we have seen in Chapter 3, compact homogeneous manifolds are well-understood. Thus it seems reasonable to study natural classes of complex manifolds, which are in a sense close to the compact ones. These classes are obtained by imposing various conditions of analytic or topological nature.

A complex manifold X with dim $X \geq 2$ is said to be *strictly pseudoconcave* if X admits an exhaustion by open subsets $X_\nu \subset X$, $\nu = 1, 2, \ldots$, with the property that $X_\nu \subset\subset X_{\nu+1}$ for each ν and there exists a strictly plurisubharmonic function ϕ_ν defined in a neighborhood W_ν of ∂X_ν, such that $X_\nu \cap W_\nu = \{x \in W_\nu \mid \phi(x) > 0\}$. One example of a strictly pseudoconcave manifold is a manifold, which is obtained from a compact one by deleting a finite number of points. A.T.Huckleberry and D.Snow gave a complete classification of strictly pseudoconcave homogeneous manifolds, see [HuSn1], [HuSn2].

Let Y be a flag manifold equivariantly embedded in $\mathbb{P}_n$. We identify $\mathbb{P}_n$ with a hyperplane in $\mathbb{P}_{n+1}$ and fix a point o in the complement $\mathbb{P}_{n+1} - \mathbb{P}_n$. Let $C(Y)$ be the cone in $\mathbb{P}_{n+1}$, which is obtained by connecting each point of Y to the origin o by a line. Then o is the only possible singularity of $C(Y)$. We refer to $C(Y)$ and $C(Y) - Y$ as a *projective* and, respectively, an *affine cone* over Y. Removing the vertex o from each of these cones, we obtain the manifolds, which are homogeneous under a complex Lie transformation group.

Let X be a non-compact strictly pseudoconcave homogeneous manifold. Then all possibilities are given as follows:

(i) X is an affine cone over a flag manifold with its vertex removed;

(ii) $X = Z - M$, where $M = K/L$ is a compact Riemanniam symmetric space of rank 1 with K the connected isometry group, Z is a non-singular equivariant completion of $M_{\mathbb{C}} = K_{\mathbb{C}}/L_{\mathbb{C}}$ (such a completion is in fact unique, Z is a flag manifold of a semisimple group S containing $K_{\mathbb{C}}$, and X is an open orbit on Z of a non-compact real form of S);

(iii) $X = \mathbb{P}_n - \overline{\mathbb{B}}_n$, where $\mathbb{B}_n$ is the unit ball in $\mathbb{C}^n \subset \mathbb{P}_n$ (the transitive group is $\mathrm{Aut}(\mathbb{B}_n) = \mathrm{PU}(n,1)$).

Another natural class, which admits a nice description in group-theoretic terms, is the class of homogeneous complex manifolds with two ends. The statement of the result is particularly simple if X is an algebraic variety homogeneous under the regular action of a connected linear algebraic group G. Then X has two ends if and only if the isotropy subgroup is the kernel of a non-trivial regular character of a parabolic subgroup $P \subset G$, see [A3]. Geometrically speaking, this means that X is a principal $\mathbb{C}^*$-bundle over a flag manifold G/P. There is a generalization of this result to homogeneous complex manifolds of the form $X = G/H$, where G is a connected complex Lie group, $H \subset G$ a closed complex Lie subgroup with finitely many connected components, see [Ho], [GiHu2].

An analytic condition which ensures that a homogeneous complex manifold X of a complex Lie group has at most two ends is that $\mathcal{O}(X) \neq \mathbb{C}$. Under this assuption X has two ends if and only if X admits a fibering with compact connected fiber over an affine homogeneous cone minus its vertex, see [Gi].

Let d_X denote the codimension of the top nonvanishing homology group of X with coefficients in $\mathbb{Z}_2$. In the preceding result one can replace the two ends condition by $d_X = 1$. The next case $d_X = 2$ is considered in [AGi]. Homogeneous complex manifolds of real Lie groups, which have more than one end and are Kähler, are classified in [GiOelRi].

2 Hypersurfaces in homogeneous complex manifolds

For a connected complex manifold X we denote by $\mathcal{M}(X)$ the field of meromorphic functions of X and by $\mathcal{H}(X)$ the set of all hypersurfaces in X. Given two points $x, y \in X$ we write $x \sim y$ if

$$x \in F \iff y \in F$$

for any $F \in \mathcal{H}(X)$. If X is homogeneous with respect to a (real) Lie group of holomorphic transformations, then this is an equivalence relation. More precisely, if $X = G/H$ then there exists a closed subgroup $J \subset G$ containing H, such that $Y = G/J$ has an invariant complex structure, the canonical mapping

$$\pi : X = G/H \to Y = G/J$$

is holomorphic, and $x \sim y$ if and only if $\pi(x) = \pi(y)$. Furthermore, $H^\circ \lhd J$ and J/H° is a complex Lie group acting holomorphically on the fiber. For the proof of this fact the reader is referred to [GR] (in the case of compact X) and to [Oel1] (in the general case).

By definition of π each hypersurface $F \in \mathcal{H}(X)$ is of the form $F = \pi^{-1}(F')$, where $F' \in \mathcal{H}(Y)$. Moreover, Y is *hypersurface separable*, i.e., if two points of Y are equivalent in the above sence then they coincide. The map $\pi : X \to Y$ is called a *hypersurface reduction* of X. H.Grauert and R.Remmert proved in [GR] that a compact hypersurface separable homogeneous complex manifold X is projective algebraic. If one drops the compactness assumption then it is still true that X is a Kähler manifold. Moreover, this result holds for homogeneous complex manifolds whose hypersurface reduction has discrete fibers. The proof uses a theorem of W.Richthofer on smoothing positive (p,p)-currents on homogeneous complex manifolds, see [Hu3].

In a similar way one can prove the existence of the *meromorphic reduction* $\pi^\star : X = G/H \to Y = G/J^\star$ of a homogeneous complex manifold X. The mapping $\pi^\star$ is uniquely defined by the property

$$\pi^\star(x) = \pi^\star(y) \iff f(x) = f(y) \quad \text{for all } f \in \mathcal{M}(X).$$

Obviously, $J^\star \supset J$ but, in general, $J^\star \neq J$. For example, J.Winkelmann showed that the universal covering X of $\mathbb{C}^2 - \mathbb{R}^2$ is hypersurface separable but all meromorphic functions on X come from $\mathbb{C}^2 - \mathbb{R}^2$.

If G is a complex Lie group then J and $J^\star$ are complex Lie subgroups. There is at least one case when the hypersurface and meromorphic reductions are well-understood. Namely, for G semisimple F.Berteloot and K.Oeljeklaus proved that $J = J^\star = \hat{H}$, where $\hat{H}$ is the Zariski closure of H, see [BerOel]. Since, as we have already noted, G/J is a Kähler manifold, this theorem is a consequence of the following fact. A complex manifold homogeneous under a semisimple complex Lie group is Kähler if and only if the isotropy subgroup is algebraic, see [Ber1]. The proof is based on a theorem about plurisubharmonic functions on $\mathrm{SL}(2,\mathbb{C}) \times (\mathbb{C}^*)^q$ invariant under cyclic subgroups of certain type, see [Ber2], [BerOel].

Let $\mathcal{H}(G)^H$ denote the set of all hypersurfaces in G, which are invariant under the right translations by H, where $H \subset G$ is an arbitrary subgroup. If H is a closed complex Lie subgroup then $\mathcal{H}(G)^H = \mathcal{H}(G/H)$. The similar notation $\mathcal{O}(G)^H$ (resp. $\mathcal{M}(G)^H$) is used for the algebra of H-invariant holomorphic functions (resp. the field of H-invariant meromorphic functions) on G. The result in [BerOel] is, in fact, stronger than the theorem stated above. Namely, if G is semisimple then $\mathcal{H}(G)^H = \mathcal{H}(G)^{\hat{H}}$ and, as a consequence, $\mathcal{M}(G)^H = \mathcal{M}(G)^{\hat{H}}$. For H Zariski dense we have $\mathcal{H}(G)^H = \emptyset$ and $\mathcal{M}(G)^H = \mathbb{C}$. This has been proved in [A4] for arithmetic discrete subgroups, e.g. $\mathrm{SL}(n, \mathbb{Z}) \subset \mathrm{SL}(n, \mathbb{C})$, and also for $G = \mathrm{SL}(2, \mathbb{C})$. The proof in the general case is due to A.T.Huckleberry and G.A.Margulis, see [HuMar].

Let H be an arbitrary subgroup of a connected complex Lie group G. Assume that H is not contained in a proper parabolic subgroup and that $\mathcal{O}(G)^H = \mathbb{C}$. It is then natural to ask, whether $\mathcal{H}(G)^H = \mathcal{H}(G)^{H \cdot G'}$, where, as usual, G' denotes the commutator subgroup of G, see [A8].

It follows from [GR] that the answer is positive if the quotient $G/\overline{H}$ is compact, where $\overline{H}$ is the closure of H (in the Euclidean topology). If G is semisimple then H is Zariski dense in G and, as we have seen above, the answer is also positive. The same is true if G is nilpotent [A8], solvable [OelRi2], and, also, if G is the direct product of a semisimple group and a solvable group [OelRi3]. In the general case the question remains open.

3 Almost homogeneous complex spaces

Let X be an irreducible normal complex space, G a connected complex Lie group holomorphically acting on X, and assume that X is almost homogeneous with respect to G. Denote by Ω the open G-orbit in X and let $E := X - \Omega$. Under certain assumptions on E one can try to classify all possible pairs (X, G).

Historically, the first theorem of this type is the following characterization of $\mathbb{P}_n$ due to E.Oeljeklaus. If X is a compact manifold and E contains an isolated point, then X is isomorphic to $\mathbb{P}_n$, see [Oel]. More generally, if X is not necessarily compact and is allowed to have normal singularities, then X is either a projective or an affine cone over a flag manifold with its vertex removed (the *cone theorem*). For complete algebraic varieties under the regular action of a linear algebraic group the cone theorem is proved in [A3]. The proof is based on the characterization of homogeneous varieties with two ends. The same method yields a description of complete almost homogeneous varieties with disconnected E. The proof of the cone theorem in the general case is given in [HuOe1].

If E is 1-dimensional, X compact and non-singular, then the complete list of all possibilities is given in [HuOe2], see Ch. II, §4, Th.1. A generalization to complex spaces can be found in [Les].

Some of the above results can be generalized to real transformation groups, see [HuOe2].

A complete normal algebraic variety X with a regular action of a connected linear algebraic group G is called a *two-orbit variety* if X is almost homogeneous and G acts transitively on E. Two-orbit varieties are classified if codim $E = 1$ (see [A7]) and if codim $E = 2$ (for reductive G; see [Feld]). A generalization of the former result to Kähler manifolds can be found in [HuSn3]. The classification in the case codim $E = 1$ is based on the following fact. Assume that G has no quotient group isomorphic to $\mathbb{C}$. Then a homogeneous variety G/H admits an equivariant completion

$$G/H \xrightarrow{\sim} \Omega \hookrightarrow X,$$

such that $E = X - \Omega$ is a homogeneous hypersurface or the disjoint union of two such hypersurfaces, if and only if the codimension of a generic K-orbit on G/H equals 1, where K is a maximal compact subgroup of G.

If G is reductive then a two-orbit variety with codim $E = 1$ or 2 is spherical (see §5.3 for the definition). It is unknown whether this is true if codim $E > 2$.

4 Symmetric spaces in the sense of A.Borel

Let X be a connected reduced complex space, G a (not necessarily connected) complex Lie group acting holomorphically and effectively on X. Then X is said to be *symmetric in the sense of Borel* or *complex-symmetric* with respect to G if for every $x \in X$ there exists an element $s_x \in G$ of order 2, such that x is an isolated fixed point of s_x (see [Bo5]). As an example one can take any Hermitian symmetric manifold X with $G = \mathrm{Aut}(X)$. A general complex-symmetric space X is not necessarily G-homogeneous. However, A.Borel proved in [Bo5] that for X compact G° acts on X with a finite number of orbits. In particular, X is almost homogeneous with respect to G°.

Though the definition is complex-analytic, the only known results are obtained under the assumption that X is a normal complete algebraic variety and G° is a linear algebraic group acting on X regularly. Namely, if G° is simple then one can use the classification of two-orbit varieties to show that X is either G°-homogeneous or isomorphic to the projective quadric $\mathbb{Q}(n)$ with $G^\circ = \mathrm{SO}(n-1)$ having on X two orbits [A9]. In the latter case X is homogeneous under a larger group. In both cases X is Hermitian symmetric. More generally, if G° is reductive then X is the direct product of a normal complex-symmetric torus embedding and a Hermitian symmetric space, see [Le].

5 Equivariant complex extensions

Let M be a real analytic manifold of dimension n. Its *complex extension* is a pair (Ω, φ), where Ω is a complex manifold of complex dimension n and φ is a real analytic isomorphism of M onto a closed totally real submanifold $\varphi(M) \subset \Omega$. If M is paracompact, then a complex extension always exists, see [WB], [Shu]. Moreover, one can choose Ω to be a Stein manifold, see [G]. Assume that we are given a real analytic action of a (real) Lie group G on M. A complex extension (Ω, φ) is said to be *G-equivariant* if G acts analytically on Ω by holomorphic transformations and $\varphi(gx) = g\varphi(x)$ for all $x \in M, g \in G$. An equivariant complex extension may not exist even if M is a homogeneous manifold, see [A12]. However, if the G-action on M is proper then an equivariant extension always exists. Moreover, one can find an equivariant extension (Ω, φ), such that the G-action on Ω is also proper, and one can choose Ω to be a Stein manifold, see [Hei2], [Ku].

Suppose (Ω, φ) is an equivariant extension of M and assume that the G-action on M is proper. Then there exist invariant domains $D \subset \Omega$ containing M, which are Stein and the action on which is proper. An explicit construction of such domains for homogeneous manifolds M can be found in [AGi], [A12].

Bibliography

[A1] D.N.AKHIEZER - On the cohomology of compact complex homogeneous spaces (I & II), Matem. Sbornik 84 (1971), 290 - 300 ; 87 (1972), 587 - 593 (Russian).

[A2] D.N.AKHIEZER - Compact complex homogeneous spaces with solvable fundamental group, Izv. Akad. Nauk SSSR, ser. mat., 38 (1974), 59 - 80 (Russian). AMS translation: Math. USSR Izvestija 8 (1974), 61 -83.

[A3] D.N.AKHIEZER - Dense orbits with two ends, Izv. Akad. Nauk SSSR, ser. mat., 41 (1977), 308 - 324 (Russian). AMS translation: Math. USSR Izvestija 11 (1977), 293 - 307.

[A4] D.N.AKHIEZER - Invariant meromorphic functions on complex semisimple Lie groups, Invent. Math. 65 (1981), 325 - 329.

[A5] D.N.AKHIEZER - Compact groups of automorphisms of Stein manifolds, C. R. Acad. Bulgare des Sci. 35 (1982), 577 - 580.

[A6] D.N.AKHIEZER - An estimate of the dimension of the automorphism group of a compact complex homogeneous space, Bull. Acad. Sci. Georgian SSR 110 (1983), 469 - 472 (Russian).

[A7] D.N.AKHIEZER - Equivariant completions of homogeneous algebraic varieties by homogeneous divisors, Ann. Glob. Anal. & Geom. 1 (1983), 49 - 78.

[A8] D.N.AKHIEZER - Invariant analytic hypersurfaces in complex nilpotent Lie groups, Ann. Glob. Anal. & Geom. 2 (1984), 129 - 140.

[A9] D.N.AKHIEZER - On algebraic varieties that are symmetric in the sense of Borel, Dokl. Akad. Nauk SSSR 279 (1984), 13 - 16 (Russian). AMS translation: Sov. Math. Dokl. 30 (1984), 579 - 582.

[A10] D.N.AKHIEZER - Actions with a finite number of orbits, Funct. Anal. Appl. 19 (1985), 1 -5 (Russian). AMS translation: Funct. Anal. Appl. 19 (1985), 1 - 4.

[A11] D.N.AKHIEZER - Homogeneous complex manifolds, in "Several complex variables IV", Sovr. Probl. Mat. 10, Moscow, 223 - 275, 1986 (Russian). English translation: Encyclopaedia of Mathematical Sciences, Volume 10, Springer, 1990, 195 - 244.

[A12] D.N.AKHIEZER - Equivariant complex extensions of homogeneous spaces, Matem. Zametki 51 (1992), 3 - 9 (Russian).

[A13] D.N.AKHIEZER - Invariant plurisubharmonic functions and geodesic convexity, Expo. Math. 11 (1993), 261 - 270.

[AGi] D.N.AKHIEZER, B.GILLIGAN - On complex homogeneous spaces with top homology in codimension two, to appear in Can. J. of Math.

[AGin] D.N.AKHIEZER, S.G.GINDIKIN - On Stein extensions of real symmetric spaces, Math. Ann. 286 (1990), 1 - 12.

[Ar] F.ARIBAUD - Une nouvelle démonstration d'un théorème de R.Bott et B.Kostant, Bull. Soc. Math. France 95 (1967), 205 - 242.

[Aus] L.AUSLANDER - On radicals of discrete subgroups of Lie groups, Amer. J. of Math. 85 (1963), 145 - 150.

[B] R.BOTT - Homogeneous vector bundles, Ann. of Math. 66 (1957), 203 - 248.

[BaOe] W.BARTH, E.OELJEKLAUS - Über die Albanese-Abbildung einer fasthomogenen Kählermannigfaltigkeit, Math. Ann. 211 (1974), 47 - 62.

[BaOt1] W.BARTH, M. OTTE - Über fast-uniforme Untergruppen komplexer Liegruppen und auflösbare komplexe Mannigfaltigkeiten, Comment. Math. Helv. 44 (1969), 269 - 281

[BaOt2] W.BARTH, M.OTTE - Invariante holomorphe Funktionen auf reduktiven Liegruppen, Math. Ann. 201 (1973), 97 - 112

[BarGr] R.G.BARTLE, L.M.GRAVES - Mappings between function spaces, Trans. Amer. Math. Soc. 72 (1952), 400 - 413.

[Be] E.BEDFORD - On the automorphism group of a Stein manifold, Math. Ann. 266 (1983), 215 - 227.

[BeDa] E.BEDFORD, J.DADOK - Bounded domains with prescribed group of automorphisms, Comment. Math. Helvetici 62 (1987), 561 - 572.

[Ber1] F.BERTELOOT - Existence d'une structure kählérienne sur les variétés homogènes semi-simples, C. R. Acad. Sci. Paris 305, Série I (1987), 809 - 812.

[Ber2] F.BERTELOOT - Fonctions plurisousharmoniques sur $\mathrm{SL}(2, \mathbb{C})$ invariantes par un sous-groupe monogène, Journ. Anal. Math. 48 (1987), 267 - 276.

[BerOel] F.BERTELOOT, K.OELJEKLAUS - Invariant plurisubharmonic functions and hypersurfaces on semi-simple complex Lie groups, Math. Ann. 281 (1988), 513 - 530.

[Bi] A.BIALYNICKI-BIRULA - On homogeneous affine spaces of linear algebraic groups, Amer. J. of Math. 85 (1963), 577 - 582.

[BiHoMo] A.BIALYNICKI-BIRULA, G.HOCHSCHILD, G.D.MOSTOW - Extensions of representations of algebraic linear groups, Amer. J. of Math. 85 (1963), 131 - 144.

[Bir] D.BIRKES - Orbits of linear algebraic groups, Ann. of Math. 93 (1971), 459 - 475.

[Bl] A.BLANCHARD - Sur les variétés analytiques complexes, Ann. Sci. École Norm. Sup. (3) 73 (1956), 157 - 202.

[BM1] S.BOCHNER, D.MONTGOMERY - Groups of differentiable and real or complex analytic transformations, Ann. of Math. 46 (1945), 685 - 694.

[BM2] S.BOCHNER, D.MONTGOMERY - Locally compact groups of differentiable transformations, Ann. of Math. 47 (1946), 639 - 653.

[BM3] S.BOCHNER, D.MONTGOMERY - Groups on analytic manifolds, Ann. of Math. 48 (1947), 659 - 669.

[Bo1] A.BOREL - Les bouts des espaces homogènes de groupes de Lie, Ann. of Math. 58 (1953), 443 - 457.

[Bo2] A.BOREL - Kählerian coset spaces of semi-simple Lie groups, Proc. Nat. Acad. Sci. USA 40 (1954), 1147 - 1151.

[Bo3] A.BOREL - Compact Clifford-Klein forms of symmetric spaces, Topology 2(1963), 111 - 122.

[Bo4] A.BOREL - *Linear algebraic groups*, Benjamin, New-York, 1969.

[Bo5] A.BOREL - Symmetric compact complex spaces, Arch. Math. 33 (1979), 49 - 56.

[Bog] F.A.BOGOMOLOV - Holomorphic tensors and vector bundles on projective varieties, Izv. Akad. Nauk SSSR, ser. mat., 42 (1978), 1227 - 1287 (Russian). AMS translation: Math. USSR Izvestija, 13 (1979), 499 - 555.

[BoHi] A.BOREL, F.HIRZEBRUCH - Characteristic classes and homogeneous spaces (I, II, & III), Amer. J. Math. 80 (1958), 458 - 538; 81 (1959), 315 - 382; 82 (1960), 491 - 504.

[BoRe] A.BOREL, R.REMMERT - Über kompakte homogene Kählersche Mannigfaltigkeiten, Math. Ann. 145 (1962), 429 - 439.

[BorSh] Z.I.BOREVICH, I.R.SHAFAREVICH - *Number theory*, Nauka, Moscow, 1964 (Russian). English translation: Pure and Appl. Math, vol. 20, Acad. Press, New York, 1966.

[Bou] N.BOURBAKI - *Groupes et algèbres de Lie*, Chap. 1 - 8. Hermann, Paris, 1960 - 1975.

[Br1] M.BRION - Quelques propriétés des espaces homogènes sphériques, Manuscr. Math. 55 (1986), 191 - 198.

[Br2] M.BRION - Classification des espaces homogènes sphériques, Comp. Math. 63 (1987), 189 - 208.

[Br3] M.BRION - Sur l'image de l'application moment, p. 177 -192 in "Séminaire d'algebre Paul Dubreil et Marie-Paule Malliavin", Lect. Notes in Math., 1296, Springer, 1987.

[Br4] M.BRION - Groupe de Picard et nombres caractéristiques des variétés sphériques, Duke Math. J. 58 (1989), 397 - 424.

[BrLuVu] M.BRION, D.LUNA, Th.VUST - Espaces homogènes sphériques, Invent. Math. 84 (1986), 617 - 632.

[C] C.CARATHÉODORY - Über die Abbildungen, die durch Systeme von analytischen Funktionen von mehreren Veränderlichen erzeugt werden, Math. Z. 34 (1932), 758 - 792.

[Ca] É.CARTAN - Sur les domaines bornés homogènes de l'espace de n variables complexes, Abh. Math. Sem. Univ. Hamburg 11 (1935), 116 - 162.

[Ca1] H.CARTAN - Les fonctions de deux variables complexes et le problème de la représentation analytique, J. Math. pures et appl. 10 (1931), 1 - 114.

[Ca2] H.CARTAN - Sur les fonctions de plusieurs variables complexes: l'itération des transformations intérieurs d'un domaine borné, Math. Z. 35 (1932), 760 - 773.

[Ca3] H.CARTAN - *Sur les groupes de transformations analytiques*, Act. Sci. Ind. 198, Hermann, Paris, 1935.

[CalEck] E.CALABI, B.ECKMANN - A class of compact, complex manifolds which are not algebraic, Ann. of Math. 58 (1953), 494 - 500.

[Ch] W.L.CHOW - On compact complex analytic varieties, Amer. J. of Math. 71 (1949), 893 - 914.

[De1] M.DEMAZURE - Une démonstration algébrique d'un théorème de Bott, Invent. Math. 5 (1968), 349 - 356.

[De2] M.DEMAZURE - A very simple proof of Bott's theorem, Invent. Math. 33 (1976), 271 - 272.

[DoNa] J.DORFMEISTER, K.NAKAJIMA - The fundamental conjecture for homogeneous Kähler manifolds, Acta Math. 161 (1988), 23 - 70.

[Feld] D.FELDMÜLLER - Two-orbit varieties with smaller orbit of codimension two, Arch. Math. 54 (1990), 582 - 593.

[Fu] H.FUJIMOTO - On the holomorphic automorphism groups of complex spaces, Nagoya Math. J. 33 (1968), 85 - 106.

[G] H.GRAUERT - On Levi's problem and the imbedding of real-analytic manifolds, Ann. of Math. 68 (1958), 460 - 472.

[Gi1] B.GILLIGAN - Ends of complex homogeneous manifolds having non-constant holomorphic functions, Arch. Math. 37 (1981), 544 - 555.

[Gi2] B.GILLIGAN - Holomorphic reductions of homogeneous spaces, p. 27 - 36 in Lect. Notes Math. 1014, Springer, 1983.

[GiHu1] B.GILLIGAN, A.T.HUCKLEBERRY - On non-compact complex nil-manifolds, Math. Ann. 238 (1978), 39 - 49.

[GiHu2] B.GILLIGAN, A.T.HUCKLEBERRY - Complex homogeneous manifolds with two ends, Mich. J. Math. 28 (1981), 183 - 196.

[GiOelRi] B.GILLIGAN, K.OELJEKLAUS, W.RICHTHOFER - Homogeneous complex manifolds with more than one end, Can. J. of Math., Vol. 41 (1989), 163 - 177.

[Gl] A.GLEASON - Groups without small subgroups, Ann. of Math. 56 (1952), 193 - 212.

[Go] R.GODEMENT - *Topologie algébrique et théorie des faisceaux*, Hermann, Paris, 1958.

[GPV1] S.G.GINDIKIN, I.I.PIATETSKY-SHAPIRO, E.B.VINBERG - On the classification and canonical realization of complex bounded homogeneous domains, Trudy Mosk. Matem. Ob. 12 (1963), 359 - 388 (Russian). AMS Translation: Trans. Moscow Math. Soc. 12 (1963), 404 - 437.

[GPV2] S.G.GINDIKIN, I.I.PIATETSKY-SHAPIRO, E.B.VINBERG - Homogeneous Kähler manifolds, p. 3 - 87 in "Geometry of homogeneous bounded domains", Roma, Edizioni Cremonese, 1968.

[GR] H.GRAUERT, R.REMMERT - Über kompakte homogene komplexe Mannigfaltigkeiten, Arch. Math. 13 (1962), 498 - 507.

[GR1] H.GRAUERT, R.REMMERT - *Analytische Stellenalgebren*, Springer, 1971.

[GR2] H.GRAUERT, R.REMMERT - *Theorie der Steinschen Räume*, Springer, 1977.

[GR3] H.GRAUERT, R.REMMERT - *Coherent analytic sheaves*, Springer, 1984.

[Gr] Ph.GRIFFITHS - Some geometric and analytic properties of homogeneous complex manifolds, Acta Math. 110 (1963), 115 - 208.

[GrHa] Ph.GRIFFIRHS, J.HARRIS - *Principles of algebraic geometry*, Wiley, New York, 1978.

[GrSch] Ph.GRIFFITHS, W.SCHMID - Locally homogeneous complex manifolds, Acta Math. 123 (1969), 253 - 302.

[Gu1] R.GUNNING - On Vitali's theorem for complex spaces with singularities, J. Math. Mech. 8 (1959), 133 - 141

[Gu2] R.GUNNING - *Introduction to holomorphic functions of several variables*, vol. I. Wadsworth, Belmont, California, 1990.

[GuRo] R.GUNNING, H.ROSSI - *Analytic functions of several complex variables*, Prentice-Hall, Englewood Cliffs, N.J., 1965.

[GV] S.G.GINDIKIN, E.B.VINBERG - Kähler manifolds admitting a transitive solvable automorphism group, Matem. Sbornik 74 (1967), 357 - 377 (Russian). AMS Translation: Matem. Sbornik 74 (1967), 333 - 351.

[H] M.W.HIRSCH - *Differential topology*, Graduate Texts in Math. 33, Springer, 1976.

[Ha] J.I.HANO - On Kählerian homogeneous spaces of unimodular Lie groups, Amer. J. of Math. 79 (1957), 885 - 900.

[Ha-Ch] HARISH-CHANDRA - Discrete series for semisimple Lie groups II, Acta Math. 116 (1966), 1 - 111.

[HaWe] F.R.HARVEY, R.O.WELLS - Zero sets of non-negative strictly plurisubharmonic functions, Math. Ann. 201 (1973), 165 - 170

[He] S. HELGASON - *Differential geometry, Lie groups and symmetric spaces*, Acad. Press, New York, 1978.

[Hei1] P.HEINZNER - Geometric invariant theory on Stein spaces, Math. Ann. 289 (1991), 631 - 662.

[Hei2] P.HEINZNER - Equivariant holomorphic extensions of real analytic manifolds, Bull. Soc. Math. France, 121 (1993), 445 - 463.

[Hi] F.HIRZEBRUCH - *Topological methods in algebraic geometry* , 3d enlarged edition, Springer, 1966.

[Ho] O.M.HOSROVJAN - On complex homogeneous spaces with two ends, p.35 - 42 in "Geometric methods in problems of analysis and algebra" (ed. A.L.Onishchik), Yaroslavl, 1978 (Russian).

[Hu1] A.T.HUCKLEBERRY - Actions of groups of holomorphic transformations, in "Several complex variables VI", Encyclopaedia of Mathematical Sciences, Volume 69, Springer, 1989, 143 - 196.

[Hu2] A.T.HUCKLEBERRY - The classification of homogeneous surfaces, Expo. Math. 4 (1986), 289 - 334.

[Hu3] A.T.HUCKLEBBERY - Subvarieties of homogeneous and almost homogeneous manifolds, to appear.

[HuLiv] A.T.HUCKLEBERRY, L.LIVORNI - A classification of complex homogeneous surfaces, Can. J. of Math. 33 (1981), 1096 - 1109.

[HuMar] A.T.HUCKLEBERRY, G.A.MARGULIS - Invariant analytic hypersurfaces, Invent. Math. 71 (1983), 235 - 240.

[HuOe1] A.T.HUCKLEBERRY, E.OELJEKLAUS - A characterization of complex homogeneous cones, Math. Z. 170 (1980), 181 - 194.

[HuOe2] A.HUCKLEBERRY, E.OELJEKLAUS - *Classification theorems for almost homogeneous spaces*, Travaux Inst. Elie Cartan 9, Univ. de Nancy, 1984.

[HuOe3] A.T.HUCKLEBERRY, E.OELJEKLAUS - On holomorphically separable complex solv-manifolds, Ann. Inst. Fourier 36 (1986), 57 - 65.

[HuSn1] A.T.HUCKLEBERRY, D.SNOW - Pseudoconcave homogeneous manifolds, Ann. Scuola Morm. Sup. Pisa, Serie IV, 7 (1980), 39 - 54.

[HuSn2] A.T.HUCKLEBERRY, D.SNOW - A classification of strictly pseudoconcave homogeneous manifolds, Ann. Scuola Norm. Sup. Pisa, Serie IV, 8 (1981), 231 - 255.

[HuSn3] A.T.HUCKLEBERRY, D.SNOW - Almost homogeneous Kähler manifolds with hypersurface orbits, Osaka J. Math. 19 (1982), 763 - 786.

[Hum1] J.E.HUMPHREYS - *Introduction to Lie algebras and representation theory,* Springer, New-York, 1972.

[Hum2] J.E.HUMPHREYS - *Linear algebraic groups*, Grad. Texts in Math. 21, Springer, 1975.

[K] K.KODAIRA - On Kähler varieties of restricted type, Ann. of Math. 60 (1954), 28 - 48.

[Ka1] W.KAUP - Infinitesimale Transformationsgruppen komplexer Räume, Math. Ann. 160 (1965), 72 - 92.

[Ka2] W.KAUP - Reelle Transformationsgruppen und invariante Metriken auf komplexen Räumen, Invent. Math. 3 (1967), 43 - 70.

[Kan] I.L.KANTOR - The cross-ratio and other invariants on homogeneous spaces with parabolic isotropy subgroups, Trudy Sem. Vect. Tens. Anal. 17 (1974), 250 - 313 (Russian).

[Ke] J.L.KELLEY - *General topology*, Van Nostrand, Princeton (N.J.), 1955.

[Ker] H.KERNER - Über die Automorphismengruppen kompakter komplexer Räume, Arch. Math. 11 (1960), 282 - 288.

[Ki] A.A.KIRILLOV - *Elements of the theory of representations*, Nauka, Moscow, 1972 (Russian). Engllish translation: Springer, 1976.

[KimVi] B.N.KIMEL'FELD, E.B.VINBERG - Homogeneous domains on flag manifolds and spherical subgroups, Funct. Anal. Appl. 12 (1978), 12 - 19 (Russian). AMS Translation: Funct. Anal. Appl. 12 (1978), 168 - 174.

[Kl] J.L.KOSZUL - Sur la forme hermitienne canonique des espaces homogènes complexes, Can. J. of Math. 7 (1955), 562 - 576.

[Kn] F.KNOP - The Luna-Vust theory of spherical embeddings, pp. 225 - 249 in "Proceedings of the Hyderabad conference on algebraic groups", Hyderabad, 1989.

[Ko] S.KOBAYASHI - *Hyperbolic manifolds and holomorphic mappings*, Marcel Dekker, New York, 1970.

[KoNo] S.KOBAYASHI, K.NOMIZU - *Foundations of differential geometry*, vol. I, II. Interscience Publ., New York, 1963, 1969.

[KoSp] K.KODAIRA, D.C.SPENCER - On deformations of complex analytic structures (I & II), Ann. of Math. 67 (1958), 328 - 466.

[Kr] M.KRÄMER - Sphärische Untergruppen in kompakten zusammenhängenden Liegruppen, Comp. Math. 38 (1979), 129 - 153.

[KSS] H.KRAFT, P.SLODOWY, T.A.SPRINGER (editors) - *Algebraische Transformationsgruppen und Invariantentheorie*, DMV Seminar, Band 13, Birkhäuser Verlag, Basel - Boston - Berlin, 1989.

[Kt] B.KOSTANT - Lie algebra cohomology and the generalized Borel-Weil theorem, Ann. of Math. 74 (1961),329 -387.

[Ku] F.KUTZSCHEBAUCH - Eigentliche Wirkungen von Liegruppen auf reell-analytischen Mannigfaltigkeiten, Schriftenreihe des Graduiertenkollegs Geometrie und Mathematische Physik, Institut für Mathematik, Ruhr-Universität Bochum, Heft 5, 1994.

[La1] M.LASSALLE - Séries de Laurent des fonctions holomorphes dans la complexification d'un espace symétrique compact, Ann. Sci. École Norm Sup. 11 (1978), 167 - 210.

[La2] M.LASSALLE - Deux généralisations du "théoreme des trois cercles" de Hadamard, Math. Ann. 249 (1980), 17 - 26.

[Le] R.LEHMANN - Complex-symmetric spaces, Ann. Inst. Fourier 39 (1989), 373 - 416.

[Les1] F.LESCURE - Élargissement du groupe d'automorphismes pour des variétés quasihomogènes, Math. Ann. 261 (1982), 455 - 462.

[Les2] F.LESCURE - Compactifications équivariantes par des courbes, Mém. Soc. Math. France, n° 26, Suppl. au Bull. de la S.M.F., Tome 115 (1987).

[Lo1] J.-J. LOEB - Action d'une forme réelle d'un groupe de Lie complexe sur les fonctions plurisousharmoniques, Ann. Inst. Fourier 35 (1985), 59 - 97.

[Lo2] J.-J. LOEB - Plurisousharmonicité et convexité sur les groupes réductifs complexes, Pub. IRMA - Lille, Vol. 2, n. VIII, 1 - 12 (1986).

[Lo3] J.-J.LOEB - Pseudo-convexité des ouverts invariants et convexité geodesique dans certains espaces symetriques, in "Séminaire d'Analyse P.Lelong - P.Dolbeault - H.Skoda (1983/1984)", pp. 172 - 190. Lecture Notes in Math. 1198, Springer, 1986.

[Lo4] J.-J. LOEB - Fonctions plurisousharmoniques invariantes sur certains espaces homogènes, C. R. Acad. Sci. Paris 312, Série I (1991), 681 - 683.

[Lu] D.LUNA - Slices étales, Bull. Soc. Math. France, Mémoire 33 (1973), 81 - 105.

[LuVu] D.LUNA, Th.VUST - Plongements d'espaces homogènes, Comm. Math. Helv. 58 (1983), 186 - 245.

[Ma] A.I.MAL'CEV - On a class of homogeneous spaces, Izv. Akad. Nauk SSSR, ser. mat., 13 (1949), 9 - 32 (Russian). English translation: Amer. Math. Soc. Transl. 39 (1951),

[Mat1] Y.MATSUSHIMA - Sur les espaces homogènes kählériens d'un groupe de Lie réductif, Nagoya Math. J. 11 (1957), 53 - 60.

[Mat2] Y.MATSUSHIMA - Espaces homogènes de Stein des groupes de Lie complexes I, Nagoya Math J. 16 (1960), 205 - 218.

[Mat3] Y.MATSUSHIMA - Espaces homogénes de Stein des groupes de Lie complexes II, Nagoya Math. J. 18 (1961), 153 - 164.

[MatMor] Y.MATSUSHIMA, A.MORIMOTO - Sur certains espaces fibrés holomorphes sur une variété de Stein, Bull. Soc. Math. France 88 (1960), 137 - 155.

[Mi] E.MICHAEL - Continuous selections I, Ann. of Math. 63 (1956), 361 - 382.

[Mik] I.V.MIKITYUK - On the integrability conditions of invariant hamiltonian systems with homogeneous configuration spaces, Matem. Sbornik 129 (1986), 514 - 534 (Russian); AMS translation: Math. USSR Sbornik 57 (1987), 527 - 546.

[Mo1] G.D.MOSTOW - Factor spaces of solvable groups, Ann. of Math. 60 (1954), 1 - 27.

[Mo2] G.D.MOSTOW - Self-adjoint groups, Ann. of Math. 62 (1955), 44 - 55.

[Mo3] G.D.MOSTOW - On the fundamental group of a homogeneous space, Ann. of Math. 66 (1957), 249 - 255.

[Mor] A.MORIMOTO - Non-compact complex Lie groups without non-constant holomorphic functions, in "Proceedings of the conference on complex analysis at the University of Minneapolis", Springer, 1965, 256 - 272.

[MZ] D.MONTGOMERY, L.ZIPPIN - *Topological transformation groups*, Wiley (Interscience), New York, 1955.

[N] R.NARASIMHAN - *Several complex variables*, Univ. of Chicago Press, 1971.

[NaOk] M.S.NARASIMHAN, K.OKAMOTO - An analogue of the Borel-Weil-Bott theorem for hermitian symmetric pairs of non-compact type, Ann. of Math., 91 (1970), 486 - 511.

[Oe1] E.OELJEKLAUS - Ein Hebbarkeitssatz für Automorphismengruppen kompakter Mannigfaltigkeiten, Math. Ann. 190 (1970), 154 - 166.

[Oe2] E.OELJEKLAUS - Fasthomogene Kählermannigfaltigkeiten mit verschwin dender erster Bettizahl, Manuscr. Math. 7 (1972), 175 - 183.

[Oel1] K.OELJEKLAUS - Hyperflächen und Geradenbündel auf homogenen komplexen Mannigfaltigkeiten. Schriftenreihe des Mathematischen Instituts der Universität Münster, Ser. 2, Heft 36, Münster 1985.

[Oel2] K.OELJEKLAUS - On the holomorphic separability of discrete quotients of complex Lie groups, Math. Z. 211 (1992), 627 - 633.

[OelRi1] K.OELJEKLAUS, W.RICHTHOFER - Homogeneous complex surfaces, Math. Ann. 268 (1984), 273 - 292.

[OelRi2] K.OELJEKLAUS, W.RICHTHOFER - On the structure of complex solvmanifolds, J. Diff. Geom. 27 (1988), 399 - 421.

[OelRi3] K.OELJEKLAUS, W.RICHTHOFER - Recent results on homogeneous complex manifolds, p. 78 - 119 in "Complex analysis III, Proc. Spec. Year, College Park (Md.), 1985 -86", Springer, Lect. Notes in Math. 1277.

[On1] A.L.ONISHCHIK - Complex envelopes of compact homogeneous spaces, Dokl. Acad. Nauk SSSR 130 (1960), 726 - 729 (Russian).

[On2] A.L.ONISHCHIK - Inclusion relations among transitive compact transformation groups, Trudy Moskov. Matem. Obshch. 11 (1962), 199 - 242 (Russian). English translation: Amer. Math. Soc. Transl. (2) 50 (1966), 5 - 58.

[On3] A.L.ONISHCHIK - Transitive compact transformation groups, Mat. Sbornik 60 (1963), 447 - 485 (Russian). English translation: Amer. Math. Soc. Transl. (2) 55 (1966), 153 - 194.

[On4] A.L.ONISHCHIK - *Topology of transitive transformation groups*, J.A.Barth, Leipzig - Berlin - Heidelberg, 1994.

[OnVi] A.L.ONISHCHIK, E.B.VINBERG - *Lie groups and algebraic groups*, Nauka, Moscow, 1988 (Russian). English translation: Springer, 1990.

[OtPo] M.OTTE, J.POTTERS - Beispiele homogener Mannigfaltigkeiten, Manuscr. Math. 10 (1973), 117 - 127.

[P] I.I.PIATETSKY-SHAPIRO - *Geometry of classical domains and theory of automorphic functions*, Nauka, Moscow, 1961 (Russian). French translation: Dunod, Paris, 1966.

[Pa] R.PALAIS - *A global formulation of the Lie theory of transformation groups*, Mem. Amer. Math. Soc. 22, Providence, Rhode Island, 1957.

[Pi1] S.I.PINCHUK - Homogeneous domains with piecewise-smooth boundaries, Matem. Zametki 32 (1982), 729 - 735 (Russian). AMS Translation: Math. Notes Acad. Sci. USSR, May 1983, 849 - 852.

[Pi2] S.I.PINCHUK - Holomorphic mappings in $\mathbb{C}^n$ and the problem of holomorphic equivalence, in "Several complex variables III", Sovr. Probl. Mat. 9, Moscow, 195 - 223, 1986 (Russian). English translation: Encyclopaedia of Mathematical Sciences, Springer (to appear).

[Ro] J.-P.ROSAY - Sur une caractérisation de la boule parmi les domaines de $\mathbb{C}^n$ par son groupe d'automorphismes, Ann. Inst. Fourier 29 (1979), 91 - 97.

[Rot] O.S.ROTHAUS - Envelopes of holomorphy of domains in complex Lie groups, in "Problems of analysis", pp. 309 - 317. Princeton: University Press, 1970.

[SaZa] R.SAERENS, W.ZAME - The isometry groups of manifolds and the automorphism groups of domains, Trans. Amer. Math. Soc. 301 (1987), 413 - 429.

[Sch1] W.SCHMID - On a conjecture of Langlands, Ann. of Math. 93 (1971), 1 - 42.

[Sch2] W.SCHMID - L^2-cohomology and discrete series, Ann. of Math. 103 (1976), 375 - 394.

[Se1] J.-P.SERRE - Représentations linéaires et espaces homogènes kähleriens des groupes de Lie compacts, Sém. Bourbaki, n. 100, Paris (1954).

[Se2] J.-P.SERRE - Faisceaux algébriques cohérents, Ann. of Math. 61 (1955), 197 - 278.

[Se3] J.-P.SERRE - *Lie algebras and Lie groups*, Benjamin, New-York, 1965.

[Se4] J.-P.SERRE - *Complex semisimple Lie algebras*, Springer, 1987.

[Sh] I.R.SHAFAREVICH (ed.) - *Algebraic surfaces*, Proc. Steklov Inst. of Math. LXXV, Nauka, Moscow, 1965 (Russian). English translation: Amer. Math. Soc., Providence, Rhode Island, 1967.

[Shi] H.SHIMA - Homogeneous Kählerian manifolds, Japan. J. of Math. 10 (1984), 71 - 98.

[Shu] H.B.SHUTRICK - Complex extensions, Quart. J. Math. Oxford (Ser. 2), 9 (1958), 189 - 201.

[Sn1] D.SNOW - Reductive group action on Stein spaces, Math. Ann. 259 (1982), 79 - 97.

[Sn2] D.SNOW - Stein quotients of connected complex Lie groups, Manuscr. Math. 50 (1985), 185 - 214.

[Sn3] D.SNOW - Cohomology of twisted holomorphic forms on Grassmann manifolds and quadric hypersurfaces, Math. Ann. 276 (1986), 159 - 176.

[Sn4] D.SNOW - Vanishing theorems on compact hermitian symmetric spaces, Math. Z., 198 (1988), 1 - 20.

[St] M.STEINSIEK - Transformation groups on homogeneous-rational manifolds, Math. Ann. 260 (1982), 423 - 435.

[Su] A.A.SUKHANOV - Description of the observable subgroups of linear algebraic groups, Matem. Sbornik 137 (179) (1988), 90 - 102 (Russian). AMS Translation: Math. USSR Sbornik, Vol. 65 (1990), 97 - 108.

[Ti1] J.TITS - *Sur certaines classes d'espaces homogénes de groupes de Lie*, Mém. Acad. Roy. Belg., 29 (3), Bruxelles, 1955.

[Ti2] J.TITS - Espaces homogènes complexes compacts, Comment. Math. Helv. 37 (1962), 111 - 120.

[Vi] E.B.VINBERG - Complexity of action of reductive groups, Funct. Anal. Appl. 20 (1986), 1 - 13 (Russian). AMS Translation: 20 (1986), 1 - 11.

[W1] J.A.WOLF - *Spaces of constant curvature*, McGraw-Hill, New York, 1967.

[W2] J.A.WOLF - The action of a real semisimple group on a complex flag manifold I. Orbit structure and holomorphic arc components. Bull. Amer. Math. Soc. 75 (1969), 1121 - 1237.

[Wa1] H.-C.WANG - Two point homogeneous spaces, Ann. of Math. 55 (1952), 177 - 191.

[Wa2] H.-C.WANG - Closed manifolds with homogeneous complex structures, Amer. J. of Math. 76 (1954), 1 - 32.

[Wa3] H.-C.WANG - Complex parallelizable manifolds, Proc. Amer. Math. Soc. 5 (1954), 771 - 776.

[War] G.WARNER - *Harmonic analysis on semi-simple groups* I, Springer, 1972.

[WB] H.WHITNEY, F.BRUHAT - Quelques propriétés fondamentales des ensembles analytiques réels, Comment. Math. Helv. 33 (1959), 132 - 160.

[Wi] J.WINKELMANN - Classification des espaces complexes homogènes de dimension 3 (I & II), C. R. Acad. Sci. Paris 306, Série I (1988), 231 - 234; 405 - 408.

[Wo] B.WONG - Characterization of the unit ball in $\mathbb{C}^n$ by its automorphism group, Invent. Math. 41 (1977), 253 - 257.

[Ya] H.YAMABE - Generalization of a theorem of Gleason, Ann. of Math. 58 (1953), 351 - 365.

Index of Notations

Index of Terminology

The Riemann-Hilbert Problem

A Publication from the Steklov Institute of Mathematics
Adviser: Sergeev, Armen.

by D. V. Anosov and A. A. Bolibruch

1994. x, 191 pages (Aspects of Mathematics, Volume E22; edited by Klas Diederich) Hardcover
ISBN 3-528-06496-X

From the contents: Introduction – Counterexample to Hilbert's 21st problem – Irreducible representations – Miscellaneous topics – The case p = 3 – Fuchsian equations.

The Riemann-Hilbert problem (Hilbert's 21st problem) belongs to the theory of linear systems of ordinary differential equations in the complex domain. The problem concerns the existence of a Fuchsian system with prescribed singularities and monodromy. Hilbert was convinced that such a system always exists. However, this turned out to be a rare case of a wrong forecast made by him. In 1989 the second author (A. B.) discovered a counterexample, thus obtaining a negative solution to Hilbert's 21st problem in its original form.

Vieweg Publishing · P.O. Box 58 29 · D-65048 Wiesbaden

vieweg

Aspekte der Mathematik

Edited by Klas Diederich

*A Publication of the Max-Planck-Institut für Mathematik, Bonn

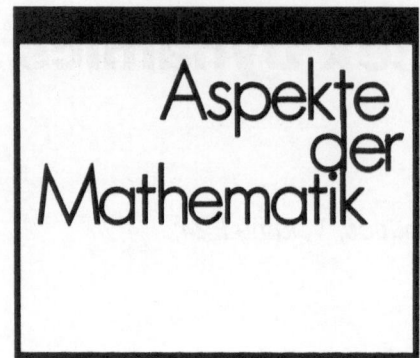